MW00710860

E. Smulders

Laundry Detergents

Description of the picture on the front cover

The pictures are described clockwise beginning at the top side.

Position: Top right side
Pearls of water gliding on a hydrophobic surface.

Position: Middle of right side
"The Lady in White" as a symbol for the brand Persil® as it was created by the Berlin artist Kurt Heiligenstaedt in 1922.

Position: Bottom right side
Cotton woven textile in plain weave seen through a microscope.

Position: Bottom in the middle
Behavior of surfactants in water (schematic drawing).
Demonstrated is the equilibrium between surfactants forming a layer on the water surface, dissolved surfactants in the water phase (left) and a micelle (right).

Position: Middle of left side
Spray drying tower for the production of detergents in powder form.

Position: Top left side
Front door and porthole of a typical European drum-type washing machine.

Position: Centre
Direction and magnitude of power vectors of water molecules depending on their location either at the interface water/air or dissolved in the water bulk phase.

Eduard Smulders

Laundry Detergents

in collaboration with
Wilfried Rähse, Wolfgang von Rybinski,
Josef Steber, Eric Sung, Frederike Wiebel

Authors:
Dr. Eduard Smulders, 40724 Hilden, Germany
Dr. Wilfried Rähse, Henkel KGaA, 40191 Düsseldorf, Germany
Dr. Wolfgang von Rybinski, Henkel KGaA, 40191 Düsseldorf, Germany
Dr. Josef Steber, Henkel KGaA, 40191 Düsseldorf, Germany
Dr. Eric Sung, 41068 Mönchengladbach, Germany
Dr. Frederike Wiebel, Henkel KGaA, 40191 Düsseldorf, Germany

Library of Congress Card No.: applied for

British Library Cataloguing-in-Publication Data:
A catalogue record for this book is available from the British Library

Die Deutsche Bibliothek – CIP-Einheitsaufnahme
A catalogue record for this book is available from Die Deutsche Bibliothek

ISBN 3-527-30520-3

Printed in the Federal Republic of Germany
Printed on acid-free paper.

Composition: Rombach GmbH, Freiburg
Printing: strauss-offsetdruck, Mörlenbach
Bookbinding: Wilhelm Osswald + Co., Neustadt

Dear Friend of Henkel

Eighteen years ago, a team of now retired colleagues from laundry detergent development, research, production and life sciences contributed the chapter on "Detergents" to "Ullmanns Enzyklopädie der technischen Chemie" (Ullmann's Encyclopedia of Industrial Chemistry). This comprehensive account of laundry washing and detergent chemistry with descriptions of production, analysis and environmental behavior met with huge interest at that time. The exclusively published offprints in German and English were therefore soon unobtainable.

In the meantime there have been so many innovations and changes in laundry product development, that an update has become necessary.

In its present form, the content is now about one third larger to accommodate many new aspects. Today, the Internet is the "library" people visit to find such reference works. The work as a whole no longer exists in book form but only selected chapters, such as this one, do.

It therefore gives us great pleasure to present you, as a Friend of Henkel, this special edition which is exclusively issued for Henkel. We hope that the content will refresh your memory and answer questions or simply serve as an information source that you will frequently and happily dip into.

We are obliged to Dr. Eduard Smulders and his team of co-authors for their valuable contributions to this monograph.

Dr. Thomas Müller-Kirschbaum
Vice President R&D / Technology Laundry and Home Care
Henkel KGaA

Contents

1. **Historical Review** 1

2. **Physical Chemistry of the Washing Process** 7
2.1. Influence of the Water 7
2.2. Types of Soil 8
2.3. The Soil Removal Process 11
2.3.1. Oily/Greasy Soil 13
2.3.2. Particulate Soil 21
2.3.3. Calcium-Containing Soil 29
2.3.4. Influence of Textile Fiber Type 33
2.4. Subsequent Processes 34
2.4.1. Dispersion and Solubilization Processes 34
2.4.2. Adsorption 35
2.4.3. Soil Antiredeposition and Soil Repellent Effects 35
2.5. Concluding Remarks 38

3. **Detergent Ingredients** 38
3.1. Surfactants 39
3.1.1. Anionic Surfactants 45
3.1.2. Nonionic Surfactants 52
3.1.2.1. Alcohol Ethoxylates (AE) 52
3.1.2.2. Alkylphenol Ethoxylates (APE) 55
3.1.2.3. Fatty Acid Alkanolamides (FAA) 55
3.1.2.4. Alkylamine Oxides 56
3.1.2.5. N-Methylglucamides (NMG) 56
3.1.2.6. Alkylpolyglycosides (APG) 56
3.1.3. Cationic Surfactants 57
3.1.4. Amphoteric Surfactants 61
3.2. Builders 61
3.2.1. Alkalies 62
3.2.2. Complexing Agents 63
3.2.3. Ion Exchangers 68
3.2.4. Builder Combinations 74
3.3. Bleaches 74
3.3.1. Bleach-Active Compounds 75
3.3.2. Bleach Activators 80
3.3.3. Bleach Catalysts 83
3.3.4. Bleach Stabilizers 83
3.4. Further Detergent Ingredients 84
3.4.1. Enzymes 84

3.4.2. Soil Antiredeposition Agents, Soil Repellent/Soil Release Agents 88
3.4.3. Foam Regulators 90
3.4.4. Corrosion Inhibitors 92
3.4.5. Fluorescent Whitening Agents 92
3.4.6. Dye Transfer Inhibitors 96
3.4.7. Fragrances 96
3.4.8. Dyes 97
3.4.9. Fillers and Formulation Aids 98

4. Household Laundry Products 98
4.1. Heavy-Duty Detergents 99
4.1.1. Conventional Powder Heavy-Duty Detergents 100
4.1.2. Compact and Supercompact Heavy-Duty Detergents 103
4.1.3. Extruded Heavy-Duty Detergents 106
4.1.4. Heavy-Duty Detergent Tablets 106
4.1.5. Color Heavy-Duty Detergents 107
4.1.6. Liquid Heavy-Duty Detergents 108
4.2. Specialty Detergents 110
4.2.1. Powder Specialty Detergents 111
4.2.2. Liquid Specialty Detergents 111
4.3. Laundry Aids 112
4.3.1. Pretreatment Aids 113
4.3.2. Boosters 114
4.3.3. Aftertreatment Aids 115
4.3.3.1. Fabric Softeners 116
4.3.3.2. Stiffeners 117
4.3.3.3. Laundry Dryer Aids 119
4.3.4. Other Laundry Aids 119
4.3.4.1. Refreshing Products for Dryer Application 119
4.3.4.2. Odor Removers for Washer Application 120

5. Industrial and Institutional Detergents 120

6. Production of Powder Detergents 122
6.1. Technology Overview 123
6.2. Manufacturing Processes 125
6.2.1. Traditional Spray-Drying Process 125
6.2.2. Superheated Steam Drying 127
6.2.3. Nontower Agglomeration Process 127
6.2.4. Nontower Compound Technology 129
6.3. Densification Processes 129
6.3.1. Dry Densification in a Mixer 130
6.3.2. Dry Densification in a Spheronizer 130
6.3.3. Dry Densification in a Roller Press 131
6.3.4. Wet Granulation 131

6.3.5. Spaghetti Extrusion 132
6.3.6. Postaddition Process 135
6.3.7. Dry Densification in a Tablet Press 135
6.4. Raw Materials 138
6.4.1. Anionic Surfactants 138
6.4.2. Nonionic Surfactants 139
6.4.3. Builders 141
6.4.4. Peroxygen Bleaches 142
6.4.5. Enzymes 143

7. Analysis of the Composition 145
7.1. Detergent Ingredients 146
7.2. Purposes of Detergent Analysis 147
7.3. Sample Preparation 147
7.4. Analytical Methods 147
7.4.1. Qualitative Analysis 147
7.4.2. Sample Preparation 148
7.4.3. Quantitative Analysis 149
7.4.4. Separation Methods 151
7.4.5. Structure Determination 152
7.4.6. Determination of Characteristic Values 153
7.4.7. Analysis Automation 153
7.5. Sources of Information 153

8. Test Methods for Laundry Detergents 154
8.1. Laboratory Methods 155
8.2. Practical Evaluation 156
8.3. Consumer Tests 157

9. Economic Aspects 157
9.1. Detergent Components 158
9.1.1. Surfactants 158
9.1.2. Builders 160
9.2. Laundry Detergents 161
9.3. Fabric Softeners 164
9.4. Other Laundry Aids 165

10. Ecology 165
10.1. Laundry, Wastewater, and the Environment 165
10.2. Contribution of Laundry to the Sewage Load 166
10.3. Detergent Laws 167
10.3.1. Development of the European Detergent Legislation 168
10.3.2. Regulatory Limitations on Anionic and Nonionic Surfactants 169
10.3.3. Primary Biodegradation Test Procedures 170
10.3.4. Regulation of Maximum Phosphate Content in Detergents 172

10.4. General Criteria for the Ecological Evaluation of Detergent Chemicals 173
10.4.1. Concept 174
10.4.2. Environmental Exposure Assessment 174
10.4.2.1. Biodegradation 175
10.4.2.2. Biodegradability Standard Test Methods 176
10.4.2.3. Supplementary Biodegradation Test Methods 178
10.4.2.4. Exposure Analysis 179
10.4.3. Assessment of Environmental Effects 181
10.4.3.1. Basic Ecotoxicity Tests 182
10.4.3.2. Subchronic and Chronic Ecotoxicity Tests 184
10.4.3.3. Biocenotic Ecotoxicity Tests 184
10.4.3.4. Bioaccumulation 184
10.4.4. Process of Environmental Risk Assessment 185
10.5. Ecological Characterization of Main Detergent Ingredients 186
10.5.1. Surfactants 186
10.5.1.1. Anionic Surfactants 186
10.5.1.2. Nonionic Surfactants 190
10.5.1.3. Cationic Surfactants 192
10.5.2. Builders 193
10.5.2.1. Zeolites 193
10.5.2.2. Polycarboxylates 194
10.5.2.3. Citrates 195
10.5.2.4. Sodium Carbonate (Soda Ash) 195
10.5.2.5. Nitrilotriacetate (NTA) 196
10.5.3. Bleaching Agents 196
10.5.3.1. Sodium Perborate 196
10.5.3.2. Sodium Percarbonate 197
10.5.3.3. Tetraacetylethylenediamine (TAED) 197
10.5.4. Auxiliary Agents 198
10.5.4.1. Phosphonates 198
10.5.4.2. EDTA 199
10.5.4.3. Enzymes 200
10.5.4.4. Optical Brighteners 200
10.5.4.5. Carboxymethyl Cellulose 201
10.5.4.6. Dye Transfer Inhibitors 201
10.5.4.7. Fragrances 201
10.5.4.8. Foam Regulators 202
10.5.4.9. Soil Repellents 202
10.5.4.10. Dyes 202
10.5.4.11. Sodium Sulfate 203

11. Toxicology 203
11.1. Detergent Ingredients 204
11.1.1. Surfactants 204
11.1.2. Builders 206
11.1.3. Bleach-Active Compounds 206
11.1.4. Auxiliary Agents 207
11.2. Finished Detergents 208
11.3. Conclusions 208

12. Textiles 209

13. Washing Machines and Wash Programs (Cycles) 220
13.1. Household Washing Machines 220
13.1.1. Classification 220
13.1.2. Operational Parameters 224
13.1.3. Wash Programs 227
13.1.3.1. Japanese Washing Machines and Washing Conditions 227
13.1.3.2. North American Washing Machines and Washing Conditions 227
13.1.3.3. European Washing Machines and Washing Conditions 228
13.1.4. Energy and Water Consumption 230
13.1.5. Construction Materials Used in Washing Machines 230
13.1.6. The Market for Washing Machines 232
13.2. Laundry Dryers 234
13.3. Washing Machines for Institutional Use 236
13.3.1. Batch-Type Machines 236
13.3.2. Continuous Batch Washers 237

14. References 238

15. Persil – Nine Decades of Research for Detergent Users 261

16. Index 267

1. Historical Review

The symbol used by the ancient Egyptians to represent a launderer was a pair of legs immersed in water. This choice was logical, because at that time the standard way to wash clothes was to tread on them. The "fullones" of ancient Rome earned their bread, too, by washing clothes with their feet. The washing process then was very simple: laundry of every kind was subjected to purely mechanical treatment consisting of beating, treading, rubbing, and similar procedures. It has long been known, however, that the washing power of water can be increased in various ways. Rainwater, for example, was found to be more suitable for washing than well water. Hot water also was found to have more washing power than cold, and certain additives seemed to improve any water's effectiveness.

Even the ancient Egyptians used soda ash as a wash additive. This was later supplemented with sodium silicate to make the water softer. These two substances also formed the basis for the first commercial detergent brand to appear on the German market, Henkel's "Bleichsoda", introduced in 1878. Its water-softening effect was a result of the precipitation of calcium and magnesium ions, and it simultaneously eliminated iron salts, which had a tendency to turn laundry yellow. Used along with soap, which had also been known since antiquity, this product prevented the formation of inactive material known as "lime soaps", and the laundry no longer suffered from a buildup of insoluble soap residues.

Soap is the oldest of the surfactants. It was known to the Sumerians by ca. 2500 B.C., although the Gauls were long credited with its discovery. For more than 3000 years, soap was regarded strictly as a cosmetic — in particular a hair pomade — and as a remedy. Only in the last 1000 years has it come to be used as a general purpose washing and laundering agent. Soap remained a luxury until practical means were discovered for producing soda ash required for saponification of fats. With the beginning of the 20th century and the introduction in Germany of the first self-acting laundry detergent (Persil, 1907), soap took its place as one ingredient in multicomponent systems for the routine washing of textiles. In these, soap was combined with so-called builders, usually sodium carbonate, sodium silicate, and sodium perborate. The new detergents were capable of sparing people the weather-dependent drudgery of bleaching their clothes on the lawn, and the enhanced performance of these new agents substantially reduced the work entailed in doing laundry by hand.

The next important development was the transition brought about by technology from the highly labor-intensive manual way of doing laundry to machine washing. This change in turn led to a need for appropriate changes in the formulation of laundry detergents. Soap, notorious for its sensitivity to water hardness, was gradually replaced by so-called synthetic surfactants with their more favorable characteristics. The term "synthetic surfactants" is erroneous when used to distinguish from soap, which itself is made through synthesis as well. The first practical substitutes for soap were fatty alcohol sulfates, discovered in Germany by BERTSCH and coworkers in 1928 [1]. The availability of synthetic alkyl sulfates based on natural fats and oils made possible the first neutral detergent for delicate fabrics: Fewa, introduced in Germany in 1932. This was followed in 1933 on the U.S. market by Dreft, a similarly conceived product. Fatty alcohol sulfates and their derivatives (alkyl ether sulfates, obtained by treating fatty alcohols with ethylene oxide and subsequent sulfation) still retain their importance in many applications, particularly in heavy-duty detergents, specialty detergents, dish-washing agents, cosmetics, and toiletries. The general acceptance worldwide of synthetic surfactants, with their reduced sensitivity to water hardness relative to soap, is a development of the 1940s. Procter & Gamble introduced the synthetic detergent Tide in the USA in 1946. By the 1950s the widespread availability of tetrapropylenebenzene-sulfonate (TPS), a product of the petrochemical industry, had largely displaced soap as the key surfactant from the detergent market in the industrialized nations. The only remaining role of soap in the industrialized countries became that of a foam regulator. The favorable economics associated with TPS, along with its desirable properties, caused this branched-chain synthetic surfactant to capture ca. 65 % of the total synthetic surfactant demand in the Western world in the late 1950s.

However, a new criterion soon appeared for surfactants, a criterion that had long been ignored: the biodegradability of the products. As a result of several unusually dry summers in Germany, especially that of 1959, water flow in many streams and rivers was severely restricted. Great masses of stable foam began to build up in the vicinity of weirs, locks, and other constructions in waterways. These were caused by insufficiently biodegraded TPS and nonylphenol ethoxylates, another class of surfactants that had been introduced at that time. The discovery that many surfactants could emerge unchanged even from a modern sewage treatment plant and thus enter surface waters led to adoption of the first German Detergent Law in 1961 [2], whose provisions took effect in 1964 [3]. Manufacturers were subsequently enjoined from marketing any detergents or cleansing agents whose biodegradability fell below 80 % in a test devised by the Detergent Commission. Initially only anionic surfactants were affected, but these were later joined by nonionic surfactants. The German precedent was soon followed by the enactment of similar legislation in such countries as France, Italy, and Japan. In the UK and the USA, the transition to biodegradable surfactants occurred as a result of voluntary agreements between industry and government.

For a long time branched alkylbenzenesulfonates and nonylphenol ethoxylates have been replaced in most countries by the much more rapidly and effectively degradable linear alkylbenzenesulfonates and long-chain alcohol ethoxylates.

Another important step in the development of laundry detergents was replacement of builders such as sodium carbonate by complexing agents. The first complexing agents that were used were of the sodium diphosphate type, but these were replaced after World War II by the more effective sodium triphosphate. Inorganic ion exchangers such as zeolite A have meanwhile replaced sodium triphosphate in many countries, particularly in those where phosphate legislation has been enacted. Table 1 summarizes the performance characteristics of detergents containing various builders and surfactants [4].

Table 1. Properties of detergents based on different surfactants and builders

Detergent	Single wash cycle performance	Soil antiredeposition capability	Deposition on fabrics and washing machines	Yellowing and bad odor
Soap – sodium carbonate – sodium silicate	poor	poor	very heavy	heavy
Synthetic nonionic surfactants – sodium carbonate – sodium silicate	fair to good	fair	heavy	none
Synthetic surfactants – sodium diphosphate	good	fair to good	heavy	none
Synthetic surfactants – sodium triphosphate	good	good	weak	none
Synthetic surfactants – sodium triphosphate – zeolite 4 A	good	good	weak	none
Synthetic surfactants – zeolite 4 A – polycarboxylates – sodium carbonate	good	good	weak	none

The last decades also saw the introduction of other ingredients for improving detergency performance, and their presence in formulations has been state of the art for modern laundry detergents. Key ingredients among these are the following:

Ion exchangers (zeolites)
Soil antiredeposition agents
Enzymes
Fluorescent whitening agents
Foam regulators
Bleach activators
Soil repellents
Polycarboxylate cobuilders

Table 2 shows highlights of the historical development of detergent ingredients and detergents with concurrent developments in textile fibers and washing machines. Comparison of the changes clarifies how closely the various participants in the washing process are tied to one another. The laundry occupies the center of the stage, and it has consistently kept pace both with detergents and with available washing processes and machines, at least to the extent that the newly developed textiles have been found to have value and market appeal. Thus, close cooperation between all manufacturers involved has become essential for the development of more and more optimized washing processes.

Table 2. Development of detergent ingredients, detergents, textile fibers, and washing equipment from 1876 to 2000

Year	Detergent ingredients	Detergents	Textile fibers	Washing equipment
1876	sodium silicate soap starch		cotton linen wool	boiler
1878	sodium carbonate sodium silicate	prewashing product and laundry softener (Henko, Henkel, Germany)		
1890			cupro	
1907	soap sodium carbonate sodium perborate sodium silicate	heavy-duty detergent (Persil, Henkel, Germany)	rayon acetate silk	wooden vat machine
1913	proteases (pancreatic enzymes)	prewashing product (Burnus, Röhm & Haas, Germany)		
1920			viscose staple fiber	metal tub agitator washing machine
1932/33	synthetic surfactants (fatty alcohol sulfates)	specialty detergent (Fewa, Henkel, Germany; Dreft, P & G, USA)		
1933	sodium diphosphate magnesium silicate	heavy-duty detergent (Persil, Henkel, Germany)		
1940	alkylsulfonates (Mersolat) antiredeposition agents (CMC[a])	heavy-duty detergent (Henkel, Unilever, Germany)		automatic agitator washing machine (Blackstone, USA)
1946	fatty alcohol sulfates alkylbenzenesulfonates sodium triphosphate	heavy-duty detergent (Tide, P & G, USA)	polyamide	automatic drum-type machine (Bendix, USA)
1948	nonionic surfactants	heavy-duty detergent (All, Monsanto, USA)		
1949	fluorescent whitening agents (optical brighteners)	rinsing additive (Sil, Henkel, Germany)		
1950	fragrances cationic surfactants	heavy-duty detergent (Dial, Armour Dial, USA) fabric softener (CPC International, USA)		
1954	anionic – nonionic combinations foam-regulators (soap)	heavy-duty detergent (Dash, P & G, USA)	polyacrylonitrile	semiautomatic drum-type machine (washing automatically, rinsing by hand, spinning separately, Europe)
1957			polyester resin-finished cotton	automatic drum-type machine (washing and spinning separately, Europe)
1959	proteases (enzymes)	prewashing detergent (Bio 40, Gebr. Schnyder, Switzerland)		

Table 2. (continued)

Year	Detergent ingredients	Detergents	Textile fibers	Washing equipment
1962	foam-regulators (behenate soap)	heavy-duty detergent (Dash, P & G, Germany)	polyester – cotton blend	fully automatic drum-type machine (washing, rinsing, and spinning automatically, Europe)
1965			polyurethane wool with reduced felting	wash dryer (washing, rinsing, spinning, and tumble drying automatically in one machine)
1966			nonwovens	
1970	fatty acid amine condensation product	detergent for delicates with fabric softening effect (Perwoll, Henkel, Germany)	resin-finished linen	
1972	bleach activators (TAGU)[b]	heavy-duty detergent (Cid, Henkel, Germany)		
	NTA[c]	heavy-duty detergent (various brands, Canada)		
1973	amylases (enzymes)	heavy-duty detergent (Mustang, Henkel, Germany)		
1974	sodium percarbonate	heavy-duty detergent (Dixan, Henkel, Cyprus; Persil, Lever, UK)		
1975	sodium citrate	heavy-duty liquid detergent (Wisk, Lever, USA)		
1976	zeolite 4 A	heavy-duty detergent (prodixan, Henkel, Germany; Tide, P & G, USA)		
	foam-regulators (silicone oils)	heavy-duty detergent (Mustang, Henkel, Germany)		
		heavy-duty liquid detergent without builders (Era, P & G, USA)		
1977	layered silicates – cationic surfactants	dual-function detergent with fabric-softening effect (Bold 3, P & G, USA)		
1978	bleach activator TAED[d]	heavy-duty detergent (Skip, Lever, France)	Dunova (modified polyacrylonitrile fiber)	microprocessor operation, electronic sensing
1981		liquid heavy-duty detergent in Europe (Liz, Henkel, Germany; Vizir, P & G, Germany		
	esterquats	fabric softeners (Minidou, Lesieur Cotelle, France)		
1982	zeolite 4 A – NTA builder systems	nonphosphate heavy-duty powder detergent, (Dixan, Henkel, Switzerland)	PTFE[e]-fiber Goretex	
1984	poly(acrylic acid), poly(acrylicacid-*co*-maleic acid)	heavy-duty detergents		

Table 2. (continued)

Year	Detergent ingredients	Detergents	Textile fibers	Washing equipment
1986			PES[f]-polyether fiber Sympatex	controlled water intake (Europe)
1987		compact heavy-duty detergent (Attack, Kao, Japan)		
	cellulase (enzyme)	compact heavy-duty detergent (Attack, Kao, Japan)		
	soil repellent	heavy-duty detergents, USA		
1988	lipase (enzyme)	compact heavy-duty detergent (Hi Top, Lion, Japan)		
1990	alkylpolyglycosides (APG)	liquid heavy-duty detergent (Persil, Henkel, Germany)		
	foam regulators (paraffins)	heavy-duty detergent (Persil, Henkel, Germany)		
1991	dye transfer inhibitor PVP[g]	heavy-duty color detergent (Persil Color, Henkel, Germany)		
1992		extruded supercompact heavy-duty detergent (Persil Megaperls, Henkel, The Netherlands, Belgium, Austria)		
1994	*N*-methylglucamide	heavy-duty detergents (Ariel, P & G, Germany; Tide, P & G, USA)		
	bleach catalyst Mn-TACN[h]	heavy-duty detergent (Omo Power, Lever, The Netherlands; Persil Power, Lever, UK)		
1995				fuzzy logic control
1996		gel-structured liquid heavy-duty detergents (Persil Gel and Dixan Gel, Henkel, Europe)		
1997-1998		heavy-duty detergent tablets (Europe: Colon, Benckiser, Spain; Persil, Lever, UK; Le Chat, Henkel, France; Dixan, Henkel, Italy; Persil, Henkel, Germany)		

[a] CMC = carboxymethyl cellulose.
[b] TAGU = tetraacetylglycoluril.
[c] NTA = nitrilotriacetic acid.
[d] TAED = tetraacetylethylenediamine
[e] PTFE = polytetrafluoroethylene
[f] PES = poly(ethylene glycol terephthalate)
[g] PVP = poly(*N*-vinylpyrrolidone)
[h] Mn – TACN = manganese – tetraazacylononane complex

2. Physical Chemistry of the Washing Process

Washing and cleaning in aqueous wash liquor is a complex process involving the cooperative interaction of numerous physical and chemical influences. In the broadest sense, washing can be defined as both the removal by water or aqueous surfactant solution of poorly soluble matter and the dissolution of water-soluble impurities from textile surfaces.

A fundamental distinction exists between the primary step, in which soil is removed from a substrate, and secondary stabilization in the wash liquor of dispersed or molecularly dissolved soil. Stabilization is necessary to prevent redeposition onto the fibers of soil that has already been removed. The terms *single* and *multiple wash cycle or multicycle performance* are used respectively in conjunction with the two phenomena.

The following components constitute a partnership in the overall washing process:

Water
Soil
Textiles
Washing equipment
Detergent

Wash performance is highly sensitive to such factors as textile properties, soil type, water quality, washing technique (amount and kind of mechanical input, time, and temperature), and detergent composition. Not all these mutually interrelated factors are amenable to random variation; indeed, they are generally restricted within rather narrow limits. Of particular importance is the composition of the detergent.

2.1. Influence of the Water

The most obvious role of water is to serve as a solvent, both for the detergent and for soluble salts within the soil. Water is also the transport medium for dispersed and colloidal soil components, however. The washing process begins with wetting and penetration of the soiled laundry by the detergent solution. Water has a very high surface tension (72 mN/m), and wetting can only take place rapidly and effectively if the surface tension is drastically reduced by surfactants to values of 30 mN/m and below. The surfactants thus become key components of any detergent. The surface tension as a function of the surfactant concentration is shown in Figure 1 for typical surfactants used in detergents [5].

Water hardness has also a significant influence on the results of the washing process. Water hardness is defined in terms of the amount of calcium and magnesium salts present, measured in millimoles per liter (mmol/L). A calcium hardness of 1 mmol/L

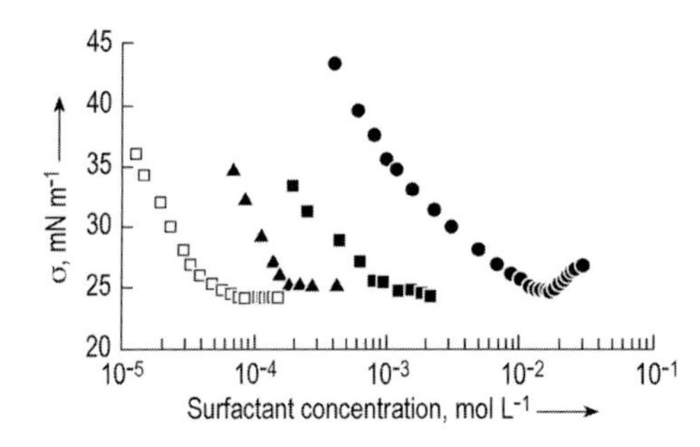

Figure 1. Surface tension as a function of concentration for different alkylglycosides at 60 °C in distilled water ● = C_8 monoglycoside; ■ = C_{10} monoglycoside; ▲ = C_{12} monoglycoside; □ = $C_{12}-C_{14}$ alkylpolyglycoside

corresponds to 40.08 mg of calcium ions per liter of water. Additional hardness data are provided in Table 3 [6], along with other measures of hardness and relationships among them.

Water hardness can vary considerably from one country to another (Table 4) and even from region to region within a country. Soft water is relatively rare in most of Europe, whereas it is common in North America, Brazil, Japan, and many other countries.

Water of poor quality can severely impair the washing process and have detrimental effects on washing machines. The calcium and magnesium ions, which are responsible for water hardness, are prone to precipitate, either in the form of carbonates or as insoluble compounds derived from ingredients present in detergents. These precipitates may cause the formation of residues on the laundry, but they can also build up as scale in the washing machine, thereby adversely affecting the function of heating coils and other machine components. A high calcium content in the water also impedes the removal of particulate soil. The presence of trace amounts of iron, copper, or manganese ions in water may also be detrimental to washing. These ions can catalyze decomposition of bleaching agents during the washing process. Complexing agents or ion exchangers are often found in detergents, and one of their functions is to bind multivalent alkaline earth and heavy-metal ions through chelation or ion exchange.

2.2. Types of Soil

One way to categorize soils is by their origin, i.e.:

Dust from the atmosphere
Bodily excretions
Impurities derived from domestic, commercial, or industrial activity

From a detergency standpoint, however, it is more appropriate to regard the principal types of soil in other ways. Thus, one can distinguish the following:

Table 3. Units currently used for expressing the hardness of water [6]

Name of unit	Definition	Symbol	Conversion factors						
			Ca^{2+}		CaO	$CaCO_3$			
			mmol/L	meq/L	°d	mg/kg *	°e	°a	°f
Millimole per liter	1 mmol of calcium ions (Ca^{2+}) in 1 L of water	mmol/L	1	2.000	5.600	100	7.020	5.8500	10.00
Milliequivalent per liter	20.04 mg of calcium ions (Ca^{2+}) in 1 L of water	meq/L	0.500	1	2.800	50	3.510	2.9250	5.00
German degree of hardness	10 mg of calcium oxide (CaO) in 1 L of water	°d	0.178	0.357	1	17.8	1.250	1.0440	1.78
Milligram per kilogram	1 mg of calcium carbonate ($CaCO_3$) in 1 L of water	mg/kg *	0.010	0.020	0.056	1	0.070	0.0585	0.10
English degree of hardness	1 grain of calcium carbonate ($CaCO_3$) in 1 gal (UK) of water	°e	0.142	0.285	0.798	14.3	1	0.8290	1.43
American degree of hardness	1 grain of calcium carbonate ($CaCO_3$) in 1 gal (US) of water	°a	0.171	0.342	0.958	17.1	1.200	1	1.71
French degree of hardness	1 mol (= 100 g) of calcium carbonate ($CaCO_3$) in 10 m^3 of water	°f	0.100	0.200	0.560	10.0	0.702	0.5850	1

* The unit "part per million" (ppm) is often used for mg/kg.

Table 4. Distribution of water hardness shown as percentage of homes affected by defined ranges of hardness [7]

Country	Range of hardness		
	0–90 ppm	90–270 ppm	>270 ppm
Japan	92	8	0
United States	60 *	35	5
Western Europe	9 **	49 **	42 **
Austria	1.8	74.7	23.5
Belgium	3.4	22.6	74
France	5	50	45
Germany	10.8	41.7	47.5
Great Britain	1	37	62
Italy	8.9	74.7	16.4
The Netherlands	5.1	76.1	18.8
Spain	33.2	24.1	42.7
Switzerland	2.8	79.7	17.5

* Including 10 % with home water softening appliances.
** Average

Water-soluble materials:

inorganic salts
sugar
urea
perspiration

Particles:

metal oxides
carbonates
silicates
humus
carbon black (soot)

Fats and oils:

animal fat
vegetable fat and oil
sebum
mineral oil and grease
wax

Proteins from the following:

blood
grass
egg
milk
keratin from skin

Carbohydrates:

starch

Bleachable dyes from:

fruit/fruit juices
vegetables
wine
coffee
tea

Soils and stains mostly consist of mixtures of the above materials, e.g., stains from food, kitchen, or cooking.

Soil removal during the washing process is enhanced by increasing the mechanical input, wash time, and temperature. For any given washing technology, however, detergency performance is dependent on specific interactions among substrate surface, soil, and detergent components. In this context, it is important to consider not only the interactions of the wash components with one another, but also interactions among the various classes of soil. The most difficult soils to remove from fabrics are pigments, such as carbon black, inorganic oxides, carbonates, and silicates. Other problematic soils include fats, waxes, higher hydrocarbons, denatured protein, and certain natural dyes. All of these are mostly present on fibers in the form of mixed soils.

The removal of soil from a surface can either be coupled with a chemical reaction or it can occur without chemical change. A redox process involving a bleach is an example of the former, in the course of which some oxidizable substance (e.g., a natural dye from tea, wine, or fruit juice) is cleaved. Enzyme-mediated decomposition of a strongly bound protein soil and its subsequent removal is another example.

In many cases, however, the soil to be removed consists of substances that are not amenable to chemical treatment, and the only alternative is to remove the soil by interfacial processes. This requirement is reflected in the composition of modern detergents. Apart from bleaches, the primary components of modern detergents are surfactants, water-soluble complexing agents, and water-insoluble ion exchangers.

2.3. The Soil Removal Process

Physical removal of soil from a surface occurs as a result of nonspecific adsorption of surfactants on the various interfaces present [8] and through specific adsorption of chelating agents on certain polar soil components [4]. In addition, an indirect effect is caused by calcium ion exchange, whereby the release of calcium ions from soil deposits and fibers causes a loosening of the structure of the residue. Compression by electrolytes of the electrical double layer at interfaces is also significant [8]. All of these effects work together to remove oily and particulate soils from textile substrates or solid

surfaces. Interfacial properties, which are modified by the adsorption of detergent components onto the various interfaces present, include the following:
Air – water interface:

surface tension
foam generation
film elasticity
film viscosity

Liquid – liquid interface:

interfacial tension
interfacial viscosity
emulsification
electric charge
active-ingredient penetration

Solid – liquid interface:

disjoining pressure
suspension stability
electric charge

Solid – solid interface:

adhesion
flocculation
heterocoagulation
sedimentation

Interfaces in multicomponent systems:

wetting
rolling-up processes

Interfacial chemical properties undergo change as detergent compounds are adsorbed, and these changes are prerequisite to effective soil removal. The significance of a given change in respect to the overall process can vary, depending on the system involved. However, in general, the greater the equilibrium adsorption of washing active substances, and the more favorable their adsorption kinetics, the better will be their detergency performance [8].

2.3.1. Oily/Greasy Soil

Rolling-up Processes. Most oily/greasy soils are liquid at wash temperatures >40 °C. The removal of these soils is mainly influenced by the wetting behavior between washing liquid, substrate, and soil, and by the interfacial tension between washing liquid and oily soil [9]. Oily soils wet most textile substrates very effectively, and they have a tendency to spread over a surface, forming more or less closed surface layers. For this reason the following observations can be regarded as equally applicable to both liquid and solid fat residues.

In the first phase of washing, textile fibers and soil must be wetted as thoroughly as possible by the wash liquor [10]. The contact angle between a solid and a drop of a liquid applied to its surface can be taken as a measure of wetting. Figure 2 schematically depicts the way in which this angle decreases with decreasing surface tension γ_L.

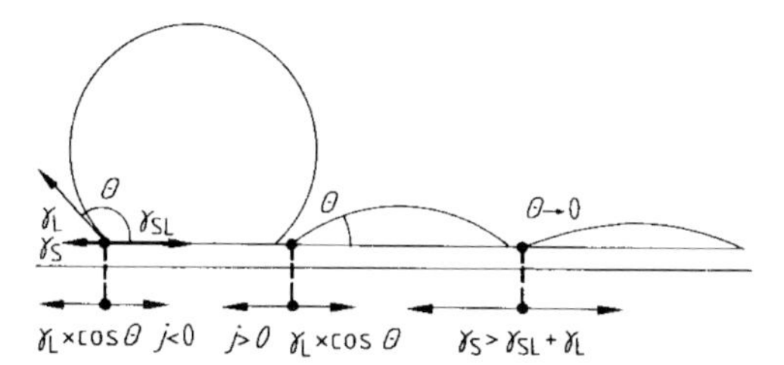

Figure 2. Schematic representation of wetting of a solid surface

To a first approximation, wetting can be described [11] by the Young equation [8]:

$$j = \gamma_S - \gamma_{SL} = \gamma_L \cos\theta \qquad (1)$$

j	= wetting tension (mN/m)
γ_S	= interfacial tension substrate/air (mN/m)
γ_{SL}	= interfacial tension substrate/liquid (mN/m)
γ_L	= interfacial tension liquid/air (mN/m)
θ	= contact angle of the wetting liquid

Total wetting of a solid is possible only if the liquid drop spreads spontaneously over the solid surface, e.g., when $\theta = 0$ and $\cos\theta = 1$. With a given solid surface possessing a low surface energy, for various liquids a linear relationship normally exists between $\cos\theta$ and surface tension. The limiting value for $\cos\theta = 1$ is a constant of the solid and is referred to as its critical surface tension, γ_c. This means that only those liquids with surface tension equal to or less than the critical surface tension of a given solid spread spontaneously, thus causing thorough wetting. Table 5 contains a collection of critical surface tension data for various synthetic materials.

Polyamide, for example, which has a γ_c of ca. 46 mN/m, is relatively easy to wet with standard commercial surfactants, whereas polytetrafluoroethylene, which has a γ_c of

Table 5. Critical surface tensions of several typical plastics [12]

Polymer	γ_c at 20 °C, mN/m
Polytetrafluoroethylene	18
Polytrifluoroethylene	22
Poly(vinyl fluoride)	28
Polyethylene	31
Polystyrene	33
Poly(vinyl alcohol)	37
Poly(vinyl chloride)	39
Poly(ethylene terephthalate)	43
Poly(hexamethylene adipamide)	46

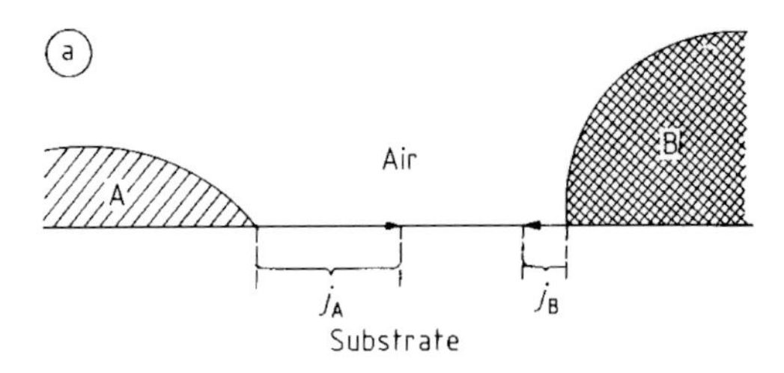

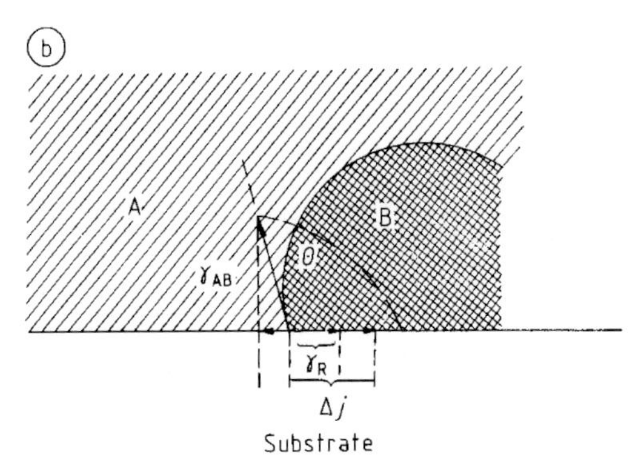

Figure 3. Two liquids on a solid surface
a) Separated; b) Overlapping
A) Wash liquor; B) Oily dirt

about 18 mN/m, is wettable only with special fluoro surfactants. This relationship simplifies the choice of a proper surfactant for a given wetting problem.

Strictly speaking, the limiting case described above, that of total wetting (Eq. 1), is only applicable if γ_{SL} is so reduced through adsorption that it approaches zero. In typical washing and cleansing operations, the situation is much more complicated, since the solid surfaces that are present tend to be irregularly covered with oily or greasy soils. Thus, the wash liquor must compete with soil in wetting the surface. Figure 3 illustrates the problem schematically.

If two drops of different liquids (e.g., wash liquor A and an oily residue B) are placed next to each other on a solid surface S, two different wetting tensions j_A and j_B act on the surface. If the two liquids come into direct contact and form a common interface, then the difference in their wetting tensions Δj, the so-called oil displacement tension,

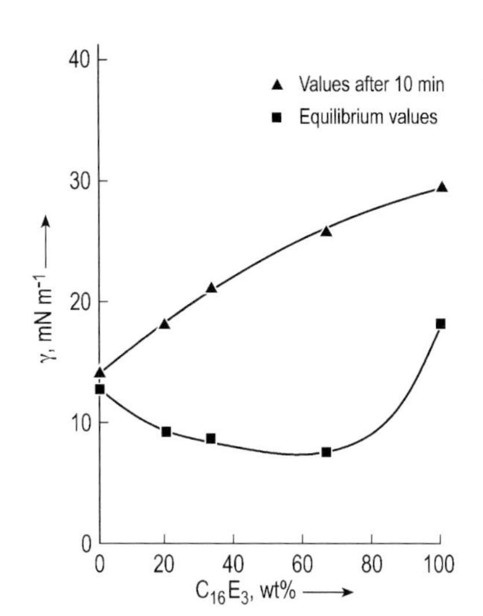

Figure 4. Interfacial tension after 10 min and after equilibrium vs. composition of the mixture (organic phase toluene, concentration 0.5 g/L, pH = 10, T = 298 K)

is effective along the line of contact. In addition, a portion of the interfacial tension γ_{AB} remains effective along the substrate surface, but with reversed sign and a magnitude of $\gamma_{AB} \cos \theta$, where θ is the contact angle in the oily phase B. The total force that results is the contact tension, γ_R, given by Equation (2):

$$\gamma_R = \Delta j + \gamma_{AB} \cos\theta \quad (2)$$

The contact tension acts in such a direction as to constrict the oil drop. As a result of adsorption of surfactants from phase A, the value of Δj is increased and that of γ_{AB} is decreased. The value of $\gamma_{AB} \cos \theta$ is negative for obtuse contact angles. From Equation (2), it can be seen that both quantities contribute to an increase in the contact tension and, thus, to better penetration of the oil drop. This complex process, in which a surface undergoes wetting first by oil and then by water, is known as "roll-up". For many cases of practical interest, roll-up is not a spontaneous phenomenon. More often, total constriction of an oil drop can be achieved only if mechanical energy is applied to the system [13]. The required energy is directly proportional to the interfacial tension γ_{AB} and decreases with increasing surfactant concentration.

All the relationships presented thus far confirm the premise that interfacial tension is the primary force resisting the removal of liquid soils, and this force must be minimized if the washing process is to be effective. One way to reduce the interfacial tension is to create appropriate mixed adsorption layers comprised of surfactants of differing constitution. For example, Figure 4 shows the interfacial tension of the system water–toluene as a function of composition for a surfactant mixture containing the anionic surfactant sodium *n*-dodecyl sulfate (SDS) and the nonionic surfactant hexadecyl triglycol ether ($C_{16}E_3$), whereby the total surfactant concentration is kept constant [14].

Even small additions of one surfactant to another may cause a significant reduction in the interfacial tension. In a broad concentration range mixtures of these surfactants

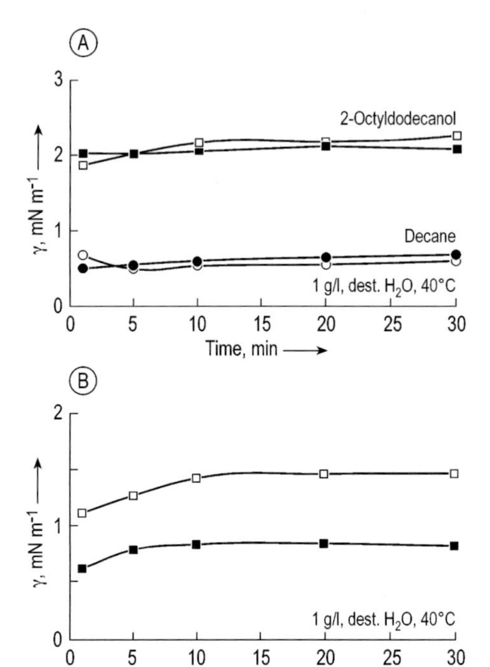

Figure 5. Dynamic interfacial tension of $C_{12/14}$ fatty alkyl sulfate (FAS) ■ and linear alkylbenzenesulfonate (LAS) □ as a function of time for decane and 2-octyldodecanol (A) and isopropyl myristate (IPM) (B)

show lower values of interfacial tension than the two surfactants alone. The presence of a small amount of nonionic surfactant in the surface layer reduces the mutual repulsion of negatively charged groups on the anionic surfactant, as a result of which adsorption increases.

Findings such as these are independent of the nature of the fiber and apply to all soils containing hydrophobic particles and oily material [15]. The oil specificity of surfactants is very important for the use in different applications [9]. Figure 5 shows the interfacial tension of (linear) alkylbenzenesulfonate (LAS) and of a fatty alcohol sulfate ($C_{12/14}$ FAS) against three different oils [16]. Both surfactants show nearly identical interfacial tension against decane and 2-octyldodecanol. In contrast the interfacial tension of FAS against isopropyl myristate is significantly lower than that of LAS. Figure 6 shows the interfacial tension of alkylpolyglycosides in dependence on the alkyl chain length. As a result, most commercial detergents contain mixtures of surfactants in carefully determined proportions. The interfacial tension in detergents is a measure of their product performance. This is shown by way of example for five different heavy-duty detergents in Figure 7 [9].

Emulsification. In addition to roll-up, the emulsification of oily liquids and greases can also play a role in the washing process, provided certain specific preconditions are met. In particular, the system must possess a sufficiently high interfacial activity so that the interfacial tension falls below $10^{-2} - 10^{-3}$ mN/m for soil to be removed by the

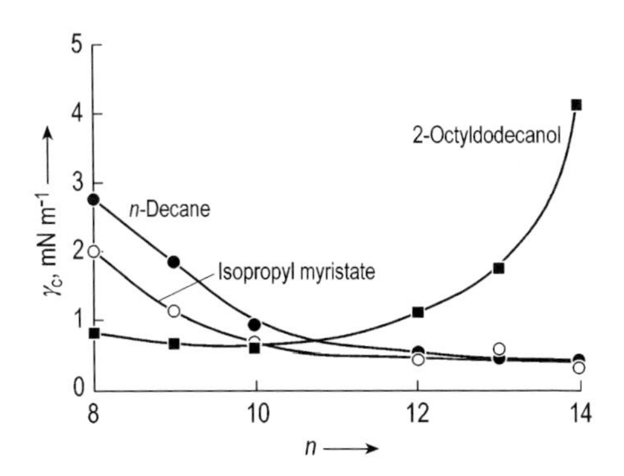

Figure 6. Plateau values of the interfacial tension γ_c in three different oil/water systems as a function of the alkyl chain length *n* of the alkylpolyglycoside surfactant at 60 °C

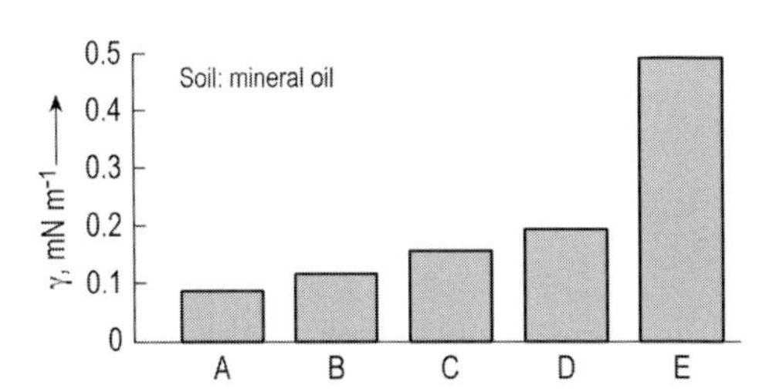

Figure 7. Interfacial tension of different detergents (A to E) against mineral oil

largely substrate-nonspecific emulsification process. However, these conditions are rarely observed in practice. As a result, emulsification is currently a major factor only in multiple wash cycle performance. In this case, the formation of largely stable emulsions inhibits the redeposition of previously removed liquid soils onto textile fibers. The emulsification of surfactant mixtures of SDS and $C_{16}E_3$ is demonstrated in Figure 8. In contrast to the interfacial tension (Fig. 4), mixtures of these surfactants show better emulsification than the individual substances in a broad concentration range.

Solubilization. Increasing surfactant concentration leads to a decrease in both surface tension and interfacial tension until the point is reached at which surfactant clusters begin to form. Above this concentration (the critical micelle concentration c_M) changes in surface and interfacial activity are only minimal. Oil-removal values also reach their upper limit at the critical micelle concentration of the surfactant employed [17]. This is illustrated in Figure 9 for the example of removing olive oil from wool.

From this behavior it can be concluded that effective washing is a result of the properties of individual surface-active ions, not of micelles. Just as with emulsification, however, micelles are able to solubilize water-insoluble materials and thereby prevent redeposition of previously removed soil in the later stages of the washing process. Values reported in the literature for critical micelle concentrations are based on pure solutions of surfactant. However, in an actual washing process, surfactants are invariably adsorbed onto a variety of surfaces, which means that the true surfactant con-

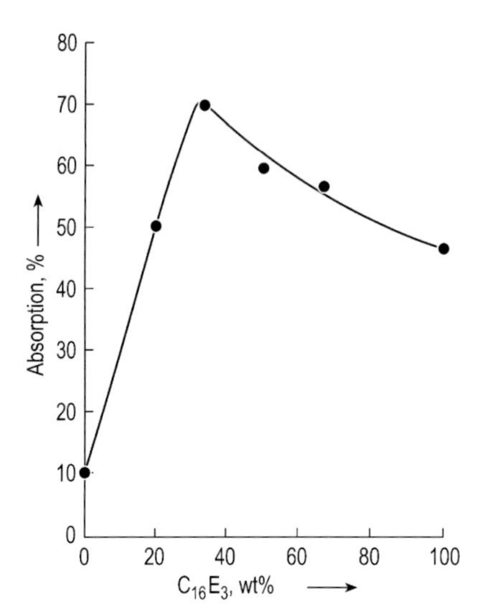

Figure 8. Emulsifying ability (% of absorption) of toluene versus composition of the mixture SDS – $C_{16}E_3$ (surfactant concentration 0.5 g/L, pH = , T = 298 K)

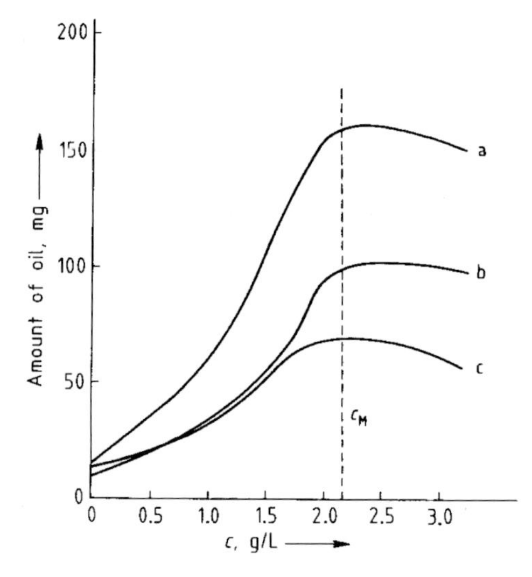

Figure 9. Removability of an olive oil – oleic acid mixture from wool, shown as a function of sodium *n*-dodecyl sulfate concentration for various degrees of oil coating (c_M=critical micelle concentration) [17] a) 4.6 % Oil coat; b) 3.3 % Oil coat; c) 2.2 % Oil coat

centration in the wash liquor is correspondingly diminished. The true surfactant concentration in solution is the only significant measure.

Phase Behavior. Penetration of individual detergent components (primarily surfactants) into an oily phase and the resulting formation of new anisotropic mixed phases can also cause a change in the water – oil interfacial tension. Development of liquid-crystalline mixed phases can be observed, for example, with olive oil – oleic acid – sodium *n*-dodecyl sulfate, and these lead to improved soil removal from textile substrates [17].

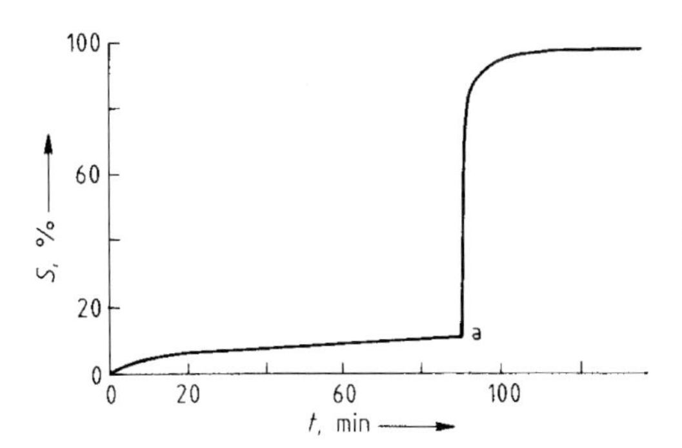

Figure 10. Washing effect S of potassium octanoate (2 wt%) on polyester fibers coated with decanol [19]
0.5 mol/L KCl added after 90 min
Prior to a: no formation of liquid-crystalline mixed phase; subsequent to a: formation of liquid-crystalline mixed phase

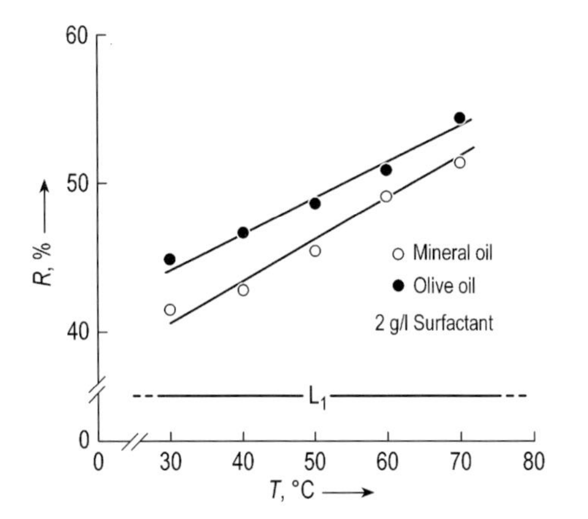

Figure 11. Phase behavior of $C_{12}E_9$ and detergency

The dramatic effect of liquid-crystalline mixed phase formation on soil removal is illustrated in Figure 10 for the model system decanol – potassium octanoate. This combination forms such phases in the presence of added electrolytes [19].

Decanol is removed completely from the fibers, provided the wash experiment is carried out under conditions leading to liquid-crystalline layers at the boundary surfaces. When this is not the case, only minimal separation of the oily residue from the polyester fibers is observed, which is a result of a slow rolling-up process.

The phase behavior of the surfactant solution has an important influence on the washing performance. Figure 11 shows the washing performance in units of reflectance versus temperature for $C_{12}E_9$ ($C_{12}E_9$ denotes a C_{12} alcohol ethoxylated with 9 mol of ethylene oxide). A micellar phase is formed over the whole temperature range, and the washing performance increases linearly. In Figure 12 the same diagram is shown for C_{12} E_4. In this system a liquid-crystalline phase appears above 45 °C. In the region of the liquid-crystalline phase, the increase in washing performance is much more pronounced than in the micellar region [18].

Specific Electrolyte Influence. In general, electrolytes are found to have only an indirect effect, and then only when anionic surfactant adsorption occurs at boundary

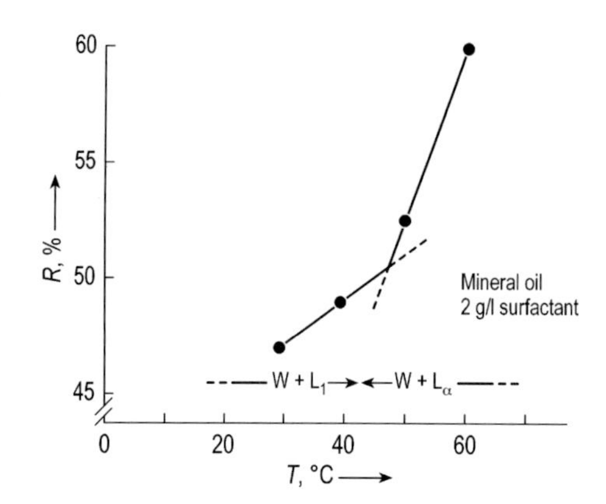

Figure 12. Phase behavior of $C_{10}E_4$ and detergency

Table 6. Interfacial tension with sebum in surfactant-free alkali and sodium triphosphate solutions at 40°C

Substance	Concentration, g/L	Interfacial tension, mN/m	pH	Water hardness, °d
NaOH	0.14	0.8	10	16
KOH	0.21	0.9	10	16
$Na_5P_3O_{10}$	2.0	0.09	9	16
$Na_5P_3O_{10}$	2.0	0.07	9	0

surfaces. An example which shows the reduction of the interfacial tension as a consequence of the addition of electrolytes is illustrated in Figure 13 for C_{16} α-ester sulfonate [20]. The addition of electrolytes causes a compression of the electrical double layer at all interfaces, which in turn causes enhanced surface adsorption of surfactants. However, if the system is devoid of surfactants, a vast difference exists between the properties of normal electrolytes, such as sodium chloride and sodium sulfate, and electrolytes prone to form complexes, including sodium triphosphate and sodium citrate.

In systems containing highly nonpolar oils, both sodium sulfate and sodium triphosphate show indifferent behavior, and they have no effect on the interfacial tension. When a small amount of oleic acid is added to the paraffin oil, however, the difference becomes clearly apparent. Although sodium sulfate has no effect on the interfacial tension, this tension is reduced dramatically by sodium triphosphate. Thus, electrolytes capable of forming complexes facilitate penetration of fatty acids through the interface, causing the liquid–liquid interfacial tension to be reduced. As a consequence, these materials are capable of activating a surfactant, an effect particularly significant for the removal of sebum, which is rich in fatty acids. Table 6 shows that the effect is not simply soap formation in an alkaline medium. The interfacial tension observed with sodium and potassium hydroxide is significantly greater than that with sodium triphosphate, even though the pH in the latter case is lower.

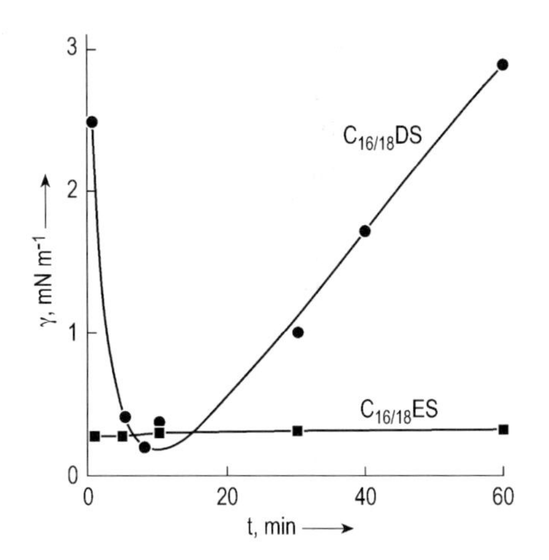

Figure 13. Dynamic interfacial tension γ between aqueous phase and mineral oil vs. time t for tallow α-ester sulfonate and tallow di-salt ($T = 60$ °C; 0.2 g/l surfactant; 4 g/l Na_2SO_4) [14]

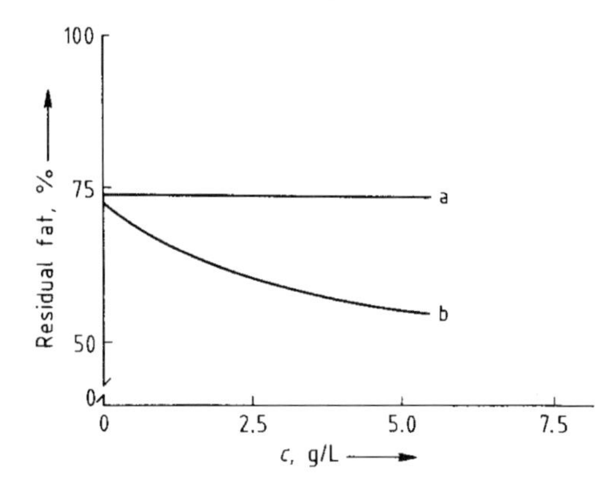

Figure 14. Removal of sebum from polyester/ cotton blend fabric studied as a function of electrolyte concentration
Sebum load: 12 g/m^2; apparatus: Launder-ometer; bath ratio: 1 : 30; wash time: 30 min; temperature: 40 °C
a) Na_2SO_4; b) $Na_5P_3O_{10}$

The influence of a change in the oil – water interfacial tension on soil removal for oily liquid soils is demonstrated by model wash experiments carried out with the system sebum – polyester/ cotton fabric (Fig. 14).

Removal of greasy soil by both water (the water value) and sodium sulfate solution is ca. 25 %. Sodium triphosphate increases the value to ca. 45 %. This clearly demonstrates the supplemental washing effect obtained when a complexing agent is introduced to reduce the interfacial tension.

2.3.2. Particulate Soil

Fundamental Principles of Adhesion and Displacement. A theoretical understanding of the interaction forces causing a solid particle to adhere to a more or less smooth surface is based on the Derjaguin – Landau – Verwey – Overbeek theory (DLVO) [21], [22]. Since this theory was developed to explain the phenomena of flocculation and

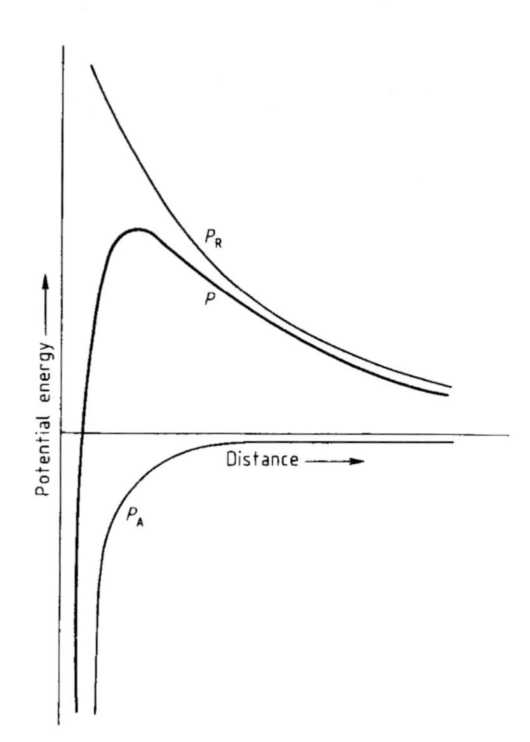

Figure 15. Calculated potential energy of attraction P_A and repulsion P_R as a function of the distance of a particle from a fabric, along with the resultant potential P; predictions based on the DLVO theory
DLVO computational parameter $z = 4$

coagulation, however, it can be applied to the washing process only in modified form [23]. A plot of potential energy as a function of the distance of a particle from a fabric shows that the potential energy passes through a maximum (Fig. 15).

The minimum in the potential energy curve corresponds to the closest possible approach, i.e., to the minimum distance that can be established between the particle and the fiber. The maximum is a measure of the potential barrier that must be overcome if the particle either is removed from the fiber or approaches the fiber from a distance. Adhering particles are more easily removed if the potential barrier is small. Conversely, a soil particle already in the wash liquor is less likely to establish renewed contact with the fabric if the potential barrier is large.

With respect to the washing process, if a particle is bound to a fabric, only a single common electrical double layer located at the overall external surface exists initially. None is present within the zone of contact. During the washing process, new diffuse double layers are created, which cause a reduction in the free energy of the system. The free energy of an electrical double layer is a function of distance and diminishes asymptotically to a limiting value that corresponds to a condition of no interaction between two double layers. Twice as much effort must be expended to bring a particle into contact with a substrate because of the presence on both surfaces of a double layer (curve $2F$ in Fig. 16). The separation of two adhering surfaces is characterized initially only by van der Waals – London attractive forces P_A and Born repulsion forces P_B, since at this point no electrical double layer exists [23]. In Figure 16, the equilibrium

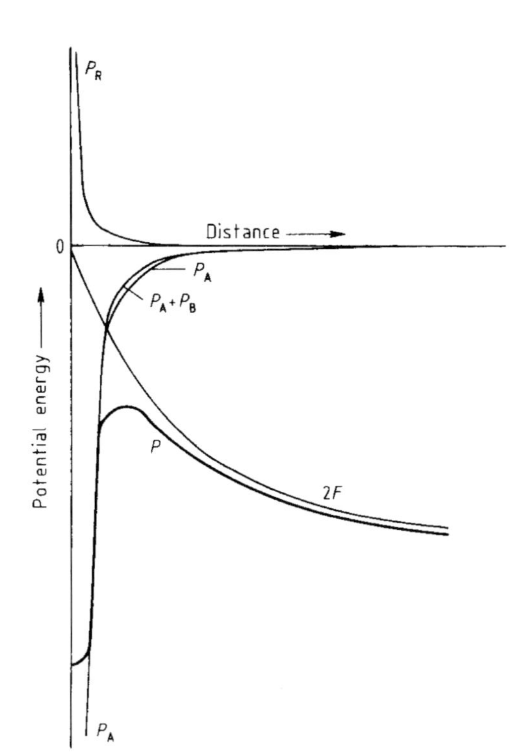

Figure 16. Potential energy diagram for the removal of an adhering particle [23]
P_B Born repulsion; P_A van der Waals attraction; F Free energy of the double layer; P Resulting potential curve
DLVO computational parameter $z = 4$

condition corresponding to the potential energy minimum has been taken as the zero point on the abscissa. With increasing distance between the particle and the contact surface, a diffuse double layer arises, which assists in the separation process by establishing an element of repulsion. Thus, the true potential curve P for the separation of an adhering particle in an electrolyte solution results from a combination of the van der Waals–Born potential and the free energy of formation of the electrical double layer. The important conclusion from Figures 15 and 16 is that an increase in the potential of the electrical double layer increases the energy barrier for particle deposition but decreases that for particle removal.

The negative influence exerted by calcium ions originating from water hardness can also be explained with the help of potential theory. According to the Schulze–Hardy rule, compression of an electrical double layer increases rapidly as the valence of a cation increases. Therefore, high concentrations of calcium might cause attractive forces to become the dominant factor, leading to significantly lower washing efficiency than would be achieved in distilled water.

Effect of Electrical Charge. The foregoing theoretical treatment offers an explanation for the behavior that is actually observed. Surface potentials cannot be measured directly. Instead, the ζ-potential or electrophoretic mobility of a particle is used as a measure of surface charge. As a rule, fibers and pigments in an aqueous medium

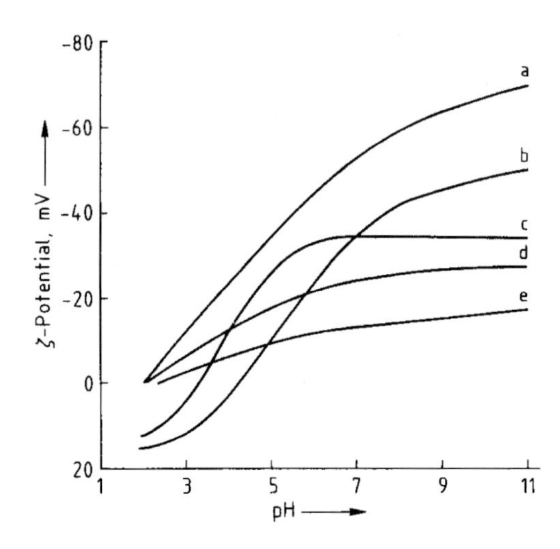

Figure 17. ζ-Potential of various fibers as a function of pH [24]
a) Wool; b) Nylon; c) Silk; d) Cotton; e) Viscose

acquire negative charges, whereby the extent of charge increases with increasing pH. This is illustrated in Figure 17, in which the ζ-potential of various fibers is taken as a measure of electrical charge and is plotted against pH [24].

Essentially similar results are obtained for all major particulate soil components. This is one of the reasons for enhancement of wash performance by mere introduction of alkali. However, repulsive forces between soil and fibers alone are insufficient to produce satisfactory washing even at high pH.

Apart from changing pH, another way to significantly alter fiber and pigment surface charges is to introduce a surfactant. The sign of the resulting charge depends on the nature of the hydrophilic group of the surfactant. Figure 18 shows the effect of several aqueous surfactant solutions on the surface potential of carbon black [25]. The surfactants chosen for the study all share the same hydrophobic component, but their hydrophilic groups vary. Electrophoretic mobility (EM) is taken as a measure of surface potential in this case.

Carbon black acquires a negative charge in water. The negative charge of pigments and fibers is further increased by adsorption of anionic surfactants. The corresponding increase in mutual repulsion is responsible for an increase in the washing effect. Dispersing power for pigments also increases for the same reason, whereas the redeposition tendency of removed soil is diminished.

In contrast to anionic surfactants, cationic surfactants reduce the magnitude of any negative surface charge. Consequently, electrostatic attraction between soil and fabric is increased. For this reason, cationic surfactants may cause a decrease in washing effect below that observed with pure water. Significant soil removal then occurs only at high surfactant concentrations, at which a complete charge reversal takes place on both fabric and soil. When the laundry is rinsed, however, a second charge reversal takes place on the fabric. At the point of electrical neutrality, redeposition of previously

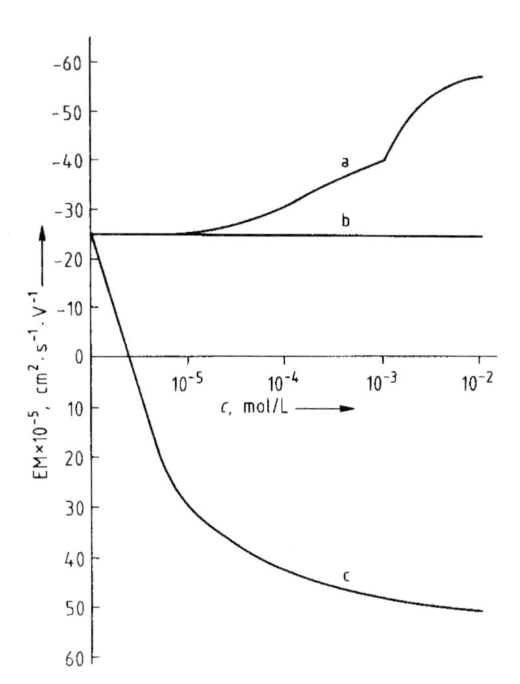

Figure 18. Electrophoretic mobility (EM) of carbon black in solutions containing various surfactants at 35 °C [25]
a) $C_{14}H_{29}OSO_3Na$ (anionic); b) $C_{14}H_{29}O(CH_2CH_2O)_9H$ (nonionic); c) $C_{14}H_{29}N(CH_3)_3Cl$ (cationic)

removed pigment soil is observed. This is why cationic surfactants are less suited for use in detergents than anionic surfactants.

Whereas surfactants are adsorbed nonspecifically at all hydrophobic surfaces, complexing agents can undergo specific attraction to surfaces that have distinct localized charges. The main process is chemisorption and is especially characteristic of metal oxides and certain fibers [4]. As shown in Figure 19, the adsorption of a complexing agent produces an effect similar to that of an anionic surfactant. The change in ζ-potential for hematite is taken as illustrative.

The specificity of adsorption of complexing agents with respect to metal oxides is so great that even displacement of anionic surfactants from surfaces with lower adsorption energies is permitted [4].

Complexing agents suppress the adsorption of anionic surfactants on metal oxides. However, adsorption is enhanced on materials such as carbon black or synthetic fibers. This effect is due to the electrolyte character of the complexing agent. The washing process generally involves removing mixed soils that consist of both hydrophilic and hydrophobic matter from fiber surfaces. For this reason, the different specificities of complexing agents and surfactants give complementary functions to these two types of material.

Adsorption Layer. In contrast to anionic and cationic surfactants, nonionic surfactants have no or little effect on surface charge [26]. Therefore, their mode of action cannot be ascribed to changes in electrical charge on pigments and fibers. Instead, the effect is related exclusively to properties of the adsorption layer. Compared to anionic

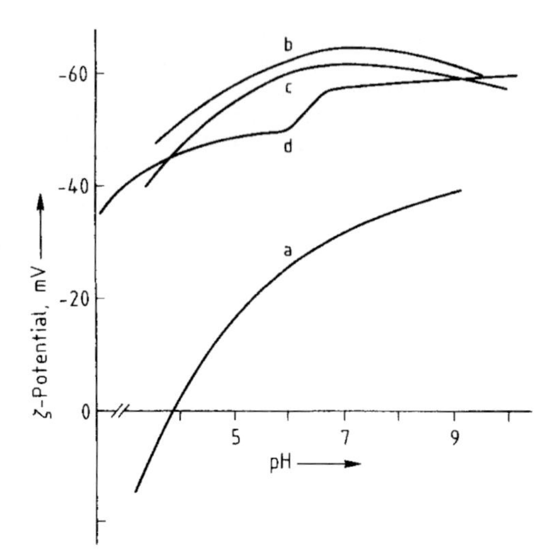

Figure 19. ζ-Potential of hematite as a function of pH at 25 °C in the presence of sodium chloride (a), sodium triphosphate (b), benzenehexacarboxylic acid (c), and 1-hydroxyethane-1,1-diphosphonic acid (d) [4]
Anion concentration: 2.5×10^{-3} mol/L

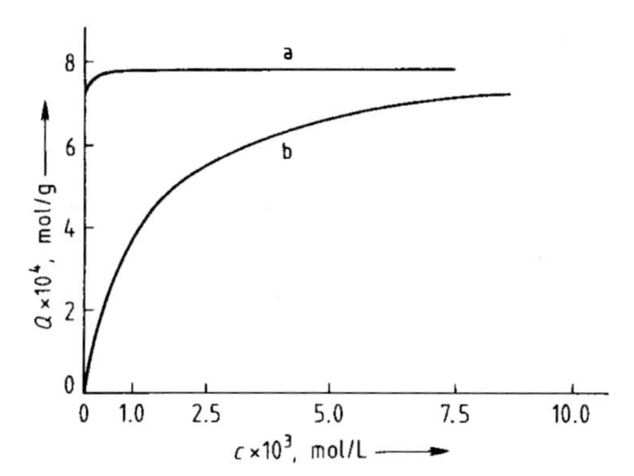

Figure 20. Adsorption isotherms for surfactants on activated carbon black [25]
25 °C, surface 1150 m^2/g (from BET measurements)
a) $C_{12}H_{25}O\ (CH_2CH_2O)_{12}H$; b) $C_{12}H_{25}OSO_3Na$

surfactants, the adsorption of nonionic surfactants is quite strong at hydrophobic, weakly polar interfaces. This is illustrated in Figure 20 by the relative adsorption of dodecyl dodecaglycol ether and *n*-dodecyl sulfate on activated carbon black [25].

In the case of *n*-dodecyl sulfate, both the hydrophilic surfactant groups and the surface have charges of the same sign. Consequently, a higher potential barrier must be overcome for its adsorption compared to the electrically neutral nonionic surfactant. With nonionic surfactants, both the adsorption equilibrium and the maximum surface coverage are displaced to relatively low concentration. The adsorption behavior of the nonionic surfactants is dependent on the structure of the molecule [28]. This is shown in Figure 21 for an alkylmonoglycoside and a fatty alcohol ethoxylate. The adsorption of the fatty alcohol ethoxylate is strongly temperature dependent in the higher concentration range and increases with increasing temperature, unlike the adsorption of the alkyl glycoside. The decisive paramenter for the temperature dependence of the adsorption of the fatty alcohol ethoxylate is apparently the cloud point (T_c = 20 °C).

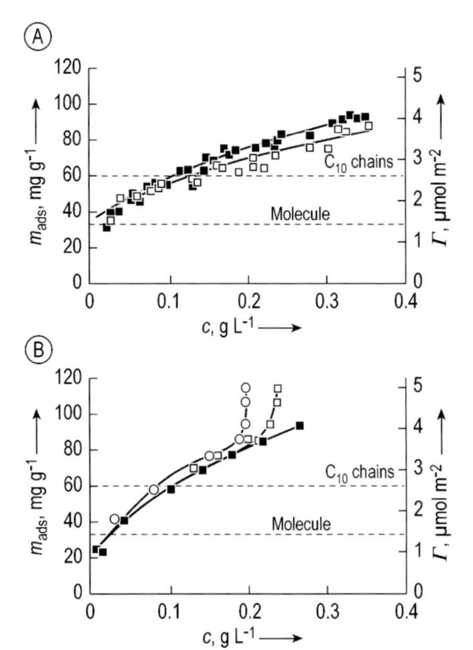

Figure 21. Adsorbed amounts m_{ads} and surface concentration Γ on graphitized carbon black as a function of the surfactant concentration cA) Alkyl monoglycoside ($C_{10}G_1$) at 22 °C (■) and 44 °C (□); B) Fatty alcohol ethoxylate ($C_{10}E_4$) at 19 °C (■), 30 °C (□), and 45 °C (○)

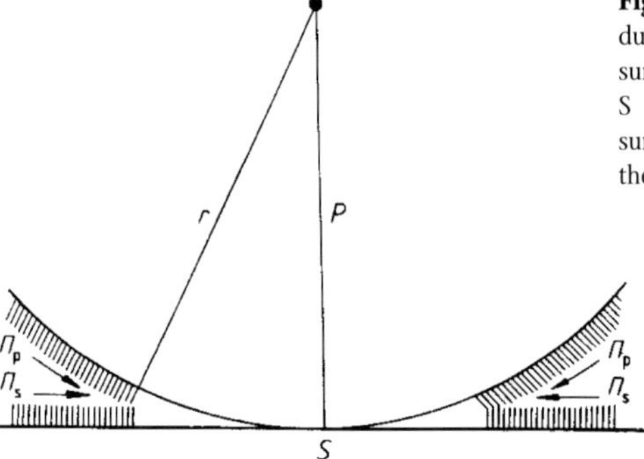

Figure 22. Schematic representation of adsorption-induced separation of a spherical particle from a hard surface
S Surface; P Particle; Π_S Splitting pressure of the surfactant layer on the surface; Π_P Splitting pressure of the surfactant layer on the particle

Figure 22 is a schematic representation of the adsorption layers on substrate and soil particles. As can be seen from the diagram, both surfactant layers advance to the point of particle – surface contact. One consequence is the development of a disjoining pressure, which leads to separation of the soil particle from the surface. This effect is obviously present with anionic surfactants as well. However, this pressure is the decisive factor with nonionic surfactants, due to the absence of any repulsive components of electrostatic origin.

Thus, hydration of hydrophilic groups is extremely important with nonionic surfactants. Adsorbed surfactant molecules are oriented such that their hydrophilic regions are directed toward the aqueous phase. Both pigment and substrate are surrounded by hydration spheres. The redeposition tendency of a soil particle is thereby reduced,

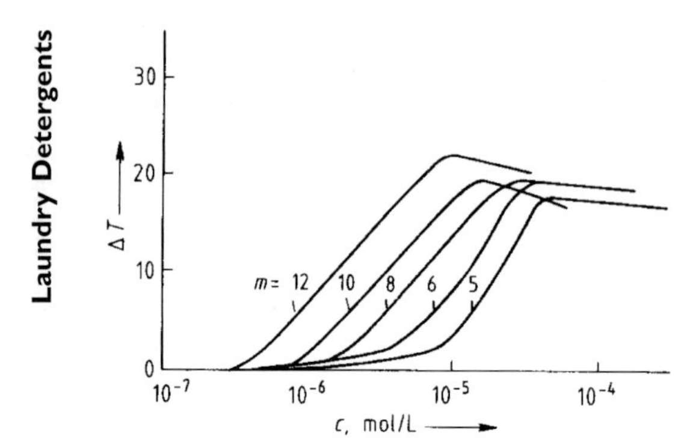

Figure 23. Reduction in the turbidity ΔT of a paraffin soil by addition of dodecyl polyglycol ethers ($C_{12}H_{25}O\ (CH_2CH_2O)_mH$) in the presence of 0.5 mol/L NaCl [29]

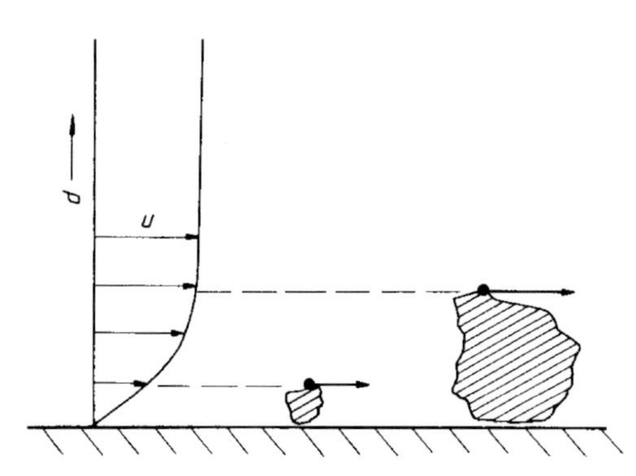

Figure 24. Schematic drawing of the effect of liquid flow on adhering particles of different size
u Flow velocity; *d* Distance from the solid surface

because the voluminous hydration sphere minimizes the effectiveness of short-range van der Waals attractive forces. This can be demonstrated by the coagulation of hydrophobic soil in the presence of nonionic surfactants containing varying numbers of ethoxy groups in their hydrophilic regions [29].

Figure 23 illustrates the stabilizing effects of dodecyl polyglycol ethers on a paraffin oil soil, where reduction in turbidity, ΔT, of the coagulated soil is plotted against the concentrations of ethers containing varying numbers of ethoxy groups. Coagulation is induced in this case by the presence of large amounts of sodium chloride, which causes compression of the electrical double layer. The decrease in turbidity becomes greater with increasing number of ethoxy groups in the molecule, and the curves are displaced in the direction of lower concentration.

Hydrodynamics. In practice, the physicochemical principles discussed thus far increase when taking advantage of hydrodynamic effects. Such hydrodynamic effects are generally strongly dependent on particle size [30]; their significance in the removal of particulate soil from fibers increases as particle size increases. Even with vigorous mechanical action, a laminar film of finite thickness in which no flow takes place is present on every surface. The flow velocity gradient increases with increasing distance from the surface, as shown schematically in Figure 24.

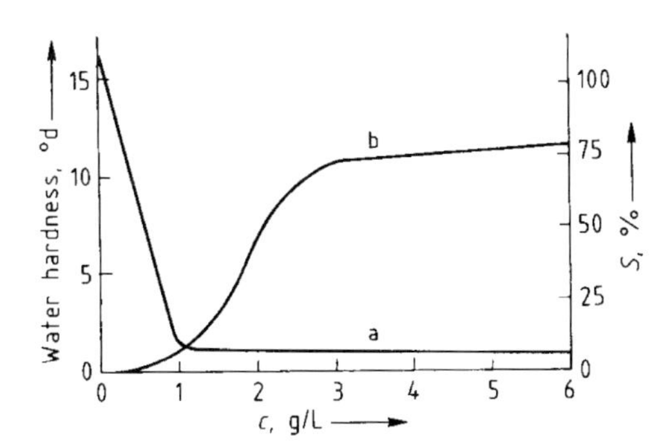

Figure 25. Water softening (a) and soil removal S from cotton soiled with a dust/sebum mixture (b) resulting from a cross-linked water-insoluble polyacrylate at 90 °C [4]

Thus, an increase in mechanical force has a significant soil-removing effect on larger particles. A high flow velocity gradient in the immediate vicinity of a surface is excluded for hydrodynamic reasons. Therefore, washing machines employ abrupt changes in direction of operation to achieve adequate turbulence near substrate surfaces. Even so, particles smaller than 0.1 μm cannot be displaced by mechanical means alone.

2.3.3. Calcium-Containing Soil

The principles discussed in Section 2.3.2 also apply without exception to calcium-containing particulate soil. However, other important mechanisms, which deserve discussion, also exist. Salts of multivalent cations are almost always present in soils and on textile fiber surfaces. Examples are calcium carbonate, calcium phosphate, and calcium stearate. Cationic bridges, which are responsible for binding soil components chemically to fibers, also frequently form. This type of linkage can be due, for example, to the carboxyl groups commonly found in cotton as a result of oxidation, but such a linkage may also arise from the presence of reactive centers associated with metal oxides or from soaps derived from sebum. Problems stem principally from poorly soluble calcium salts, whose solubility is further diminished as the water hardness of the wash liquor increases. On the other hand, their solubility in distilled water is higher because of displacement of the solubility equilibrium. When calcium salts are dissolved from a multilayered soil deposit, cavities remain in the structure. These cavities loosen the deposit and facilitate its removal from the surface. Thus, one task of detergent components is to create the highest possible calcium ion concentration gradient between soil and aqueous phase during the washing process.

Figure 25 illustrates this principle, taking the wash effectiveness of a water-insoluble cross-linked polyacrylate as an example [4]. Apparently, virtually no wash effectiveness is achieved with an ion exchanger at a concentration sufficient to eliminate most of the water hardness; an effect is observed only at higher concentrations of ion exchanger. This phenomenon can be greatly accelerated by additionally introducing an appropriate amount of water-soluble complexing agent.

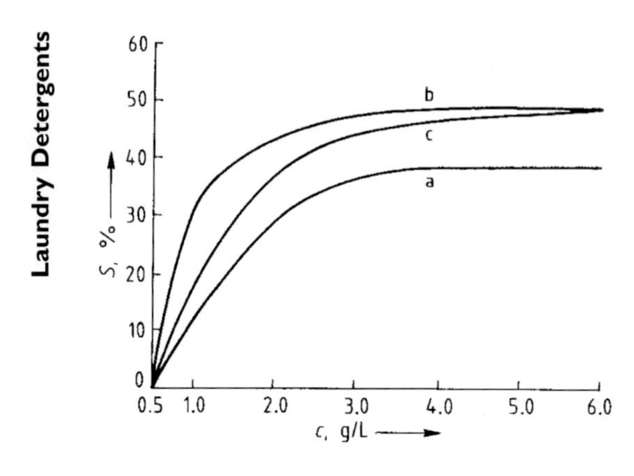

Figure 26. Comparison of soil removal S of zeolite A (a), sodium triphosphate (b), and a mixture of the two builders in the ratio zeolite A: sodium triphosphate = 9 : 1 (c); results obtained for non-resin-finished cotton tested in a Launder-ometer
Wash time: 30 min with heating; temperature: 90 °C; water hardness: 16 °d (285 ppm)

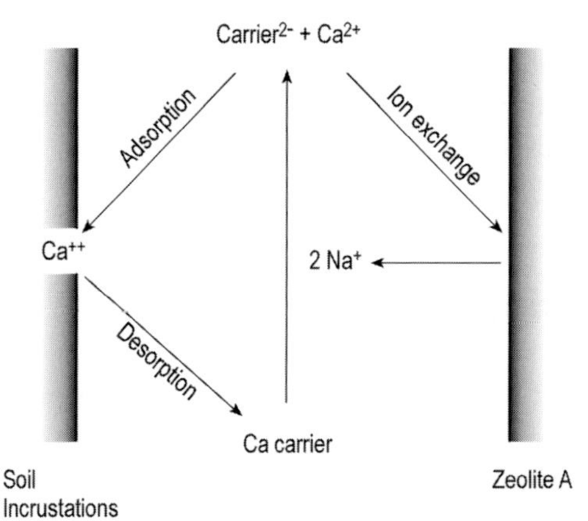

Figure 27. Mechanistic scheme for the carrier effect [24]

Before slightly soluble cations can be dissolved from soil and fibers, adsorption of the complexing agent takes place on the surface, particularly in those areas that contain multivalent cations. In the course of subsequent desorption of water-soluble multivalent cation complexes, many of the soil – fiber bonds are broken, leading to a marked enhancement of the washing effect. Removal of cations from soil and fibers by adsorption – desorption processes and displacement of solubility equilibria are the most important phenomena that accompany the use of complexing agents and ion exchangers in the washing process.

The mechanisms of action for complexing agents and water-insoluble ion exchangers are different. The two complement each other in their respective roles. Figure 26 shows the way in which a small amount of a water-soluble complexing agent can increase the washing effectiveness of the water-insoluble ion exchanger zeolite A. The effect results from an increase in the rate of dissolution of divalent ions from soil and fibers. The mode of action is depicted schematically in Figure 27. The water-soluble complexing agent serves as a carrier that transports calcium from the precipitate into the water-

Table 7. Wash experiments in a calcium-free system [31]

Soil	Wash medium*	Remission, %
80.2 % Osmosed kaolin, 16.5 % carbon black, 3.3 % black iron oxide	H_2O	65.5
	H_2O + 2 g zeolite 4 A/L	66.0
89.7 % Osmosed kaolin, 5.9 % carbon black, 2.9 % black iron oxide, 1.5 % yellow iron oxide	H_2O	59.5
	H_2O + 2 g zeolite 4 A/L	59.0

* Water is distilled.

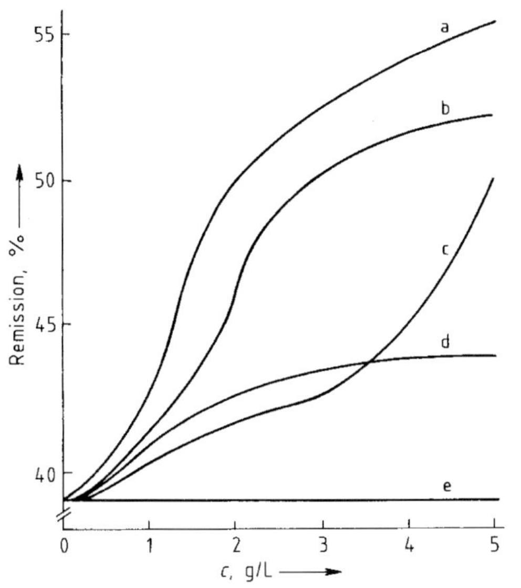

Figure 28. Soil removal of various zeolites [32]
a) Zeolite A; b) Faujasite; c) Desmine; d) Sodalite; e) Analcime
Apparatus: Launder-ometer; water hardness: 16 °d; temperature: 95 °C; wash time: 30 min with heating; fabric: non-resin-finished cotton

insoluble ion exchanger. The process is based on successive adsorption, desorption, and dissociation, and it accelerates the delivery of free calcium ions into solution.

The effectiveness of complexing agents and ion exchangers is related to the presence of calcium in the system, as is evident from Table 7. Two different soils and the cotton yarn to be studied were decalcified prior to applying artificial soil. No washing effect due to zeolite A could be observed within the method's limits of accuracy. This result can be taken as an indirect proof of the importance of dissolution of calcium from soil and fibers during the washing process.

The concentration of complexing agent and the temperature are generally the decisive factors in removing multivalent metal ions by a water-soluble complexing agent; the binding ability diminishes with increasing temperature.

Essentially the same relationships also apply when using a water-insoluble ion exchanger, although the temperature effect is usually reversed [32], [33].

Figure 28 shows the wash effectiveness of various zeolites in water with a hardness of 16 °d (285 ppm $CaCO_3$) as a function of concentration. It is apparent that the best results are achieved with zeolite A and the poorest with analcime.

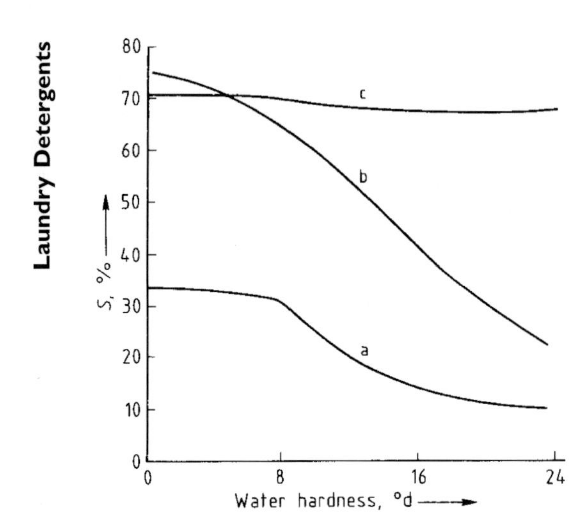

Figure 29. Soil removal *S* from wool as a function of water hardness at 30 °C [4]
a) Sodium alkylbenzenesulfonate (0.5 g/L);
b) Sodium alkylbenzenesulfonate (0.5 g/L) together with sodium sulfate (1.5 g/L); c) Sodium alkylbenzenesulfonate (0.5 g/L) together with sodium triphosphate (1.5 g/L)

In addition to the previously described reasons for removing divalent alkaline earth ions, their interaction with other detergent components must also be considered. For example, soaps form poorly soluble salts with calcium, as do many synthetic surfactants, and these can be deposited on fibers. This phenomenon is extremely common with detergents in which soap is the key surfactant, when no strongly complexing agents such as sodium triphosphate are present. Precipitation of relatively insoluble surfactant calcium salts has the additional disadvantage that it causes a severe reduction in the active surfactant concentration and, thus, to generally deteriorated conditions for soil removal.

Figure 29 illustrates how surfactants, salts, and complexing agents complement each other in the removal of soil. The negative influence of calcium is eliminated by sodium triphosphate or other complexing agents. The magnitude of the indirect counterion effect of sodium ions is made clear when sodium sulfate is added to alkylbenzenesulfonate (curve b).

Polycarboxylates are widely used in detergents as cobuilders [34], [77] – [79]. They are able to retard precipitation of sparingly soluble calcium salts such as calcium carbonate and calcium phosphate (*threshold effect*) when used in small concentrations. As anionic polyelectrolytes they bind cations (counterion condensation), whereby multivalent cations are strongly preferred. Polycarboxylates can also disperse solids in aqueous solution. Both dispersion and threshold effect result from adsorption of the polymer on the surface of soil and $CaCO_3$ particles, respectively. Stabilization of sparingly soluble salts such as $CaCO_3$ in a colloidal state by polycarboxylates occurs at substoichiometric concentrations of the cobuilder in the washing liquor, which is an advantage compared to ion excahnge or complexation. Thus, small amounts of threshold-active compounds can be used as cobuilders even in laundry detergents with a high content of sodium carbonate. The effect, however, is strongly dependent on the washing conditions, such as temperature, soda ash, and cobuilder concentration. Figure 30 illustrates the range

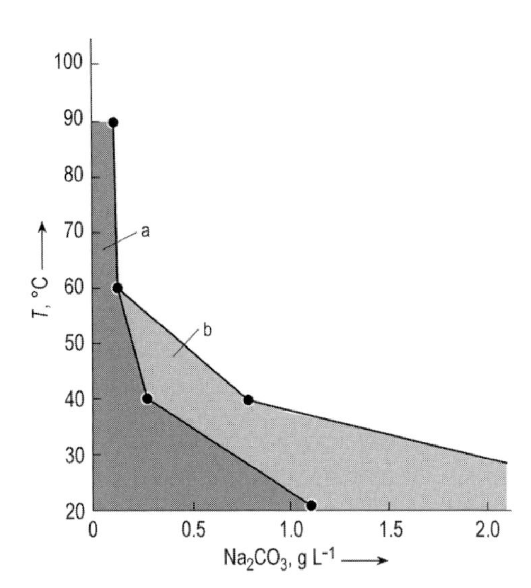

Figure 30. Precipitation inhibition of $CaCO_3$ by AC (threshold effect) as a function of temperature and soda ash concentration (3.04×10^{-3} mol/L Ca^{2+})a) 105 mg/L AC; b) 210 mg/L AC

of effectiveness of a polycarboxylate (AC) in a carbonate-containing system for typical European conditions of water hardness (3.04×10^{-3} mol/L Ca^{2+}) [35]. The appearance of $CaCO_3$ particles larger than ca. 0.2 μm within 30 min was taken as an indicator of the threshold effect. The results show that polycarboxylates are no longer threshold-active above 40 °C. This holds even more for higher carbonate concentrations, i.e., with detergents containing soda ash as the sole builder.

For zeolite A and soda ash containing products the participation of zeolite A in the elimination of calcium ions during the washing process must be taken into account. In contrast to the results obtained in the absence of zeolite A, the precipitation in the presence of zeolite A and the polycarboxylate AC is negligibly low at temperatures above 40 °C. These results can be explained by the binding of calcium ions by zeolite A and by AC in its water-soluble form. This is possible because the calcium ion concentration of the water is reduced by zeolite A. Thus, Ca^{2+} is no longer in excess with respect to AC, and the formation of an insoluble calcium salt of polycarboxylates, which decreases the threshold activity, is no longer possible. According to this mechanism, polymers with high carboxylate content and relatively high molecular weight should be used in combination with zeolite A.

2.3.4. Influence of Textile Fiber Type

The ability of a detergent to remove soil depends not only on the foregoing factors, but also on the type of textile substrate. Textile fibers that have a high calcium content at their surface (e.g., cotton) behave very differently from synthetic fibers with a low calcium content. The type of fiber has a dramatic influence on the degree of hydrophobicity/hydrophilicity, the wettability, and the extent of soil removal. Figure 31

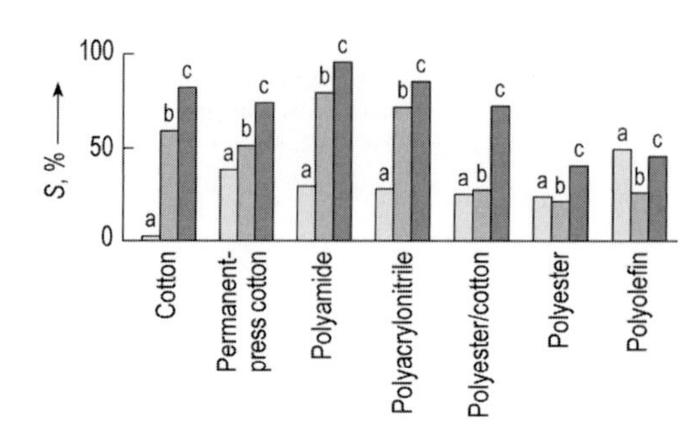

Figure 31. Influence of textile fibers on soil removal S [4]
Detergents: a) 1 g/L Alkylbenzenesulfonate + 2 g/L sodium sulfate; b) 2 g/L sodium triphosphate; c) 1 g/L alkylbenzenesulfonate + 2 g/L sodium triphosphate

clearly demonstrates in terms of soil removal the differing effects that complexing agents and surfactants have on a series of fibers. The anionic surfactant selected in this case for study (alkylbenzenesulfonate) has virtually no effect on cotton, whereas even by itself sodium triphosphate, as an example of a complexing agent, brings about substantially improved soil removal. Permanent press cotton also shows a greater effect for sodium triphosphate than for surfactants alone. The effect of sodium triphosphate is particularly pronounced with the relatively hydrophilic synthetic fibers polyamide and polyacrylonitrile. This behavior changes when more hydrophobic textile fibers are used. An effect is still apparent with polyester/cotton and polyester, but it is not greater than that of surfactants alone. At the same time, overall wash performance decreases. In the case of the very hydrophobic polyolefin fibers, the wash effectiveness of the surfactant is substantially greater than that of the complexing agent. These examples show the synergistic way in which surfactants and complexing agents or ion exchangers complement each other, not only in the case of mixed soils, but also with regard to removing soil from different fibers.

2.4. Subsequent Processes

After soil has been removed, it must be stabilized in the wash liquor, and redeposition of the removed soil must be prevented. This property of a wash liquor is known as its *soil antiredeposition capability*. Several mechanisms play a role in ensuring good antiredeposition characteristics.

2.4.1. Dispersion and Solubilization Processes

The most important factor in this context is the dispersion process. Nonspecific adsorption of surfactants and specific adsorption of complexing agents cause liquid soils to be emulsified and solid soils to be suspended or dispersed. Poorly soluble substances are solubilized by surfactant micelles as molecular dispersions. These mechanisms have been discussed in detail in Sections 2.3.1 and 2.3.2.

2.4.2. Adsorption

Supplementary adsorption effects play an important role when zeolites are used in detergents. Such effects are not observed with detergents containing only the complexing agent sodium triphosphate.

Zeolites can significantly enhance washing effectiveness by serving as competitive substrates for the adsorption of molecularly dispersed soluble substances and colloidal particles. Their presence is particularly advantageous under extreme conditions involving large quantities of particulate matter. Adsorption and heterocoagulation of the soil by zeolites substantially reduce redeposition of soil on the fabric, leading to a significant increase in whiteness maintenance.

2.4.3. Soil Antiredeposition and Soil Repellent Effects

Soils and detergents also contain natural and synthetic macromolecules in addition to low molecular mass compounds. Proteins from blood and protein-containing food, for example, can be adsorbed onto textile fibers and must by some means be desorbed during the wash process.

Detergents often contain polymeric soil antiredeposition agents, whose role is to be adsorbed onto the substrate, thereby creating a protective layer that sterically inhibits redeposition of previously removed particulate soil. The desorption of proteins and the adsorption of antiredeposition agent in a single process represent competing phenomena, so that careful selection of the proper antiredeposition agent is required.

In a surfactant-free environment, the adsorption of most macromolecules onto a solid surface is effectively an irreversible process. The reason for this irreversibility is the vast number of points of contact that exist between macromolecule and substrate. From a statistical standpoint, the number of bonds between the two is so large that adherence is ensured regardless of the strength of the bonds. Most of the observed binding results from weak hydrophobic interactions.

In a multicomponent system containing both surfactants and macromolecules, competitive adsorption is possible at the substrate surface. This permits the surfactant to successively destroy the individual points of contact binding the macromolecule, thereby displacing it from the interface. This mechanism is most often observed for nonionic surfactants. Anionic surfactants are often capable of forming polymer–surfactant complexes, in the course of which the conformation of the macromolecule is altered in such a way as to reduce the extent of its attraction to the interface.

Figure 32 depicts the adsorption behavior of gelatin on glass both in the presence and in the absence of sodium *n*-dodecyl sulfate. The two adsorption isotherms are very different; in the presence of a surfactant, adsorption of the macromolecule is virtually eliminated. Even preadsorbed gelatin is desorbed from the surface by subsequent

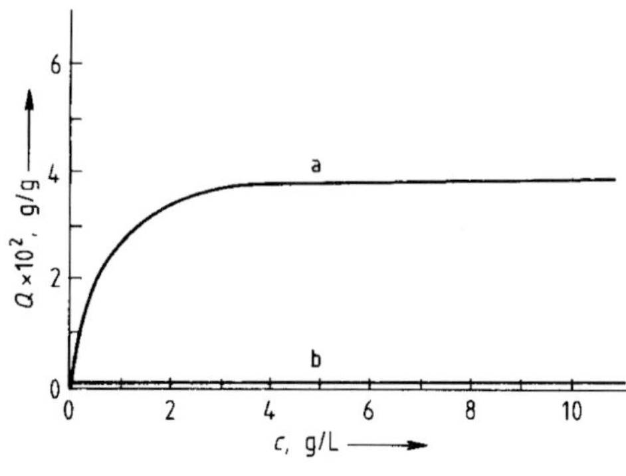

Figure 32. Adsorption of gelatin on powdered glass (temperature 25 °C)
a) Without sodium dodecyl sulfate; b) With sodium dodecyl sulfate (gelatin – surfactant mass ratio 1 : 1.44)

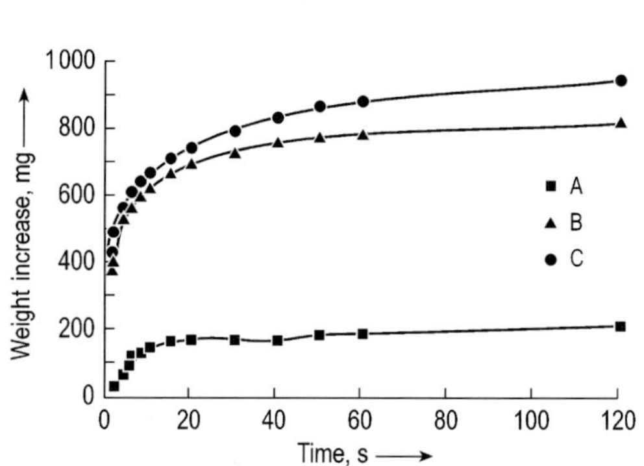

Figure 33. Kinetics of wetting of polyester fibers by three different detergents. The increase in weight of the fibers during wetting by pure water is plotted as a function of the wetting time [9]

surfactant addition. This competitive adsorption phenomenon involving surfactants and macromolecules is of great importance in soil removal, as is the formation of macromolecule – surfactant complexes. Both significantly impair the desired adsorption of polymeric antiredeposition agents.

The adsorption of antiredeposition agents is usually a selective process dependent on the chemical constitutions of both fiber and polymer. For example, the soil antiredeposition effect of carboxymethyl cellulose is rather limited on hydrophilic fibers such as cotton. Cellulose ethers, such as methylhydroxypropyl cellulose, or polymers from terephthalic acid and polyethylene glycol (soil repellent) are effective especially with more hydrophobic fibers such as polyester. Therefore, combinations of several antiredeposition agents often must be used to ensure satisfactory results with mixed laundry. In this case the absolute amount of adsorbed substance is not the determining factor, but rather the extent to which adsorption confers hydrophilic characteristics, i.e., the change in surface characteristics relative to untreated fiber surfaces. This can be characterized either by measuring the kinetics of wetting (Fig. 33) or by observing the resulting differences in the wetting tension with respect to pure water (cf. Eqs. 1 and 2). Figure 33 reveals clear differences in the time-dependent increase in weight when prewashed polyester is immersed in water [9]. The polyester fibers had previously been washed with different detergent formulations. Table 8 shows that carboxymethyl cellulose, a frequently used antiredeposition agent for cotton, has no effect on polyester.

Table 8. Comparison of changes in polyester wetting tensions with changes in remission for a heavy-duty detergent (I) and a low-temperature heavy-duty detergent (II) for polyester

Detergent I,	Detergent II,	Antiredeposition agent (in % detergent charge)			Change in wetting tension*		Change in remission**	
		Sodium carboxy-methyl cellulose	Methylhydroxy-propyl cellulose	Hydroxyethyl cellulose	I Δj_v,	II Δj_v,	I ΔR,	II ΔR,
g/L	g/L				mN/m	mN/m	%	%
7.4	4.5	0	0	0	0	–1		
0	0	0.5	0	0	1	1		
7.4	4.5	0.5	0	0	2	0	3	–1
0	0	0	0.5	0	22	22		
7.4	4.5	0	0.5	0	14	14	17	16
0	0	0	0	0.5	22	21		
7.4	4.5	0	0	0.5	3	2	5	0

* Δj_v = calculated for advancing contact angle conditions.
** ΔR = measured after three washes.

By contrast, methylhydroxypropyl cellulose and soil repellents cause the polyester surface to become considerably more hydrophilic. It is noteworthy that the effects are retained with both detergents, albeit to a somewhat reduced extent. Only with these formulations is a significant increase in the soil antiredeposition effect observed, as evidenced by the changes in percentage remission. Hydroxyethyl cellulose can be regarded as a representative of numerous polymers which, though they are readily adsorbed from aqueous solutions and are capable of showing considerable antiredeposition activity in pure water, nonetheless lose most of their effectiveness in a detergent solution as a result of competitive adsorption and displacement by surfactants.

2.5. Concluding Remarks

To simplify a physicochemcial approach of the washing process, it has been necessary to treat separately each of the several phenomena involved and to isolate them from one another. In any real washing process, the various mechanisms are all at work more or less simultaneously. Thus, these mechanisms affect one another in a mutually supportive and additive way. Investigations on the physical chemistry of washing have made possible a rather thorough understanding of the process. Despite the complexity of the washing phenomenon and the continued presence of certain unanswered questions, increasing knowledge of physicochemical correlations has exerted a major influence on product development.

3. Detergent Ingredients

Detergents for household and institutional use are complex formulations containing up to more than 25 different ingredients. These can be categorized into the following major groups:

Surfactants
Builders
Bleaching agents
Auxiliary agents (additives)

Each individual component of a detergent has its own very specific functions in the washing process. To some extent they have synergistic effects on one another. In addition to the above ingredients, certain additives are made necessary for production reasons, whereas other materials may be added to improve product appearance.

3.1. Surfactants

Surfactants constitute the most important group of detergent components, and they are present in all types of detergents. Generally, surfactants are water-soluble surface-active agents comprised of a hydrophobic portion (usually a long alkyl chain) attached to hydrophilic or solubility-enhancing functional groups.

A surfactant can be grouped in one of four classes, depending on what charge is present in the chain-carrying portion of the molecule after dissociation in aqueous solution:

Anionic surfactants
Nonionic surfactants
Cationic surfactants
Amphoteric surfactants

Table 9 [36] provides an overview of these classes.

In general, both adsorption and detergency performance increase with increasing chain length. For example, ionic surfactants bearing *n*-alkyl groups show a linear relationship between the number of carbon atoms in the surfactant molecule and the logarithm of the amount of surfactant adsorbed on activated carbon or kaolin clay.

The structure of the hydrophobic residue also has a significant effect on surfactant properties. Surfactants with little branching in their alkyl chains generally show a good cleaning effect but relatively poor wetting characteristics, whereas more highly branched surfactants are good wetting agents but have unsatisfactory detergency performance. For compounds containing an equal number of carbon atoms in their hydrophobic chains, wetting power increases markedly as the hydrophilic groups move to the center of the chain or as branching increases. Simultaneously a decrease in adsorption and detergency performance occurs (Figs. 34 and 35).

The changes with respect to adsorption, wetting, and detergency performance that result from varying the degree of branching are far more significant for ionic surfactants than for nonionic surfactants. In the case of anionic surfactants, losses in detergency performance caused by increased branching can be compensated for to some extent, provided the overall number of carbon atoms is increased appropriately.

Household washing of textiles normally poses few situations requiring extraordinary wetting power. If problems arise, they can usually be overcome by increasing the wash time or the amount of detergent used. Most important is the effectiveness of the rolling-up process. Optical microscopy has revealed that oily and greasy soil tends to reside in more or less evenly distributed layers on the surface of fibers. These layers are gradually constricted by the action of a surfactant and its associated spreading pressure, with the layers ultimately being reduced to drops resting loosely on the fibers, which are then easily detached. Thus, oily residue as a fiber wetting agent is replaced by aqueous wash solution.

Table 9. Surfactants of various ionic nature [36]

Surfactant	Formula	Electrolytic dissociation	Ionic nature
Alkyl poly(ethylene glycol) ethers	$RO{-}(CH_2{-}CH_2{-}O)_nH$	no	nonionic
Alkylsulfonates	$R{-}SO_3^-\ Na^+$	yes	anionic
Dialkyldimethylammonium chlorides	$[H_3C{-}N^+R_2{-}CH_3]\ Cl^-$	yes	cationic
Betaines	$R{-}N^+(CH_3)_2{-}CH_2{-}C(=O){-}O^-$		amphoteric

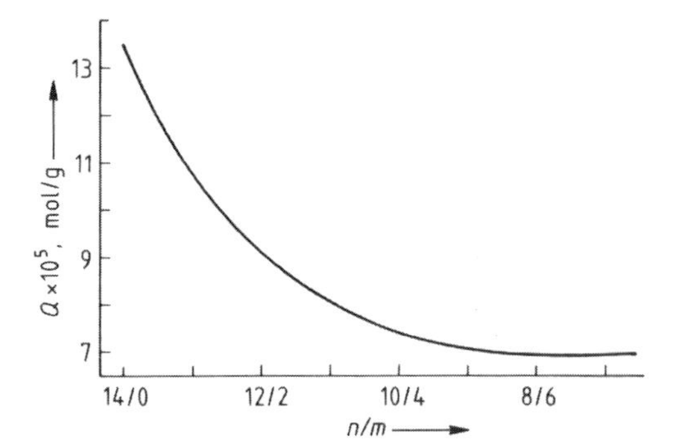

Figure 34. Decrease in adsorbed amount Q at equilibrium with increased branching of the hydrophobic residue [37]
Adsorbent: activated carbon M; amount of adsorbent: 0.050 g; particle diameter: 0.084 cm; surfactant:

NaO_3SO–CH_2–$CH(C_nH_{2n+1})(C_mH_{2m+1})$

surfactant concentration: 1×10^{-4} mol/L; temperature: 25 °C

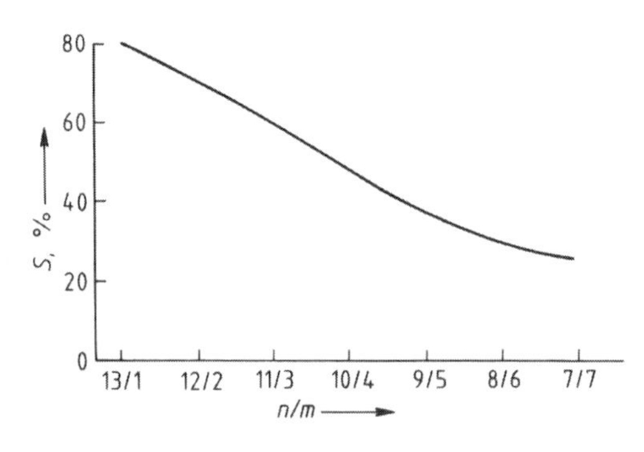

Figure 35. Decrease in soil removal S from soiled cotton as a function of increased branching in the hydrophobic residue [37]
Surfactant:

NaO_3SO–CH_2–$CH(C_nH_{2n+1})(C_mH_{2m+1})$

temperature: 90 °C; bath ratio: 1 : 12.5; water hardness: 16 °d (285 ppm); surfactant concentration: 2.91×10^{-3} mol/L

The number of types of surfactants suitable for use in laundry products has increased considerably in the past 50 years. The principal criteria for judging surfactant suitability apart from performance are toxicological and ecological characteristics. Cationic and nonionic surfactants have come to play an increasingly important role along with their anionic counterparts (cf. Table 10). Despite the wide choice of possibilities, however, only few surfactants account for the major share of the market, partly as a result of economic factors [27], [38].

Anionic surfactants are the most common ingredients in detergents designed for laundry, dishwashing, and general cleansing. Nonionic surfactants such as alcohol ethoxylates have acquired great importance during the last decades. Cationic surfactant use is largely restricted to fabric softeners because of the fundamental incompatibility of these materials with anionic surfactants and their poor cleaning efficiency. Amphoteric surfactants still lack a significant place in the market. Around the world a remarkable variability in the types of surfactants employed in products for similar purposes can be seen. The reasons are to be found in the variations in the kinds of fabric encountered worldwide, the diversity in washing machine technology, and different regional habits for fabric use and care (Fig. 36).

Wash technology has been the subject of major changes and developments during the past 50 years. The textile market has also changed, with synthetic fibers playing an increasingly large role. The wide variety of textile fibers and fiber blends and their differing characteristics have forced manufacturers to devise surfactants with a broad

Table 10. Key surfactants

Structure		Chemical name	Acronym
Anionic surfactants			
$R-CH_2-C(=O)ONa$	$R = C_{10-16}$	soaps	
$R-C_6H_4-SO_3Na$	$R = C_{10-13}$	alkylbenzenesulfonates	LAS
$R^1-CH(R^2)-SO_3Na$	$R^1 + R^2 = C_{11-17}$	alkanesulfonates	SAS
$CH_3-(CH_2)_m-CH=CH-(CH_2)_n-SO_3Na$ + $R-CH_2-CH(OH)-(CH_2)_n-SO_3Na$	$n + m = 9 - 15$ $n = 0,1,2\ldots$ $m = 1,2,3\ldots$ $R = C_{7-13}$ $x = 1,2,3$	α-olefinsulfonates	AOS
$R-CH(SO_3Na)-C(=O)OCH_3$	$R = C_{14-16}$	α-sulfo fatty acid methyl esters	MES
$R-CH_2-O-SO_3Na$	$R = C_{11-17}$	alcohol sulfates, alkyl sulfates (linear and branched)	AS
$R^1-CH(R^2)-CH_2-O-(CH_2-CH_2-O)_n-SO_3Na$	a) $R^1 = H$ $R^2 = C_{10-12}$ b) $R^1 + R^2 = C_{11-13}$ $R^1 = H, C_1, C_2\ldots$ $n = 1 - 4$	alkyl ether sulfates a) fatty alcohol ether sulfates b) oxo alcohol ether sulfates	AES
Cationic surfactants			
$[R^2-N^+(R^1)(R^3)-R^4]\ Cl^-$	$R^1, R^2 = C_{16-18}$ $R^3, R^4 = C_1$	quaternary ammonium compounds tetraalkylammonium chloride	
Nonionic surfactants			
$R^1-CH(R^2)-CH_2-O-(CH_2-CH_2-O)_n-H$	a) $R^1 = H$ $R^2 = C_{6-16}$ b) $R^1 + R^2 = C_{7-13}$ $R^1 = H, C_1, C_2\ldots$ $n = 3 - 15$	alcohol ethoxylates a) fatty alcohol ethoxylates b) oxo alcohol ethoxylates	AE
$R-C_6H_4-O-(CH_2-CH_2-O)_nH$	$R = C_{8-12}$ $n = 5 - 10$	alkylphenol ethoxylates	APE

Table 10. (continued)

Structure	Chemical name	Acronym
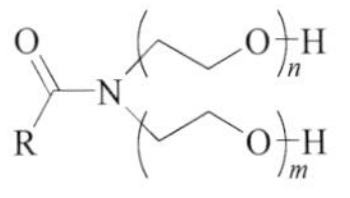 $R = C_{11-17}$, $n = 1, 2$ $m = 0, 1$	fatty acid alkanolamides	FAA
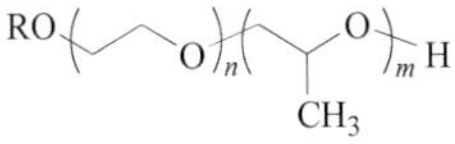 $R = C_{8-18}$, $n = 3-6$ $m = 3-6$	alcohol alkoxylates (EO/PO adducts)	
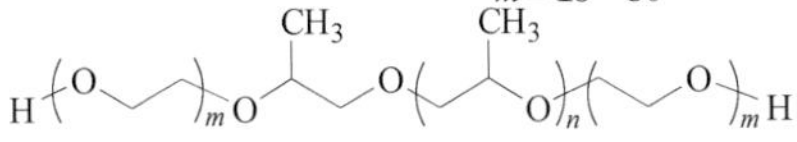 $n = 2-60$, $m = 15-80$	ethylene oxide – propylene oxide block polymers	EPE
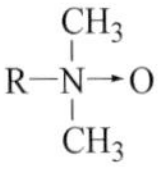 $R = C_{12-18}$	alkyldimethylamine oxides	
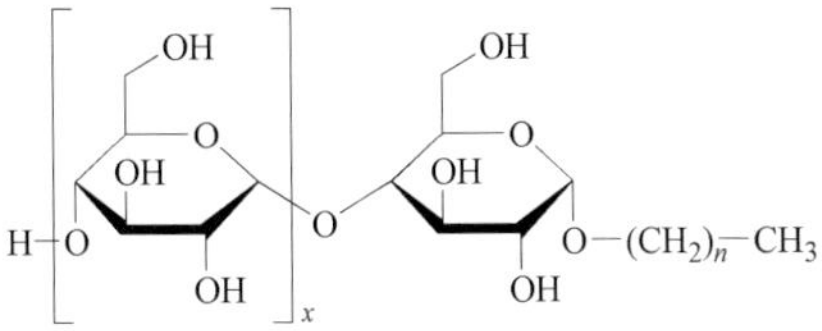 $x = 0-2$, $n = 11-17$	alkyl polyglycosides	APG
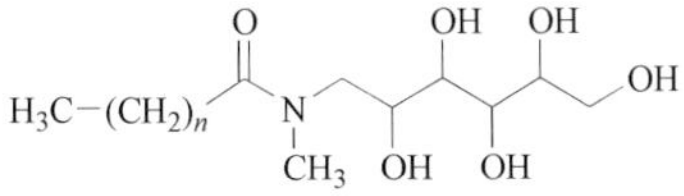 $n = 10-16$	fatty acid *N*-methylglucamides	NMG
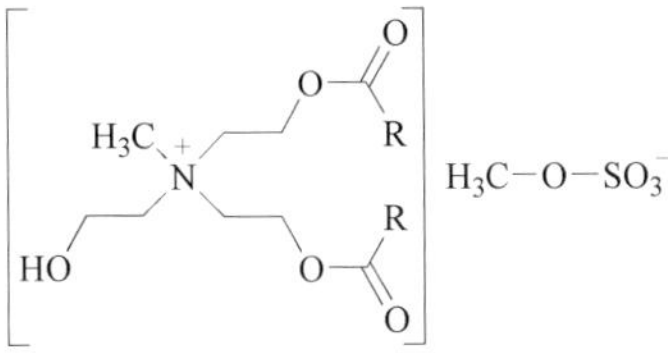 $H_3C-O-SO_3^-$	*N*-methyl-*N*,*N*-bis[2-($C_{16/18}$-acyloxy)ethyl]-*N*-(2-hydroxyethyl)ammonium-methosulfate	EQ

Table 10. (continued)

Structure	Chemical name	Acronym
$[(H_3C)_2N^+(CH_3)CH_2CH(OC(O)R)CH_2OC(O)R]\ Cl^-$	*N,N,N*-trimethyl-*N*-[1,2-dl($C_{16/18}$-acyl-oxy)propyl]-ammonium-chloride	DEQ
$(H_3C)_2N^+(R)CH_2COO^-$	alkyl betaines	

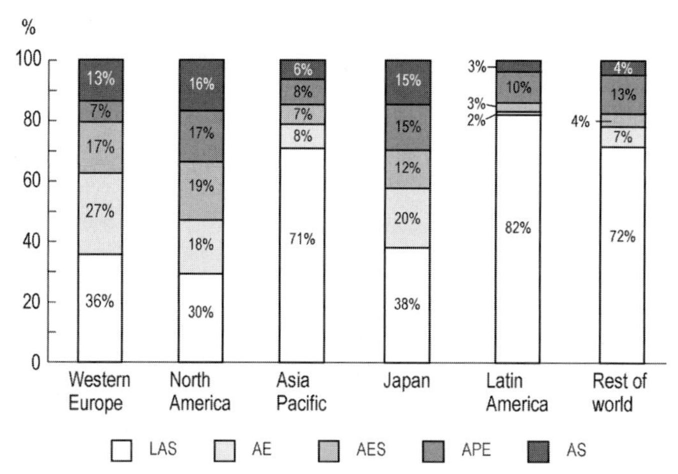

Figure 36. 1995 Global use pattern of major surfactants [39]

performance range. No single surfactant is capable of fulfilling all demands in an optimal way; consequently, the trend has been increasingly toward using surfactant blends, in which the characteristics of each component are intended to supplement those of the others. The widening scope of the demand for surfactants used in detergents not only relates to performance, but encompasses toxicological, ecological, and economic considerations as well. Surfactants suitable for detergent use are expected to demonstrate the following characteristics [40], [41]:

Specific adsorption
Excellent soil removal
Low sensitivity to water hardness
Good dispersion properties
Good soil antiredeposition capability
High solubility

Sufficient wetting power
Desirable foam characteristics
Neutral odor
Low intrinsic color
Sufficient storage stability
Good handling characteristics
Low toxicity to humans
Favorable environmental behavior
Assured raw material supply
Competitive costs

3.1.1. Anionic Surfactants

Most detergents contain larger amounts of anionic surfactants than nonionic surfactants. The following discussion deals with anionic surfactants that are in widespread use and/or have favorable characteristics. Anionic surfactants and their manufacturing processes are amply described in the literature [184].

Soap. Soap is no longer as important in many parts of the world as it was before the existence of mass-produced synthetic surfactants. Although soap powders for washing once contained as much as 40 % soap as their sole surfactant, powder detergents have since the 1950s been formulated with mixtures of far more effective surfactants in considerably smaller proportion. A further reason for the decreasing use of soap in laundry detergents is its sensitivity to water hardness, manifested through inactivation due to the formation of lime soap, which tends to accumulate on fabrics and washing machine parts. Such accumulation reduces the absorbency of fabrics and their permeability to air, and eventually through "aging" causes laundry to become discolored and to develop malodors. The primary function that remains for soap currently is as a foam regulator in laundry detergents in Europe. Nonetheless, soap has remained worldwide the largest surfactant by volume. It is still the surfactant of choice in many countries with low gross national product (cf. Chap. 9) (cf. Section 3.4.3).

Alkylbenzenesulfonates (LAS and TPS). Until the mid-1960s, this largest class of synthetic surfactants was most prominently represented by tetrapropylenebenzenesulfonate (TPS):

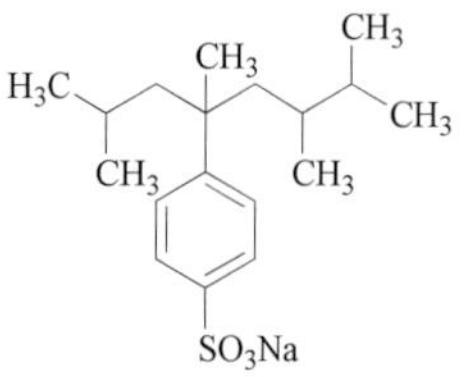

CH_3 CH_3 H_3C CH_3 CH_3 CH_3

SO_3Na

In the 1950s TPS had largely replaced soap as an active ingredient in laundry detergents in Europe, the USA, and Japan. It was later found, however, that the branched side-chain present in TPS prevents the compound from undergoing sufficient biodegradation. Thus, means were developed to replace it by more readily biodegradable straight-chain homologues. Since that time, favorable economic circumstances and good performance characteristics have permitted straight-chain or linear alkylbenzenesulfonates (LAS) to take the lead among laundry detergent surfactants in Europe, the Americas, and Asia. Nevertheless, a few countries remain in which TPS continues to be used in detergents (see Section 10.3.1). The large-scale manufacturing methods for LAS have been reviewed [48].

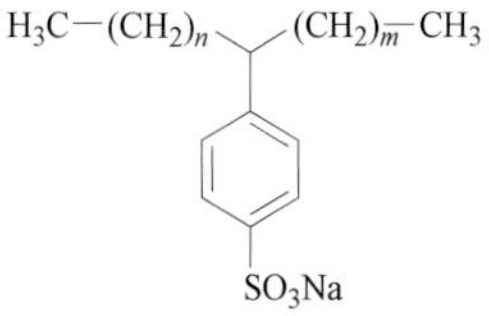

$H_3C-(CH_2)_n$ $(CH_2)_m-CH_3$

SO_3Na

$n + m = 7–10$
LAS

Apart from their very good detergency performance, LAS have interesting foaming characteristics, which are of great significance to their use in detergents. Their foaming ability is high, and the foam that is produced is readily stabilized by foam stabilizers, an important factor in many parts of the world. At the same time, however, LAS can be controlled easily by foam regulators, and this is significant with respect to detergents for the European market, where horizontal-axis drum-type washing machines are common.

As a result of their high solubility, LAS are also frequently employed in formulations for liquid detergents. However, LAS are sensitive to water hardness: the detergency performance of LAS diminishes as the hardness of the water increases. The relationship between water hardness and performance for a series of surfactants in the absence of builders or sequestrants is well demonstrated by soil removal from wool, as illustrated in Figure 37.

The decline in detergency performance with increasing water hardness is most dramatic with soap. Sensitivity to water hardness largely disappears in built formulations (zeolite A, triphosphate, soda ash, etc.) because of sequestration, binding, or ion exchange of the water hardness ions (cf. Chap. 2, Fig. 29). Figure 38 illustrates wool

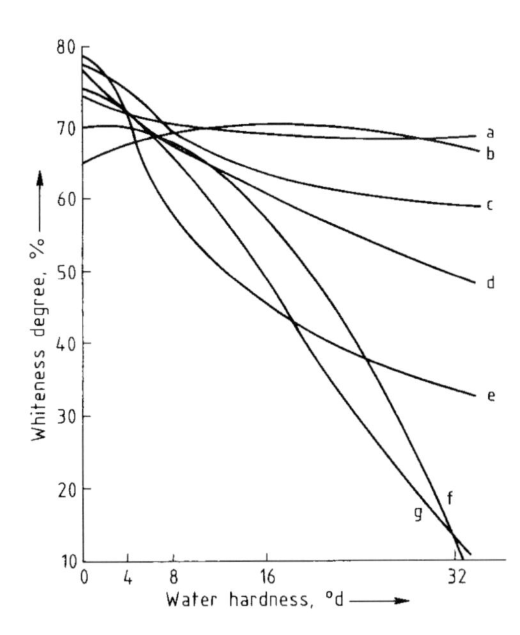

Figure 37. Detergency performance on wool by various surfactants as a function of water hardness [42]
Time: 15 min; temperature: 30 °C; bath ratio: 1 : 50; concentration: 0.5 g/L surfactant + 1.5 g/L sodium sulfate
a) Nonylphenol 9 EO; b) C_{12-14} Alcohol 2 EO sulfates; c) C_{15-18} α-Olefinsulfonates; d) C_{16-18} α-Sulfo fatty acid esters; e) C_{12-18} Alcohol sulfates; f) C_{10-13} Alkylbenzenesulfonates; g) C_{13-18} Alkanesulfonates

wash performance for various readily accessible anionic surfactants in the absence of complexing agents, plotted as a function of surfactant concentration. It can be seen that products with a lower sensitivity to hardness display only a slight advantage.

Secondary Alkanesulfonates (SAS).

R^2

R^1 — SO_3Na

$R^1 + R^2 = C_{11-17}$

Sodium alkanesulfonates (SAS) have been known as commercial surfactants since the 1940s. Their large-scale production began in the late 1960s. Sodium alkanesulfonates are still valued as specialty anionic surfactants for consumer products. Their 1998 demand amounted to only 60×10^3 t. SAS feature high solubility, fast wetting properties, chemical stability to alkali, acids, and strong oxidants including chlorine. Sodium alkanesulfonates are produced by photochemical sulfoxidation or sulfochlorination of suitable $C_{12}-C_{18}$ paraffin cuts. Manufacturing, properties, applications, and characteristics of SAS have been reviewed [43]–[46]. Sodium alkanesulfonates are the compounds that closely resemble LAS in terms of solubility, solubilizing properties, wetting power and other detergency properties. Therefore, SAS can largely be substituted for LAS in most formulations.

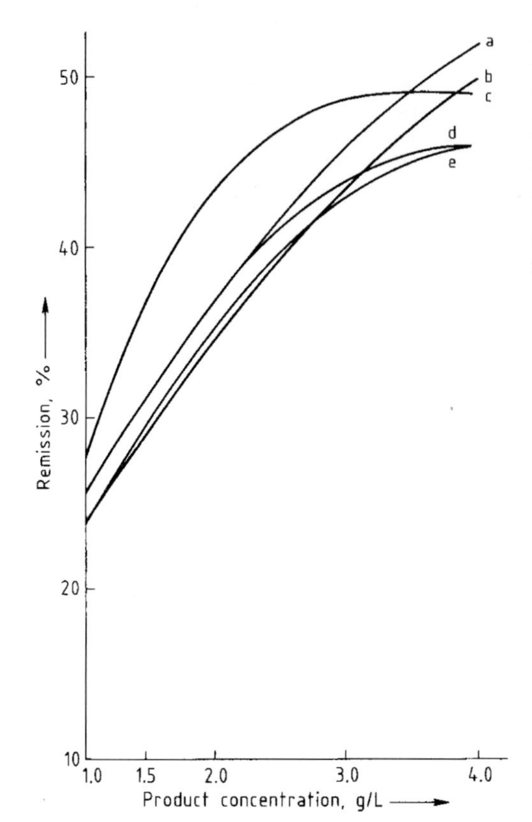

Figure 38. Detergency performance on wool of various synthetic anionic surfactants as a function of concentration [40]
Product: 25 % surfactant + 75 % sodium sulfate; temperature: 30 °C; wash time: 15 min; bath ratio: 1 : 30; water hardness: 16 °d (285 ppm $CaCO_3$)
a) C_{12-14} Alcohol 2 EO sulfates; b) C_{15-18} α-Olefinsulfonates; c) C_{16-18} α-Sulfo fatty acid esters; d) C_{13-18} Alkanesulfonates; e) Alkylbenzenesulfonates

Alkanesulfonates are completely insensitive to hydrolysis even at extreme pH values, a result of the presence of the stable carbon – sulfur bond.

The water hardness sensitivity and foaming characteristics of SAS largely resemble those of LAS, as discussed above.

α-Olefinsulfonates (AOS).

$R^1–CH_2–CH=CH–(CH_2)_n–SO_3Na$ — Alkenesulfonates

$R^2–CH_2–CH(OH)–(CH_2)_m–SO_3Na$ — Hydroxyalkanesulfonates

$R^1 = C_8 - C_{12}$ $\quad n = 1, 2, 3$

$R^2 = C_7 - C_{13}$ $\quad m = 1, 2, 3$

α-Olefinsulfonates (AOS) are produced commercially starting from α-olefins. Alkaline hydrolysis of the sultone intermediate results in ca. 60 – 65 % alkenesulfonates and ca. 35 – 40 % hydroxyalkanesulfonates. Because of the use of olefinic precursors, these mixtures are customarily called α-olefinsulfonates.

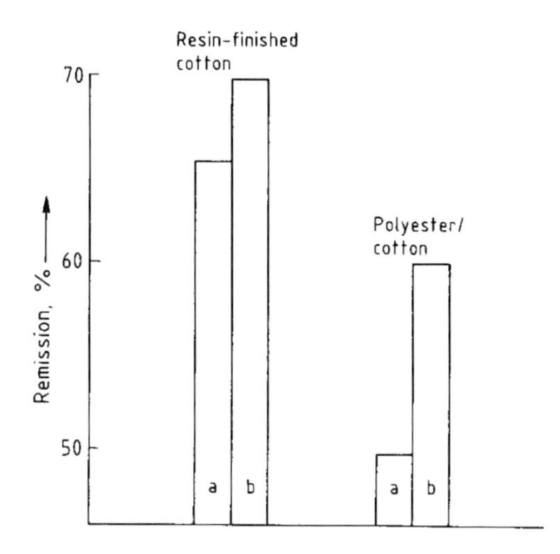

Figure 39. Performance of various olefinsulfonates [47]
Water hardness: 16 °d (285 ppm $CaCO_3$); wash time: 30 min; wash temperature: 60 °C
a) 0.75 g/L C_{15-18} Internal olefinsulfonates + 2.0 g/L sodium triphosphate; b) 0.75 g/L C_{15-18} α-Olefinsulfonates + 2.0 g/L sodium triphosphate

In contrast to LAS and SAS, AOS show little sensitivity to water hardness (cf. Fig. 37). This apparent advantage is only of significance in a few very special applications, however. Depending on chain length, AOS may cause foaming problems in drum-type washing machines, which requires the addition of special foam regulators. Since this problem does not arise with washing machines of the type used in Japan, AOS have long been an important component of some Japanese detergents.

In addition to α-olefinsulfonates, sulfonates prepared from olefins with internal or central double bonds and from vinylidene olefins also exist. These types of olefinsulfonates are unsuitable for detergent use, however, because of their poor performance characteristics. In Figure 39, a C_{15-18} α-olefinsulfonate is compared with a C_{15-18} internal olefinsulfonate, the inferior performance of which is clearly apparent. The reason for the difference is the fact that AOS have their hydrophilic group in a terminal position, whereas with the internal olefinsulfonates, additional isomers can arise in which hydrophilic groups are distributed throughout the entire hydrophobic chain. This has the same consequences as branching (cf. Section 3.1). However, olefinsulfonates prepared from internal olefins generally show very good textile wetting characteristics.

α-Sulfo Fatty Acid Esters (MES)

O
R OCH$_3$
SO$_3$Na

R = C_{12-16}
α-Sulfo fatty acid methyl esters

Another class of anionic surfactants is the α-sulfo fatty acid esters, particularly the methyl derivatives. This group of surfactants is also called methyl ester sulfonates (MES). Methyl ester sulfonates are derived from a variety of methyl ester feedstocks from renewable resources, such as coconut, palm kernel, palm stearin, beef tallow, lard, and soy. Good detergency performance is attained with products having rather long hydrophobic residues (e.g., palmitic and stearic acid derivatives) [48]. The sensitivity of MES to water hardness is small relative to that of LAS and SAS, more closely resembling that of AOS. One of the interesting detergency properties of α-sulfo fatty acid methyl esters is their exceptional dispersion power with respect to lime soap.

Special attention is to be paid to the manufacturing of finished laundry products based on MES. Insufficient storage stability of MES has so far impeded their large-scale use in detergents; they have only been used in a few Japanese detergents [49].

Alkyl Sulfates (AS).

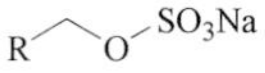

R–CH_2–O–SO_3Na

$R = C_{11-17}$

Alkyl sulfates (AS), also known as alcohol sulfates, achieved prominence as early as the 1930s in detergents for delicate fabrics in Germany and the USA and as components of textile auxiliaries. Their availability resulted from the development by SCHRAUTH to produce primary fatty alcohols by high-pressure hydrogenation of fatty acids and their methyl esters. Alkyl sulfates are produced either from natural fatty alcohols derived from palm oil, palm kernel oil, or coconut oil, or from oxo alcohols, i.e., of petrochemical origin [38]. They are characterized by desirable detergency properties, and have found increasing application not only in specialty products, but also in heavy-duty detergents, especially in those concentrated products produced by nontower technology (cf. Section 6.2.3).

Alcohol sulfates derived from natural, renewable feedstocks have been playing a steadily increasing role as surfactants in laundry detergents since the 1980s. They have found increasing application at the expense of LAS [27], [38], [50]–[53].

Alkyl Ether Sulfates (AES).

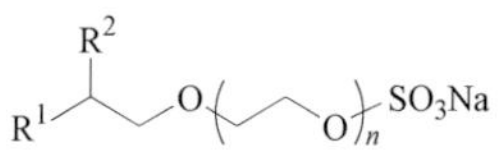

R^1–CH(R^2)–CH_2–O–(CH_2CH_2O)$_n$–SO_3Na — Alkyl ether sulfates

1. $R^1 = H$, $R^2 = C_{10-12}$ — Fatty alcohol ether sulfates
2. $R^1 + R^2 = C_{11-13}$ — Oxo alcohol ether sulfates
   $R^1 = H, C_1, C_2 \ldots$
   $n = 1–2$

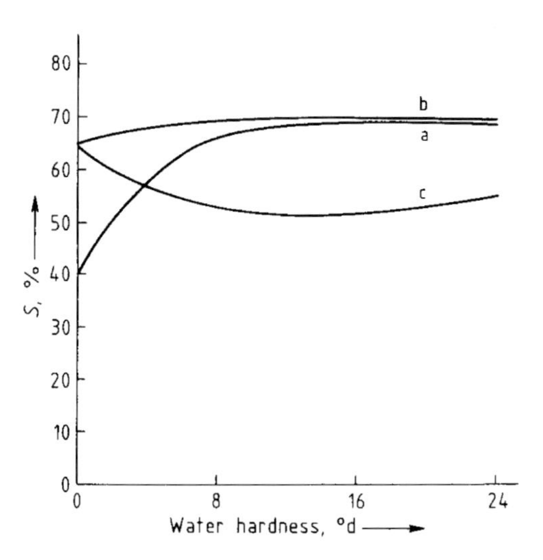

Figure 40. Soil removal S from wool as a function of water hardness at 30 °C with 0.5 g/L sodium C_{12-14} *n*-alkyl diethylene glycol ether sulfates containing no added electrolyte (a), with 1.5 g/L sodium sulfate (b), and with 1.5 g/L sodium triphosphate (c) [4]

Alkyl ether sulfates (AES), also known as alcohol ether sulfates, are obtained by ethoxylation and subsequent sulfation of alcohols derived from natural feedstocks or of synthetic alcohols. They exhibit the following unique characteristics relative to alkyl sulfates:

Low sensitivity to water hardness (cf. Fig. 37)
High solubility
Good storage stability at low temperature in liquid formulations

Commercial AES consist of alkyl ether sulfates and alkyl sulfates as the main components. Unsulfated alcohols, alcohol ethoxylates, inorganic salts, and poly(ethylene oxide) sulfates are contained as byproducts [38]. Manufacturing, properties, and analysics of AES have been reviewed [54].

Alcohol ether sulfates that are least sensitive to water hardness, e.g., sodium C_{12-14} *n*-alkyl diethylene glycol ether sulfates, actually demonstrate increased detergency performance (e.g., on wool) as the hardness increases. This is a result of the positive electrolyte effects attributable to calcium/magnesium ions. Addition of sodium sulfate produces a slight improvement of detergency performance only in regions of low water hardness. However, detergency performance declines in the presence of sodium triphosphate due to calcium/magnesium sequestration (cf. Fig. 40).

Binding of alkaline earth ions can occur not only through complexation, but also as a result of ion exchange. Calcium ions can be exchanged particularly effectively and with favorable kinetics by using zeolite A. The properties discussed above in the context of calcium sequestration can be equally well described in terms of ion exchange.

From the foregoing, it follows that if one were to undertake the development of, for example, a wool detergent, an important consideration would be the careful choice of supplemental salts; i.e., the presence of complexing agents or ion exchangers is not always advantageous in a detergent formulation containing surfactants if these are insensitive to water hardness. On the other hand, a surfactant that is sensitive to water hardness is not necessarily inferior, provided it is combined with the proper complexing agents or ion exchangers. The situation can be somewhat more complicated, however, if a detergent has to be developed for fibers whose nature requires the presence of complexing agents for proper washing.

Alkyl ether sulfates are very intensively foaming compounds that are well suited for use in high-foam detergents for vertical-axis washing machines, but are less directly applicable to detergents for horizontal-axis drum-type machines. Because of their specific properties, alkyl ether sulfates are preferred constituents of detergents for delicate or wool washables, as well as foam baths, hair shampoos, and manual dish-washing agents. The optimal carbon chain length has been established to be C_{12-14} with ca. 2 mol of ethylene oxide.

Alcohol ether sulfates have achieved some importance in the U.S. and Japanese markets. This is because their critical micelle concentration is considerably lower than that for LAS, resulting in very satisfactory washing power even at the low detergent concentrations typical in these countries. In Europe, the use of alcohol ether sulfates has so far been largely restricted to specialty detergents.

Alkyl ether sulfates were formerly prepared exclusively by ethoxylation and sulfation of native fatty alcohols. For a long time, synthetic alcohols have also been used, particularly the partially branched oxo alcohols. Oxo alcohol ether sulfates have good wetting properties comparable to those of straight-chain native fatty alcohol ether sulfates, whereas their detergency performance is inferior to the latter [55].

3.1.2. Nonionic Surfactants

3.1.2.1. Alcohol Ethoxylates (AE)

Nonionic surfactants of the alcohol ethoxylate (AE) type do not dissociate in aqueous solution. Some of their properties can be singled out for special attention:

The absence of electrostatic interactions
Behavior with respect to electrolytes
The possibility of favorable adjustment of hydrophilic–lipophilic balance values (HLB)
Anomalous solubility in water

Adsorption phenomena involving nonionic surfactants can be explained on the basis of hydrophobic interactions, in some cases coupled with steric effects. Electrolytes have no direct influence on adsorption with nonionic surfactants. Nonionic surfactant detergency performance is adversely affected by the presence of polyvalent cations. These cations cause reduced negative ζ-potentials of fiber surface and soil, leading to reduced repulsion and correspondingly poor soil removal.

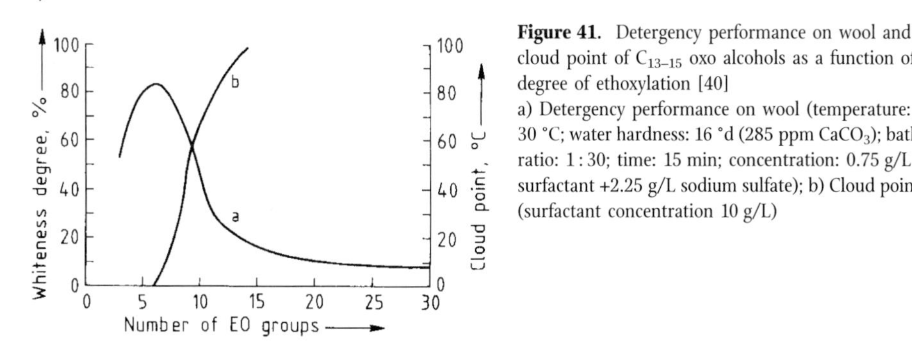

Figure 41. Detergency performance on wool and cloud point of C_{13-15} oxo alcohols as a function of degree of ethoxylation [40]
a) Detergency performance on wool (temperature: 30 °C; water hardness: 16 °d (285 ppm $CaCO_3$); bath ratio: 1 : 30; time: 15 min; concentration: 0.75 g/L surfactant +2.25 g/L sodium sulfate); b) Cloud point (surfactant concentration 10 g/L)

An important advantage of nonionic surfactants of the alcohol ethoxylate type is the fact that a proper balance (HLB) can be easily achieved between the hydrophobic and hydrophilic portions of the nonionic surfactants. For example, the hydrophilic portion of the molecule can be extended gradually by stepwise addition of ethylene oxide moieties. This leads to a stepwise increase in hydration and corresponding successive increase in solubility. On the other hand, with ionic surfactants, the presence of even one ionic group makes such a strong contribution to hydrophilic character that the introduction of further ionic groups totally eliminates the possibility of any equilibrium relationship with respect to the hydrophobic portion. The result is a rapid deterioration of typical surfactant characteristics. For a surfactant bearing two strong ionic hydrophilic groups to show washing activity, a considerably longer alkyl chain ($>C_{20}$) would be necessary.

Nonionic surfactants with a given hydrophobic residue can be adjusted to have optimal properties for various substrates with respect to adsorption and detergency performance simply by changing the degree of ethoxylation. Performance shows an initial increase with an increasing degree of ethoxylation, but a point is then reached after which it declines markedly (Fig. 41). Textile wetting power often decreases only at very high degrees of ethoxylation, whereas the wetting power for hard hydrophobic surfaces continues to climb as the number of ethoxy groups increases.

Alcohol ethoxylates show a solubility anomaly: when they are heated in aqueous solution, suddenly turbidity appears, usually at a relatively precise temperature. This is caused by a separation of the alcohol ethoxylate solution into two phases, one with a high water content and one with a low water content. The corresponding characteristic temperature for a given surfactant is known as its cloud point. The cloud point moves to a higher temperature as the number of ethoxy groups increases. If the cloud point is not greatly exceeded, then the largely aqueous phases and largely surfactant phases form an emulsion. For a given surfactant, adsorption (hence, detergency performance) decreases when the cloud point is surpassed to a significant extent. The main reason for this behavior is the reduced solubility of the alcohol ethoxylates, which are then expelled from the aqueous phase. Nevertheless, nonionic surfactants with a cloud point somewhat below the application temperature commonly show better performance than those whose cloud point is higher (Fig. 41). Thus, the wash temperature is a significant

Table 11. Detergency performance (in % remission) of alcohol ethoxylates with comparable cloud points [42]

Surfactant *	Cloud point, °C	Cotton 60 °C	Cotton 90 °C	Permanent-press cotton 60 °C	Permanent-press cotton 90 °C	Polyester/cotton 60 °C	Polyester/cotton 90 °C
C_{11-15} *sec*-Alcohol 9 EO	59	71	53	69	55	65	48
C_{9-11} Oxo alcohol 7 EO	61	57	44	66	52	53	43
Oleyl/cetyl alcohol 10 EO **	89	60	68	70	72	57	67
C_{13-15} Oxo alcohol 11 EO	88	54	69	67	69	53	58

* Surfactant concentration: 0.75 g/L; water hardness: 16 °d (285 ppm); wash time: 30 min.
** Iodine number 45.

factor in determining an optimal degree of ethoxylation (Table 11). The table shows that the best detergency performance is obtained when the temperature is maintained near the cloud point.

The cloud point can be greatly reduced by addition of several grams of electrolytes per liter, depending on the surfactant. However, everything said in this section is strictly applicable only to systems comprised of pure nonionic surfactants. For binary mixtures of nonionic and ionic surfactants, it is important to recognize that even a small amount of ionic surfactant can cause the cloud point to rise more or less dramatically.

The share of nonionic surfactants in overall surfactant production and use has been increasing since the 1970s. The major contributors to this increase have been fatty alcohol, oxo alcohol, and secondary alcohol ethoxylates, all of which are obtained by reaction of the corresponding alcohols with ethylene oxide. By varying the carbon chain length and the degree of ethoxylation, these nonionic surfactants can be tailor-made with respect to the washing temperature. For these reasons, the increased use of nonionic surfactants has been partly concomitant with the trend to wash at lower temperature and with changes in the production shares of different fibers [27], [56].

The reasons for the increased use of nonionic surfactants are found in their favorable detergency properties, particularly with respect to synthetics, and the decrease in wash temperatures observed since the 1970s.

Favorable detergency properties of nonionic surfactants derive largely from the following factors:

Low critical micelle concentration (c_M)
Very good detergency performance
Excellent soil antiredeposition characteristics with synthetic fibers

The low c_M values of nonionic surfactants indicate that they display high detergency performance even at relatively low concentrations. Table 12 provides data illustrating the low c_M of these compounds relative to that of anionic surfactants.

Table 12. Critical micelle concentration c_M of various surfactants [42]

Surfactant	c_M, g/L
LAS (C_{10-13} alkyl)	0.65
C_{12-17} Alkanesulfonates	0.35
C_{15-18} α-Olefinsulfonates	0.30
C_{12-14} Fatty alcohol 2 EO sulfates	0.30
Nonylphenol 9 EO	0.049
Oleyl/cetyl alcohol 10 EO *	0.035

* Iodine number 45.

3.1.2.2. Alkylphenol Ethoxylates (APE)

Alkylphenol ethoxylates (APE), are based on *p*-octyl-, nonyl-, and dodecylphenol poly(ethylene glycol) ethers. They achieved early success due to their exceptional detergency properties, particularly their oil and fat removal characteristics.

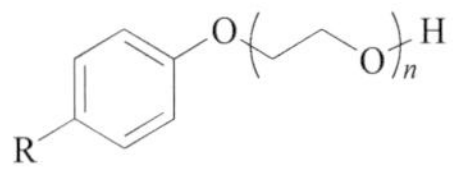

$R = C_{8-12}$ $n = 5–10$

The usage of APE has greatly declined, however, especially in Europe since 1986, due to their environmental characteristics, particularly the extent of their biodegradability and the fish toxicity of certain metabolites resulting from partial biodegradation (see Chap. 10).

3.1.2.3. Fatty Acid Alkanolamides (FAA)

Fatty acid alkanolamides are ethanolamides of fatty acids with the following structure:

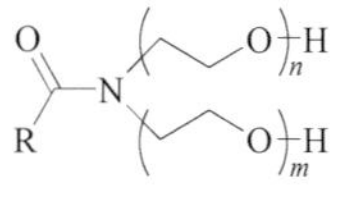

$R = C_{11-17}$ $n = 1, 2$ $m = 0, 1$

Fatty acid alkanolamides find only little application in laundry detergents. Their most important feature is foam boosting, i.e., adding desired stability to the foam produced by detergents prone to heavy foaming. This property is not desirable for horizontal-axis drum-type washing machines employed, e.g., in Europe. Nevertheless, small amounts of FAA as cosurfactants are capable of enhancing the soil removal properties of the classical detergent components at low washing temperatures.

3.1.2.4. Alkylamine Oxides

The alkylamine oxides are produced by oxidation of tertiary amines with hydrogen peroxide. They are compounds that exhibit cationic behavior in acidic conditions (pH < 3), but they behave as nonionic surfactants under neutral or alkaline conditions. For this reason they are included in the nonionic surfactant category.

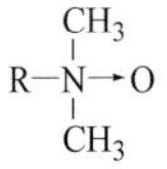

$$R{-}\overset{\displaystyle CH_3}{\underset{\displaystyle CH_3}{N}}{\rightarrow}O$$

$R = C_{12-16}$

Compounds in this class have been known since 1934 and were described as detergent components in a patent issued to IG-Farbenfabriken. Combinations of alkylbenzenesulfonates and specific amine oxides are reputed to be especially gentle to the skin. Despite good detergency properties, however, they are rarely included in laundry detergent fromulations. The reasons for this include high cost, low thermal stability, and high foam stability.

3.1.2.5. *N*-Methylglucamides (NMG)

N-Methylglucamides are a new type of nonionic surfactants, that has been introduced in detergents in the 1990s. *N*-Methylglucamides derive their hydrophilic character from the hydroxyl groups of sugar or starch glucose moieties. These surfactants are obtained by reacting sugars with methylamine and subsequent acylation with fatty acid. They have increasingly been used as co-surfactants in powder and liquid detergent formulations.

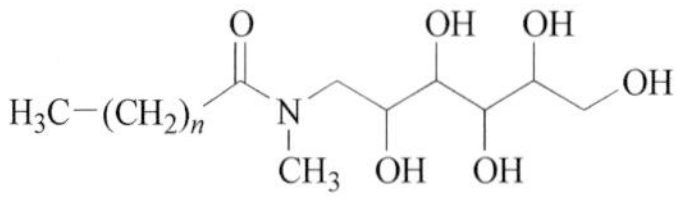

$$H_3C{-}(CH_2)_n{-}C({=}O){-}N(CH_3){-}CH_2{-}CH(OH){-}CH(OH){-}CH(OH){-}CH(OH){-}CH_2OH$$

3.1.2.6. Alkylpolyglycosides (APG)

Alkylpolyglycosides are produced by the dehydration – condensation reaction of alcohols with glucose. Alkylpolyglycosides have distinct lathering characteristics, especially in combination with anionic surfactants. Due to their good foaming properties

APG are predominantly used in dishwashing detergents, liquid detergents, and specialty detergents for fine fabrics.

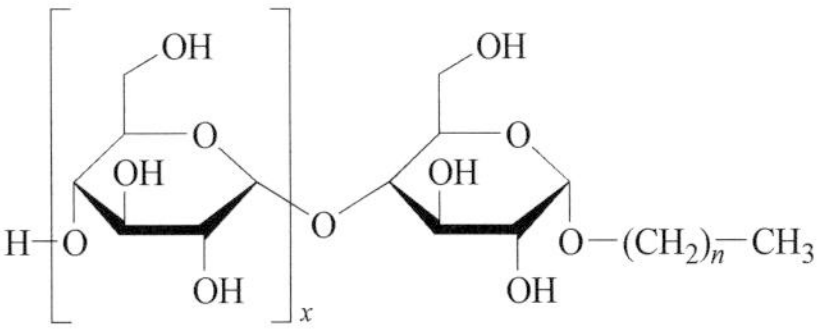

$H-[O-C_6H_{10}O_5]_x-O-(CH_2)_n-CH_3$

Excellent skin compatibility is one of the outstanding characteristics of alkylpolyglycosides. Many APG compounds with special properties can be synthesized by varying the alkyl chain length and the degree of glucose polymerization [57]. Since APG are completely based on natural resources, they biodegrade easily and rapidly to carbon dioxide and water only.

3.1.3. Cationic Surfactants

Long-chain cationic surfactants such as distearyldimethylammonium chloride (DSDMAC) exhibit extraordinarily high sorption power with respect to a wide variety of surfaces [58], [59], [62], [63]. They are very strongly absorbed to the surface of natural fibers, such as cotton, wool, and linen. Adsorption on synthetic fibers is much less pronounced. Figure 42 shows this behavior for several textile fibers. Adsorption rises steeply at low surfactant concentrations, followed by rapid saturation as the concentration increases. This behavior suggests complete coverage of boundary surfaces.

At the same time, cationic surfactants display behavior opposite to that of anionic surfactants as regards charge relationships on solids. Since the surfactant molecules bear a positive charge, their adsorption reduces the negative ζ-potential of solids present in aqueous solution, thereby reducing mutual repulsions, including that between soil and fibers. Use of higher surfactant concentrations causes charge reversal; thus, solid particles become positively charged, resulting again in repulsion. Soil removal can be achieved if adequate amounts of cationic surfactants are present and if their alkyl chains are somewhat longer than those of comparable anionic surfactants. This fact has little practical significance, however, since the subsequent rinse and dilution processes cause charge reversal in the direction of negative ζ-potentials, whereby a large fraction of the previously removed soil is once more attracted to the fibers. Therefore, cationic surfactants are employed only for the purpose of achieving certain special effects, which include applications in rinse-cycle fabric softeners, antistatic agents, and microbicides.

Mixtures made up of equivalent amounts of anionic surfactant and cationic surfactant remain virtually unadsorbed on surfaces and thus display no washing effect. Reactions between anionic and cationic surfactants produce neutral salts with extremely low water solubility. Regarding the washing process, these salts behave like

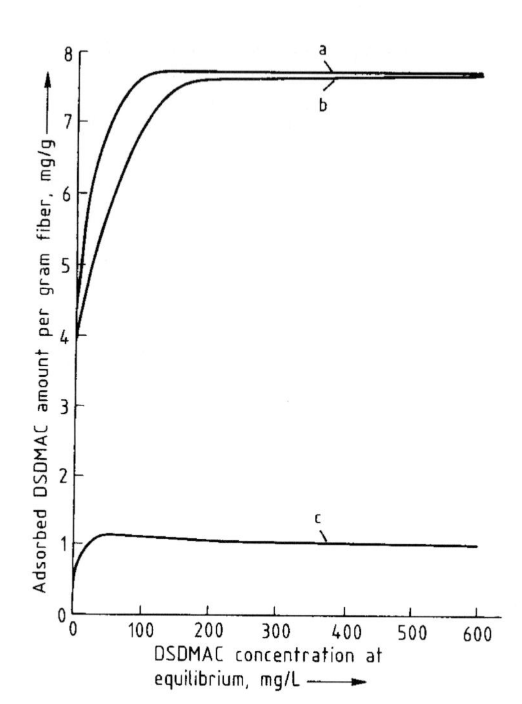

Figure 42. Adsorption isotherms of distearyldimethylammonium chloride (DSDMAC) on wool (a), cotton (b), and polyacrylonitrile (c) [58] Temperature: 23 °C; time: 20 min; bath ratio: 1 : 10

an additional load of greasy soil. On the other hand, addition of small amounts of certain specific cationic surfactants to an anionic surfactant — or even a nonionic surfactant — can enhance detergency performance.

Nonionic surfactants are more tolerant of the presence of cationic surfactants than anionic surfactants. Mixtures of the two are sometimes used in specialty detergents intended to have a wash-cycle fabric-softening effect. In such cases, it must be taken into account that adsorption of the cationic surfactant can be greatly reduced by the presence of the nonionic surfactant, depending on the concentration of the latter, a phenomenon that can have a negative influence on the fabric-softening characteristics (Fig. 43).

Dialkyldimethylammonium Chlorides. The first surfactant developed in this category was distearyldimethylammonium chloride (DSDMAC), introduced in 1949 as a fabric softener for cotton diapers and presented to the U.S. market a year later as a laundry rinse-cycle fabric softener.

$$\left[H_3C-\overset{\displaystyle R}{\underset{\displaystyle R}{\overset{|}{\underset{|}{N^+}}}}-CH_3 \right] Cl^-$$

$R = C_{16-18}$

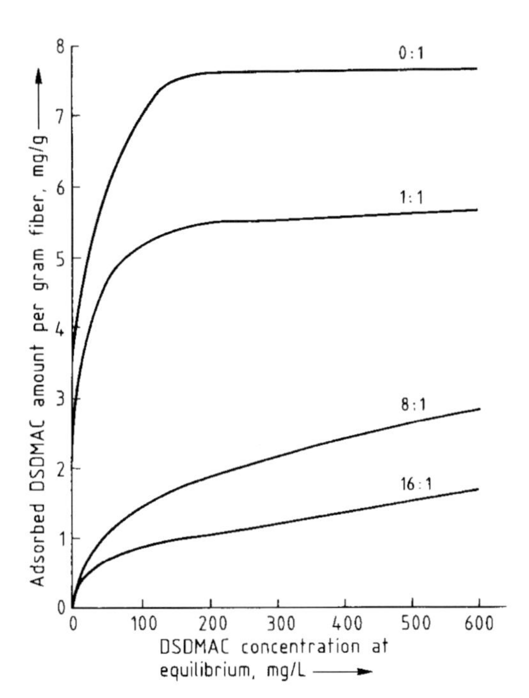

Figure 43. Adsorption isotherms of distearyldimethylammonium chloride (DSDMAC) on cotton as a function of the alkyl polyglycol ether : DSDMAC ratio at equilibrium [58]
Temperature: 23 °C; time: 20 min; bath ratio: 1 : 10

Only in the mid-1960s did these surfactants begin to have a major impact as laundry aftertreatment aids on the U.S. and Western European markets (especially in the Federal Republic of Germany).

Imidazolinium salts such as 1-(alkylamidoethyl)-2-alkyl-3-methylimidazolinium methylsulfate have achieved a place as rinse-cycle softening agents, although not nearly as significant as that of DSDMAC.

$H_3CO-SO_3^-$

R = C_{16-18}

Alkyldimethylbenzylammonium Chlorides. Compounds of the alkyldimethylbenzylammonium chloride type show only limited fabric softening character, but they have been used in laundry disinfecting agents as a result of their activity toward gram-positive and gram-negative bacteria.

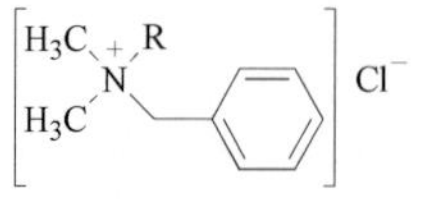

$R = C_{8-18}$

The high adsorption capability of alkyldimethylbenzylammonium chloride also leads to applications as antistatic agents in laundry aftertreatment products.

Esterquats (EQ). In the new generation of fabric softeners that came up in the 1980s/1990s, DSDMAC has been largely replaced by esterquats, e.g., products with the following structures [60], [61]:

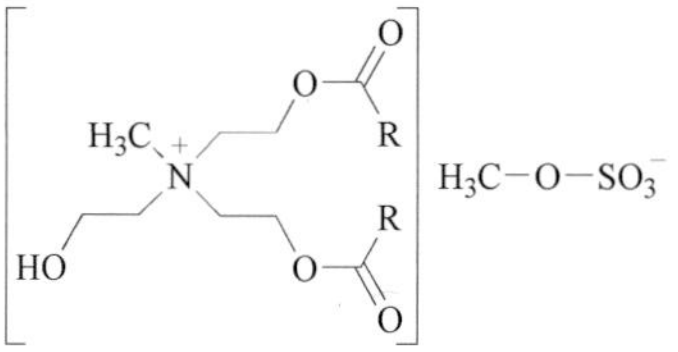

N-Methyl-*N,N*-bis[2-($C_{16/18}$-acyloxy)ethyl]-
N-(2-hydroxyethyl)ammonium-methosulfate
EQ (**Ester** **q**uat)

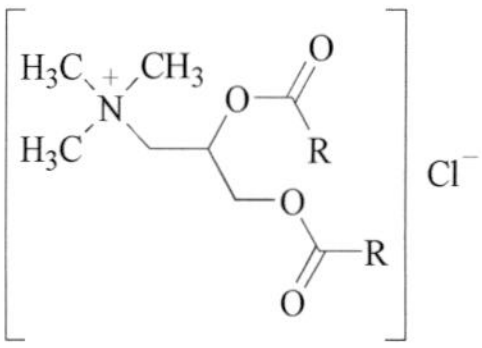

N,N,N-Trimethyl-*N*-[1,2-di($C_{16/18}$-acyloxy)
propyl] ammonium chloride
DEQ (**Diester** **q**uat)

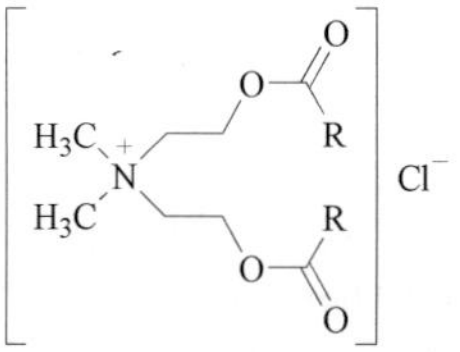

Diethyl**e**ster **dim**ethyl **a**mmonium **c**hloride (DEEDMAC)
R = Tallow fatty acid

Due to their ester bonds, which are potential breaking points, esterquats are readily biodegradable in contrast to DSDMAC, which is only poorly biodegradable (see Chap. 10). Esterquats have favorable ecotoxicological and toxicological properties. All available data proves that esterquats present no hazard for living organisms.

Esterquats have similar physicochemical and chemical properties as DSDMAC [60], [61]. It may be assumed that esterquats exhibit adsorption characteristics that are very similar to those of DSDMAC, as shown in Figure 42.

3.1.4. Amphoteric Surfactants

Compounds of the alkylbetaine or alkylsulfobetaine type possess both anionic and cationic groups in the same molecule even in aqueous solution. Despite what could be seen in some respects as excellent detergency properties, these surfactants are only rarely employed in laundry detergents, primarily for cost reasons. They are mainly applied in manual dishwashing products.

$(H_3C)_2N^+(R)-CH_2-COO^-$

R = C_{12-18} Alkylbetaines

$(H_3C)_2N^+(R)-CH_2CH_2CH_2-SO_3^-$

R = C_{12-18} Alkylsulfobetaines

3.2. Builders

Detergent builders play a central role in the course of the washing process [64]. Their function is largely that of supporting detergent action and of water softening, i.e., eliminating calcium and magnesium ions, which arise from the water and from soil.

The category of builders is predomintantly comprised of several types of materials: specific precipitating alkaline materials such as sodium carbonate and sodium silicate; complexing agents like sodium triphosphate or nitrilotriacetic acid (NTA); and ion exchangers, such as water-soluble polycarboxylic acids and zeolites (e.g., zeolite A).

Detergent builders must fulfill a number of criteria [4]:

1) Elimination of alkaline earth ions originating from
   water
   textiles
   soil
2) Soil and stain removal
   high specific detergency performance for particulate soil and fats
   distinct detergency performance on specific textile fibers
   enhancement of surfactant properties
   dispersion of soil in detergent solutions
   favorable influence on foam characteristics
3) Multiple wash cycle performance
   good soil antiredeposition capability
   prevention of lime deposits on textiles and washing machine parts
   anticorrosion properties
4) Handling properties
   sufficient chemical stability
   no hygroscopicity
   good color and odor qualities
   compatibility with other detergent ingredients
   good storage stability
   assured raw material basis
5) Human toxicological safety assurance
6) Environmental properties
   response to deactivation by biological degradation, adsorption, or other mechanisms
   no adverse effects on the biological systems found in sewage plants and surface water
   no uncontrolled accumulation
   no heavy-metal ion remobilization
   no eutrophication
   no detrimental effects on drinking water quality
7) Cost effectiveness

3.2.1. Alkalies

Alkalies such as potash and soda ash have been used to enhance the washing effectiveness of water since antiquity. Their activity is based on the fact that soil and fibers become more negatively charged as the pH value increases, resulting in increased mutual repulsion. Alkali also precipitates ions that contribute to water hardness.

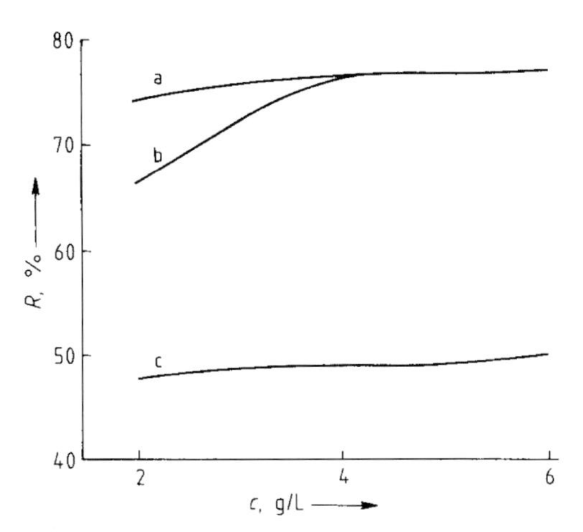

Figure 44. Comparison of soil removal (given as remission R in %) on permanent-press cotton at 0 °d and 90 °C for sodium triphosphate- and sodium carbonate-containing detergents, formulated on the basis of
a) 40 % Sodium triphosphate; b) 20 % Sodium triphosphate + 20 % sodium sulfate; c) 40 % Sodium carbonate

At the beginning of the 20th century, the principal ingredients (apart from soap) of all detergents were soda ash and silicate, which often made up nearly 50 % of the formulation of a powder detergent. These detergent ingredients were partly replaced during the 1930s by sodium monophosphate and diphosphate. All these builders removed water hardness by precipitation.

Modern builders no longer precipitate hardness; instead, water hardness is removed by complexation (sequestration) or ion exchange. Even commercial laundries using soft water have converted to detergents that are low in soda ash and contain the complexing agent sodium triphosphate. The latter has many valuable properties, among which are better detergency performance than soda ash (Fig. 44) [65].

3.2.2. Complexing Agents

Soda ash induces precipitation of calcium and magnesium salts from wash water. This can lead to formation of lime deposits on both laundry and washing machine parts. By contrast, sequestering agents form stable, water-soluble complexes with alkaline-earth ions, as well as with traces of heavy-metal present in water. Often the resulting complexes are chelates (Fig. 45). Sodium triphosphate forms a stable water-soluble complex with calcium when their stoichiometric ratio is 1 : 1 (Fig. 45). However, when a substoichiometric quantity of sodium triphosphate is present, the water-insoluble dicalcium triphosphate is formed, which precipitates on textile fibers and washing machine parts. Its precipitation can be impeded by adding small amounts of hydroxyethanediphosphonate and/or special polycarboxylates to the detergent formulation [78].

Temperature and complexing agent concentration are generally the decisive factors in successful elimination of polyvalent metal ions. Table 13 shows the calcium binding capacity of various sequestering agents as a function of temperature. Calcium binding

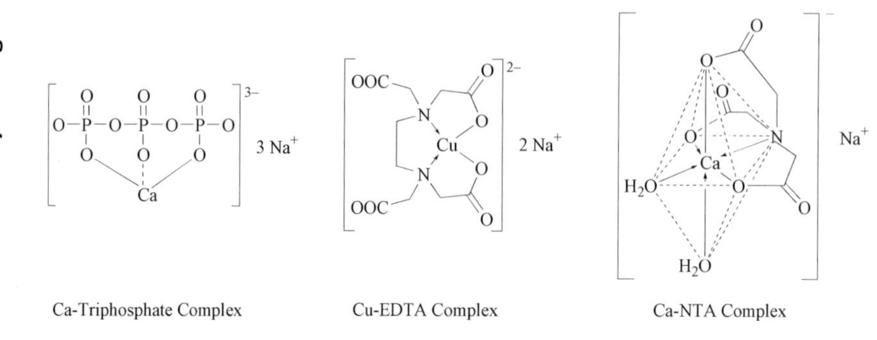

Figure 45. Metal complexes

capacity is here to be viewed as a quantitative measure of the stoichiometry of the resulting complexes. For most sequestering agents, this capacity decreases markedly with increasing temperature.

Regarding stability, the data presented show only that the stability constants (which are a function of the method of study employed, e.g., the dissolving power for freshly precipitated calcium carbonate) exceed a specific value that is dependent on the solubility of the calcium carbonate. Extremely high stability constants or stabilities, which would result in very low calcium ion concentrations, are not required; indeed, they are normally undesirable. It is important that those salts present to the greatest extent, such as calcium carbonate or others with even higher solubility products, be prevented from precipitating during the washing process. Less soluble calcium salts normally play a minor role.

Relative to other polyvalent ions, sequestration of alkaline earth ions is of primary concern because these ions are likely present in high concentration in tap water. Nevertheless, heavy-metal ions must also be eliminated because even in trace amounts their presence can have a negative effect on the washing process. For this reason, low concentrations of selective complexing agents are generally added as well, usually specific phosphonic acids, e.g., sodium hydroxyethanediphosphonate or sodium diethylenetriaminepentakis(methylenephosphonate).

If a sequestering agent is present in less than stoichiometric amounts relative to polyvalent metal ions, precipitation of carbonates and insoluble salts of the sequestering agent with the ions causing the water hardness usually results. Even with adequate amounts of sequestering agent, this effect is important because dilution during the rinse cycle can cause sequestrant concentration to drop below the necessary value, thereby permitting undesirable precipitates to form. These can build up on both fabric and washing machine components, leading eventually to serious accumulations. The problem can be particularly severe if conditions permit large crystals to form as a result of seed crystals located on fabric or machine components.

Table 13. Calcium binding capacity of selected sequestering agents [4]

Structure	Chemical name	Calcium binding capacity, mg CaO/g	
		20 °C	90 °C
$NaO-P(=O)(ONa)-O-P(=O)(ONa)-ONa$	sodium diphosphate	114	28
$NaO-P(=O)(ONa)-O-P(=O)(ONa)-O-P(=O)(ONa)-ONa$	sodium triphosphate	158	113
$HO-P(=O)(OH)-C(OH)(CH_3)-P(=O)(OH)-OH$	1-hydroxyethane-1,1-diphosphonic acid	394	378
$N(CH_2-PO_3H_2)_3$	nitrilotrimethylenephosphonic acid	224	224
$N(CH_2-C(=O)-OH)_3$	nitrilotriacetic acid	285	202
$N(CH_2-C(=O)-OH)_2(CH_2-CH_2-OH)$	*N*-(2-hydroxyethyl)iminodiacetic acid	145	91
$(HO-C(=O)-H_2C)_2N-(CH_2)_2-N(CH_2-C(=O)-OH)_2$	ethylenediaminetetraacetic acid	219	154
cyclopentane ring bearing four $-C(=O)-OH$ groups (H on each ring carbon)	1,2,3,4-cyclopentanetetracarboxylic acid	280	235
$HO(O=)C-H_2C-C(OH)(C(=O)OH)-CH_2-C(=O)OH$	citric acid	195	30
$(HO-C(=O))_2CH-O-CH_2-C(=O)-OH$	*O*-carboxymethyltartronic acid	247	123
$HO-C(=O)-CH_2-CH(O-CH_2-C(=O)-OH)-C(=O)-OH$	*O*-carboxymethyloxysuccinic acid	368	54

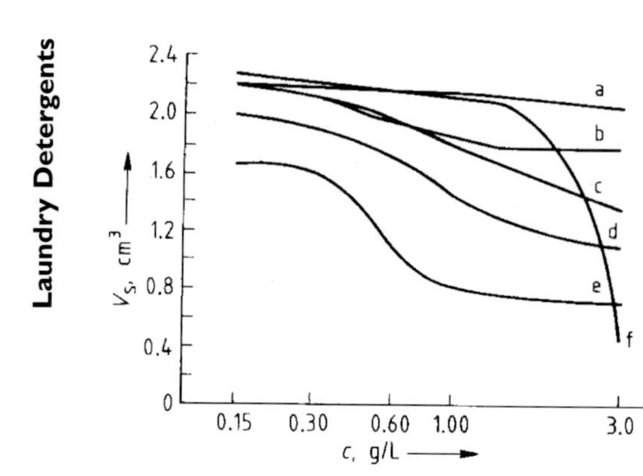

Figure 46. Sediment volume V_s of kaolin clay as a function of active ingredient concentration for water of hardness 16 °d (285 ppm $CaCO_3$) [4]
a) Sodium sulfate; b) C_{16-18} *n*-Alkyl decaglycol ethers; c) Hydroxyethyliminodiacetic acid, sodium salt; d) Nitrilotriacetic acid, sodium salt; e) Sodium triphosphate; f) Sodium *n*-dodecyl sulfate

There are a number of compounds that even in substoichiometric amounts are capable of retarding, hindering, or otherwise interfering with precipitation of insoluble salts. In some cases their action induces salts to precipitate in amorphous form, thereby strongly reducing the tendency toward formation of crystals such as calcite, whose sharp edges can be damaging to fabrics. Sodium polycarboxylates and sodium hydroxyethanediphosphonate are frequently used additives in modern detergents, which strongly exhibit the latter property even at low concentrations; this is known as a *threshold effect* [78].

Despite the many desirable properties shown by sodium triphosphate in the washing process, its continued use has been the subject of an international debate between industry, governments, and water authorities in Europe, the USA, Japan, and other regions for many years. The problem has been that sodium triphosphate is a contributor to eutrophication of standing or slowly flowing surface waters; that is, it may lead to overfertilization, which in turn encourages extreme algal growth and adversely affects marine organisms. Recognition of the problem led to an intense worldwide search for suitable replacements in the 1960s and 1970s (see also Chap. 10). Developments have been concentrated not only on sequestering agents, but also on ion exchangers, since these are capable of binding polyvalent metal ions [66] – [68]. Most of the promising substitutes examined so far were organic compounds, primarily those derived from raw materials produced by the petrochemical industry. However, few of these substances were available in large quantity. The cost for commercial production on the necessary scale of $> 10^6$ t/a was in many cases prohibitive.

Apart from eliminating cations and achieving good soil and stain removal, other important factors in the washing process are dispersion of soil and prevention of soil redeposition (cf. Chap. 2). Because of the presence of distinct localized charges, sequestering agents are readily adsorbed onto particulate soil. As a result, these compounds often act as effective dispersing agents for such soils.

Figures 46 and 47 show the relationship between dispersing agent concentration and sediment volume for two structurally different soil pigments where the sediment volume is understood as an indicator of the effectiveness of the dispersion process.

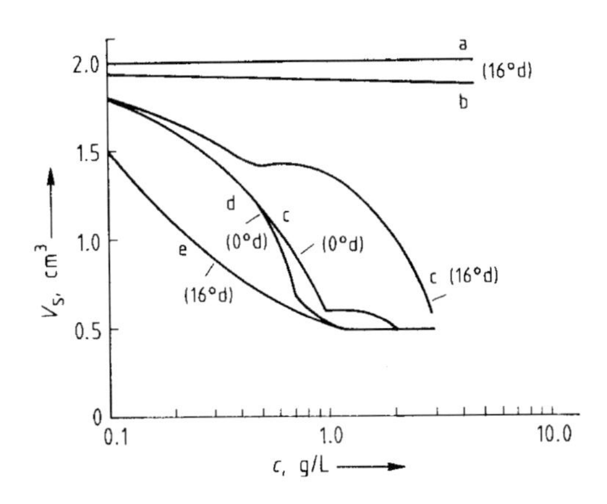

Figure 47. Sediment volume V_s of graphite as a function of active ingredient concentration [4]
a) Sodium sulfate; b) Sodium triphosphate;
c) Sodium *n*-dodecyl sulfate; d) Sodium *n*-dodecyl sulfate + 1.5 g/L sodium sulfate; e) Sodium *n*-dodecyl sulfate + 1.5 g/L sodium triphosphate

Voluminous sediments are formed in the presence of poor dispersing agents; adsorption is insufficient to produce an adequate negative surface charge, and the result is coagulation. Figure 46 shows that the extent of dispersion, in this case of kaolin clay, depends on the specificity with which a sequestering agent is adsorbed. Virtually no effect is produced by sodium sulfate (a) and a nonionic surfactant (b). Anionic surfactants begin to show stabilization only at high concentrations that are rarely used in practice (f).

With graphite, the results are exactly opposite to those observed with kaolin clay (Fig. 47). Sodium triphosphate alone shows no effect and is thus analogous to sodium sulfate (curves a and b), whereas sodium *n*-dodecyl sulfate has a very significant stabilizing effect. Because *n*-dodecyl sulfate is sensitive to water hardness, the dispersing effect increases with diminishing hardness; hence, the sediment volume decreases (both curves c: 16 and 0 °d). Although neither electrolytes nor sequestering agents alone are capable of stabilizing the particulate soil graphite, they are capable of exerting a positive influence, either indirectly through an electrolyte effect (curve d) or in hard water as a result of fixation of alkaline-earth ions (curve e). With sodium triphosphate, sequestration is accompanied not only by an electrolyte effect, but also by a pH effect. An increase in the hydroxide ion concentration and accompanying hydroxide ion adsorption causes increased electrostatic repulsion and, hence, better dispersion (reduced sediment volume). The indirect effect of sequestering agents on the dispersion of hydrophobic particulate soil applies also in principle to emulsification of water-insoluble greasy soil.

A number of additional factors must be considered when selecting a sequestering agent for detergent use. Several important criteria have been listed in Section 3.2. The extent to which evaluation of a sequestering agent can vary depends on how the several criteria are weighted, as is apparent from Table 14.

Table 14. Evaluation of sequestering agents according to various criteria

Ingredient (as sodium salt)	Calcium sequestration	Detergency performance	Incrustation on fabrics and washing machines	Hygroscopicity
Sodium diphosphate	low	very good	very heavy	insignificant
Sodium triphosphate	adequate	very good	little	insignificant
1-Hydroxyethane-1,1-diphosphonic acid	very high	very good	very little	hygroscopic
Nitrilotrimethylenephosphonic acid	high	good	little	very hygroscopic
Nitrilotriacetic acid	high	very good	very little	hygroscopic
N-(2-hydroxyethyl)iminodiacetic acid	less adequate	good		very hygroscopic
Ethylenediaminetetraacetic acid	high	good	very little	very hygroscopic
1,2,3,4-Cyclopentanetetracarboxylic acid	high	fair	little	hygroscopic
Citric acid	low	fairly low	little	insignificant
O-Carboxymethyltartronic acid	high	good	little	hygroscopic
O-Carboxymethyloxysuccinic acid	less adequate	fair		hygroscopic

3.2.3. Ion Exchangers

The disruptive effect of polyvalent metal ions can be reduced not only by the use of low molecular mass sequestering agents, but also by ion exchangers [67]. Table 15 shows that ion exchangers generally have a high binding capacity for calcium but that this usually decreases with increasing temperature.

In the early years the idea of introducing water-insoluble substances into detergents had never been seriously investigated. This was primarily because the known materials lacked sufficient calcium binding ability, were available only in forms with unsuitable particle structure, and were deemed impractical for economic reasons. Success was first achieved through research in the field of sodium aluminum silicates. Systematic investigation revealed that, among the many known types of sodium aluminum silicates, those with a regular crystalline form were appropriate for use in the washing process. One particular modification proved especially applicable and economically interesting: zeolite A [31]–[33], [70]–[74], manufactured under the registered name Sasil® (**s**odium **a**luminum **sil**icate) [66]–[68]. The ion-exchange behavior of this particular water-insoluble crystalline sodium aluminum silicate (Table 15) depends on ionic size and on the state of hydration of the ions. In addition to Ca and Mg, exchange also takes place with, e.g., Pb, Cu, Ag, Cd, Zn, and Hg ions. The elimination of calcium ions — and to a lesser extent magnesium ions — is of greatest importance for the washing process, but ion exchange capacity for heavy metal ions is also important from an ecological standpoint.

Ion exchange is dependent not only on ionic size, but also on concentration, time, temperature, and pH value. Figure 48 shows the exchange kinetics as a function of concentration and time. Calcium ions are exchanged very rapidly. The process occurs somewhat more slowly with magnesium, although exchange becomes more rapid at higher temperature. These results can be explained by the larger hydration shell of

Table 15. Calcium binding capacity of selected ion exchangers [69]

Formula	Chemical name	Calcium binding capacity**, mg CaO/g 20 °C	90 °C
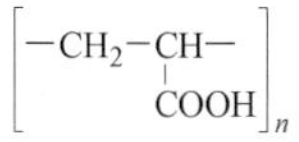 $[-CH_2-CH(COOH)-]_n$	Poly(acrylic acid)	310	260
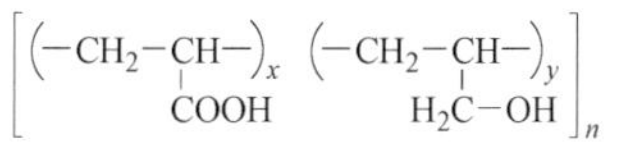 $[(-CH_2-CH(COOH)-)_x\ (-CH_2-CH(H_2C-OH)-)_y]_n$	Poly(acrylic acid-*co*-allyl alcohol)	250	140
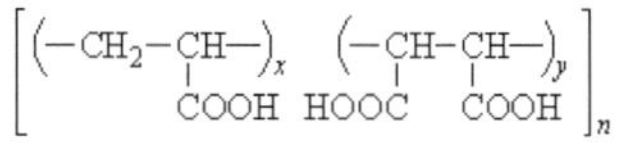 $[(-CH_2-CH(COOH)-)_x\ (-CH(HOOC)-CH(COOH)-)_y]_n$	Poly(acrylic acid-*co*-maleic acid)	330	260
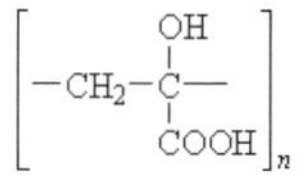 $[-CH_2-C(OH)(COOH)-]_n$	Poly(α-hydroxyacrylic acid)	300	240
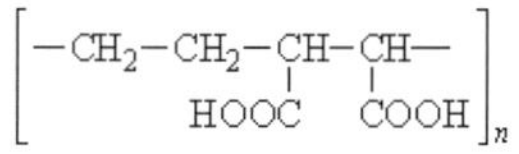 $[-CH_2-CH_2-CH(HOOC)-CH(COOH)-]_n$	Poly(tetramethylene-1,2-dicarboxylic acid)	240	240
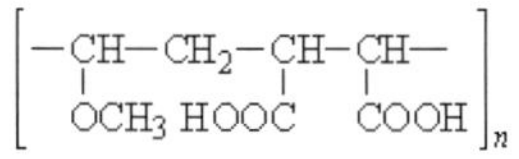 $[-CH(OCH_3)-CH_2-CH(HOOC)-CH(COOH)-]_n$	Poly(4-methoxytetramethylene-1,2-dicarboxylic acid)	430	330
$xNa_2O \cdot Al_2O_3 \cdot ySiO_2 \cdot zH_2O$	Sodium aluminium silicate (Zeolite A)	165	190

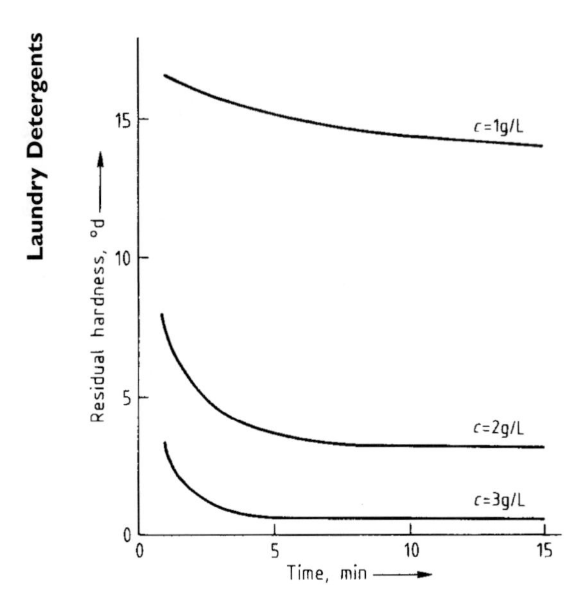

Figure 48. Kinetics of calcium binding to zeolite A [75]
Initial hardness: 30 °d (530 ppm $CaCO_3$); temperature: 25 °C

magnesium ions that impedes exchange at low temperature but is largely destroyed at higher temperature.

The fact that ion exchange occurs in a heterogeneous phase and that adsorption–desorption phenomena are absent means it is advantageous to employ combinations of sodium aluminum silicates with water-soluble sequestering agents. The latter have the ability to remove polyvalent cations, especially calcium and magnesium, from a solid surface, transport them through an aqueous medium, and then release them to the ion exchanger. This property is best described as a carrier effect consisting of the following steps (cf. also Fig. 27):

1) Sorption of the carrier onto the fiber – soil interface
2) Complexation of calcium and magnesium ions
3) Transfer from the solid fiber – soil interface through the wash liquor to the sodium aluminum silicate – water interface
4) Dissociation of the complex into carrier and calcium and magnesium ions
5) Ion exchange of calcium and magnesium ions for sodium ions in the sodium aluminum silicate crystalites
6) Renewed sorption of the carrier on the fiber – soil interface and further complexation with calcium and magnesium ions

If the properties of sodium triphosphate (a water-soluble sequestering agent) are compared with those of sodium aluminum silicate (zeolite A), a number of similarities become apparent, as do typical differences [67], [75]:

1) *Characteristics of sodium triphosphate:*
   a) complex formation with polyvalent ions
   b) alkaline reaction
   c) specific adsorption on particulate soil and fibers
   d) specific electrostatic charging of particulate soil and fibers, dissolution of polyvalent ions from soil and fibers
   e) nonspecific electrolyte effect
2) *Characteristics of sodium aluminum silicate:*
   a) binding capability for polyvalent ions through ion exchange
   b) alkaline reaction
   c) adsorption of molecularly dispersed substances
   d) heterocoagulation with particulate soil
   e) crystallization surface for poorly soluble compounds

The different properties of sodium triphosphate and sodium aluminum silicates can be explained on the basis of their differing solubilities and the divergent ways they eliminate ions contributing to water hardness. In the case of sodium triphosphate, calcium binding occurs by chelation, whereas with sodium aluminum silicates, calcium binding is a result of ion exchange [75].

Ca^{2+} binding by chelation (sodium triphosphate):

$$\left[O{-}\overset{O}{\underset{O}{P}}{-}O{-}\overset{O}{\underset{O}{P}}{-}O{-}\overset{O}{\underset{O}{P}}{-}O \right]^{5-} 5\,Na^{+} \xrightarrow{Ca^{2+}} \left[O{-}\overset{O}{\underset{O}{P}}{-}O{-}\overset{O}{\underset{O}{P}}{-}O{-}\overset{O}{\underset{O}{P}}{-}O \ \ (\text{lower O atoms coordinated to } Ca) \right]^{3-} 3\,Na^{+}$$

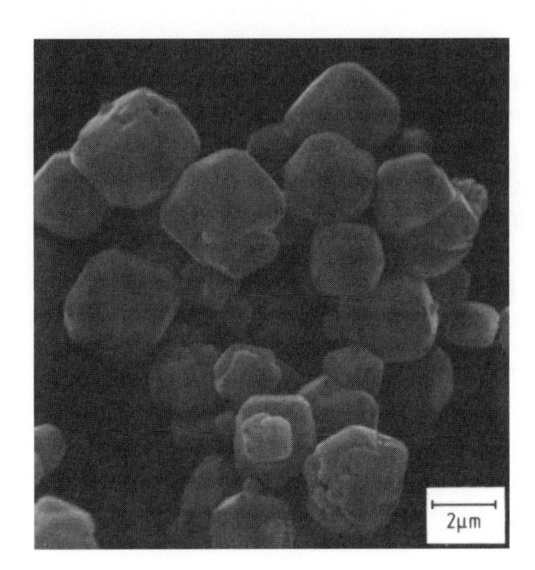

Figure 49. Scanning electron micrograph of zeolite A [71]

Ca^{2+} binding by ion exchange (zeolite A):

$$\left[\begin{array}{c} | \quad\quad | \\ O \quad\quad O \\ | \quad\quad |^{-} \\ -O-Si-O-Al-O- \\ | \quad\quad | \\ O \;\; 2\,Na^{+} \;\; O \\ |^{-} \quad\quad | \\ -O-Al-O-Si-O- \\ | \quad\quad | \\ O \quad\quad O \\ | \quad\quad | \end{array}\right] \xrightarrow{Ca^{2+}} \left[\begin{array}{c} | \quad\quad | \\ O \quad\quad O \\ | \quad\quad |^{-} \\ -O-Si-O-Al-O- \\ | \quad\quad | \\ O \;\; Ca^{2+} \;\; O \\ |^{-} \quad\quad | \\ -O-Al-O-Si-O- \\ | \quad\quad | \\ O \quad\quad O \\ | \quad\quad | \end{array}\right] + 2\,Na^{+}$$

Additional differences result from the exceptional adsorption capability of the triphosphate anion.

The binary builder system sodium triphosphate – zeolite A assures outstanding multiple wash cycle performance; that is, incrustations resulting from precipitation of calcium and magnesium phosphates are largely avoided. Replacement of sodium triphosphate by zeolite A in a detergent leads to some deterioration in removal of some soils and stains. By reoptimization of the detergent formulation with respect to surfactants, bleach system, enzymes, etc., this effect can be easily compensated for.

Alternative ion exchangers like certain water-soluble polycarboxylic acids have also achieved great importance for phosphate-free and low-phosphate detergents (see Section 3.2.4).

Zeolite A, as it is mass produced for use in detergents, appears as cubic crystals with rounded corners and edges. These crystalline particles show a tendency to form agglomerates [67], [71] (Fig. 49).

The average diameter of a zeolite A particle is ca. 4 µm. A diameter distribution curve obtained by differentiation of a residue summation curve (determined by using a

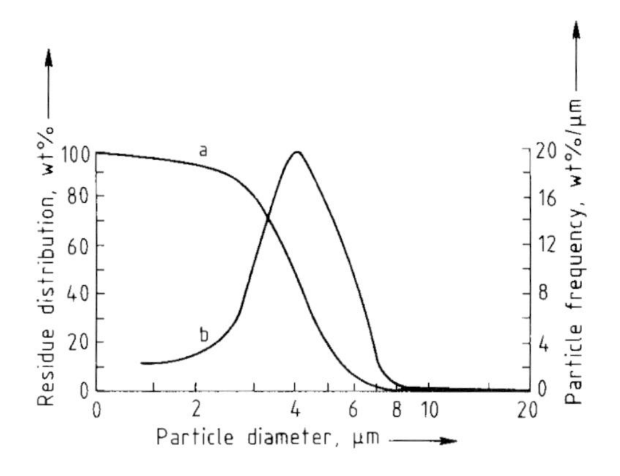

Figure 50. Residue summation curve and particle diameter distribution curve for zeolite A [71]
Coulter counter, 50 µm cell
a) Residue distribution; b) Particle frequency

Table 16. Comparison of various ion exchangers [4]

Substance (as sodium salt)	Calcium sequestration	Detergency performance	Incrustation on fabrics and washing machines	Hygroscopicity	Biodegradability*
Poly(acrylic acid)	very high	good	little	very hygroscopic	minimal
Poly(α-hydroxyacrylic acid)	high	very good	little	hygroscopic	minimal
Poly(acrylic acid-*co*-allyl alcohol)	high	fairly good	little	very hygroscopic	minimal
Poly(4-methoxytetramethylene-1,2-dicarboxylic acid)	high	good	little	very hygroscopic	minimal
Poly(tetramethylene-1,2-dicarboxylic acid)	high	good	little	very hygroscopic	minimal
Poly(acrylic acid-*co*-maleic acid)	high	good	little	hygroscopic	minimal
Sodium aluminum silicate (zeolite A)	high	good**	little	minimal	irrelevant

* Under the conditions of the Closed Bottle Test [250].
** In combination with appropriate soluble sequestering agents or soluble ion exchangers.

Coulter counter) shows quite a sharp maximum at the corresponding abscissa value (Fig. 50).

Zeolite A is specially optimized for use in detergents. Its unique particle form, cubes with rounded corners, protects fibers against damage, and the tiny particles with their narrow range of particle sizes ensures that deposition on laundry is practically nonexistent [67].

Table 16 summarizes a number of important criteria for the evaluation of water-soluble and water-insoluble ion exchangers.

3.2.4. Builder Combinations

Since the beginning of the 1980s, the ternary builder combination zeolite A – polycarboxylate – soda ash has been increasingly used in nonphosphate detergents all over the world [76]. Polycarboxylates — polyacrylate and poly(acrylate-*co*-maleinate) — have an outstanding function in this ternary builder system in that they effectively prevent lime and soil depositions on laundry and washing machines. The mechanism of action is based on the *threshold effect* (see Section 3.2.2) [77], [78]. Further ternary builder systems such as sodium triphosphate – sodium carbonate – sodium polycarboxylate are also frequently applied, of course only in those global regions where phosphate-built laundry detergents are common. The use of substoichiometric quantities of high molecular mass polymers or copolymers of acrylic and maleic acid in these formulations enables a reduction in the content of sodium triphosphate and its partial substitution by the more cost effective sodium carbonate. This is due to the ability of the polycarboxylates to retard or inhibit crystal growth of sparingly soluble salts, such as calcium phosphate and calcium carbonate. Increasing charge density of the polycarboxylic acid (i.e., maleate content in the copolymer) increases its sequestration capacity [78], [79].

Special crystalline and amorphous silicates, sodium carbonate – silicate compounds, and zeolite P are also used as alternatives to zeolite A. These alternatives are found in some European, U.S., and Japanese nonphosphate detergent formulations [80], [81]. Zeolite P was introduced as a builder in laundry detergents in 1994 [82], [83]. It has the chemical formula $Na_2O \cdot Al_2O_3 \cdot (2-5)\ SiO_2 \cdot x\ H_2O$ and is also known under the trade names zeolite MAP, Doucil A 24, and Wessalith NaP. Zeolite P has the advantage of substantially stronger calcium binding than zeolite A. Furthermore, zeolite P exhibits a superior uptake for nonionic surfactants due to is smaller average particle size and, hence, larger surface per unit weight than zeolite A [167], [200] – [202]. This property can be essential for special detergent manufacturing technologies.

Unlike zeolites, silicate-based builders are water-soluble and contribute to the alkalinity of the washing liquor. They are generally better compatible with sensitive bleach components such as percarbonate. Crystalline layered disilicate $Na_2Si_2O_5$ and sodium carbonate – sodium silicate are most suitable for use as detergent builders. They are commercially available under the tradenames SKS-6 and Nabion 15, respectively. Due to their superior efficiency in preventing buildup of lime deposits, they can replace zeolite A in nonphosphate detergents [80], [81], [194].

3.3. Bleaches

The term bleach can be taken in the widest sense to include the induction of any change toward a lighter shade in the color of an object. Physically, this implies an increase in the reflectance of visible light at the expense of absorption.

Bleaching effects can occur through mechanical, physical, and/or chemical means, specifically through change or removal of dyes and soil adhering to the bleached object. In the washing process, all of these processes occur in parallel, but to varying extents. The relative significance of each is determined in part by the nature of the soil and dyes present. Mechanical/physical mechanisms are effective primarily for the removal of particulate and greasy soil. Chemical bleaching is employed for the removal of colored nonwashable soils and stains adhering to fibers and is accomplished by oxidative or reductive decomposition of chromophoric systems. Only oxidative bleaches are used in laundry products to a great extent; many soils commonly encountered in everyday laundry contain compounds which, if bleached reductively, become colorless but may later return to their colored forms as a result of subsequent atmospheric oxidation. Nevertheless, this generalization does not rule out the use of special reductive bleaches (e.g., $NaHSO_3$, $Na_2S_2O_4$) to treat specific types of discoloration occurring in either household or institutional settings.

The extent of the bleaching effect that can be achieved is dependent on a number of factors, including the type of bleach, its oxidation potential and concentration, as well as the residence time in the washing or rinsing process, the wash temperature, the type of soil to be bleached, and the nature of the fabric.

Bleachable soils encountered in household and institutional laundry consist of a broad spectrum of diverse materials which are generally of vegetable origin and contain primarily polyphenolic compounds. These include the anthocyanin dyes derived from, e.g., cherries, blueberries, and currants, and curcuma dyes from curry and mustard. The brown tannins found in, e.g., fruit, tea, and wine stains arise from condensation of polyphenols with proteins. Other brown organic polymers include the humic acids present, for example, in coffee, tea, and cocoa. The green dye chlorophyll and the red betanin from beets are pyrrole derivatives, as are the urobilin and urobilinogen dyes derived from degradation of hemoglobin and discharged in urine. Carrot and tomato stains contain carotenoid dyes. Dyes of commercial origin such as those found in cosmetics, hair coloring agents, and ink are also important. Blood is also a bleachable soil, but its removal can sometimes present problems.

Two procedures have attained major importance in oxidative bleaching during the washing and rinsing processes: peroxide bleaching and hypochlorite bleaching. The relative extent of their application varies, depending heavily on laundering habits in the different global regions.

3.3.1. Bleach-Active Compounds

The dominant bleaches in Europe and many other regions of the world are of the *peroxide* variety. Hydrogen peroxide is converted by alkaline medium to the active intermediate hydrogen peroxide anion according to the following equation:

$$H_2O_2 + OH^- \rightleftharpoons H_2O + HO_2^-$$

The perhydroxyl anion oxidizes bleachable soils and stains. The usual sources of hydrogen peroxide are inorganic peroxides and peroxohydrates. The most frequently encountered source is sodium perborate (sodium peroxoborate tetrahydrate, $NaBO_3 \cdot 4\,H_2O$) [84], which in crystalline form contains the peroxodiborate anion:

$$\left[\begin{array}{ccccc} HO & & O{-}O & & OH \\ & B & & B & \\ HO & & O{-}O & & OH \end{array}\right]^{2-} 2\,Na^+ \cdot 6\,H_2O$$

Peroxodiborate is hydrolyzed in water to form hydrogen peroxide. Sodium perborate monohydrate and sodium percarbonate, which dissolve more quickly than perborate tetrahydrate, have been increasingly used either as separate color-safe bleach or in bleach-containing detergents at the expense of sodium perborate tetrahydrate. The monohydrate version has an improved storage stability as compared with the tetrahydrate version. Sodium perborate monohydrate is the preferred ingredient for laundry products in countries having high ambient temperatures because, contrary to perborate tetrahydrate, the monhydrate does not cause caking of powder products upon storing at elevated temperatures for prolonged periods. In contrast to what its name would suggest, sodium perborate monohydrate contains no water of crystallization. Its molecular structure is [85]:

$$\left[\begin{array}{ccccc} HO & & O{-}O & & OH \\ & B & & B & \\ HO & & O{-}O & & OH \end{array}\right]^{2-} 2\,Na^+$$

On a per weight basis the monohydrate has a considerably higher content of active oxygen than the tetrahydrate. For this reason, the monohydrate has become the favorite form of perborate in compact detergents.

The use of sodium percarbonate has gained importance in those countries in which boron is either banned or restricted for environmental and regulatory reasons, or negatively discussed. In contrast to sodium perborate, sodium percarbonate is a true peroxohydrate ($Na_2CO_3 \cdot 1.5\,H_2O_2$). In order to achieve a good storage stability for the bleach, coated or stabilized percarbonate must be used.

The salts of peroxomono- and peroxodisulfuric acid and peroxomono- and peroxodiphosphoric acid are not significant as detergent bleaches. This results largely from their insufficient bleaching power in wash liquor, either because they are insufficiently hydrolyzed to hydrogen peroxide in an alkaline medium or because their oxidation potential is too low.

Comparison shows that sodium perborate and sodium percarbonate have the most preferred prerequisites for use as a detergent bleach additive. Oxygen bleaches have been reviewed [85] – [88].

The concentration of bleach-active hydrogen peroxide anion increases with pH value and temperature. Sodium perborate exhibits significantly less bleaching efficiency at temperatures below 60 °C. Even at low temperatures (i.e., where the reaction equilibrium is unfavorable), hydrogen peroxide anions are present in the wash water, but show

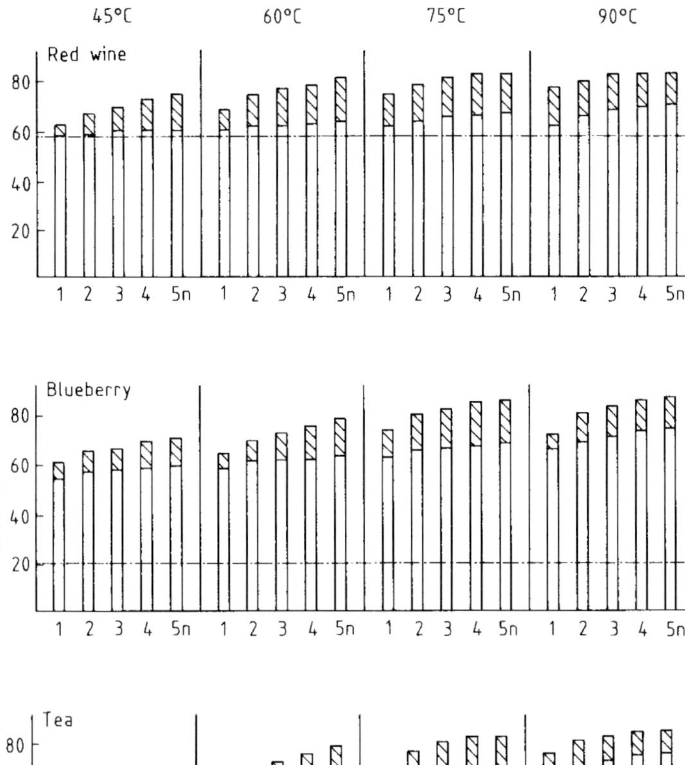

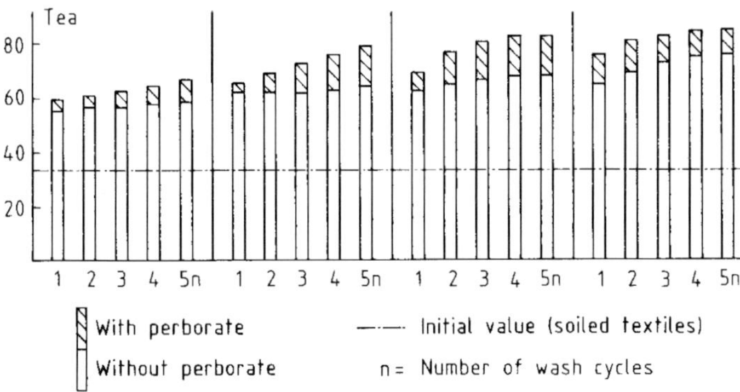

Figure 51. Bleach performance of sodium perborate vs. temperature [89]
Initial perborate tetrahydrate concentration 1.5 g/L = 150 mg of active oxygen/L

only modest bleaching power (Fig. 51). The bleaching effect also increases markedly with increasing perborate concentration (Fig. 52) and time.

Many attempts were made in the 1980s to improve the performance of a detergent by using organic peroxy acids, for example, monoperoxyphthalic acid and diperoxydodecanedioic acid (DPDDA) salts, as bleach components:

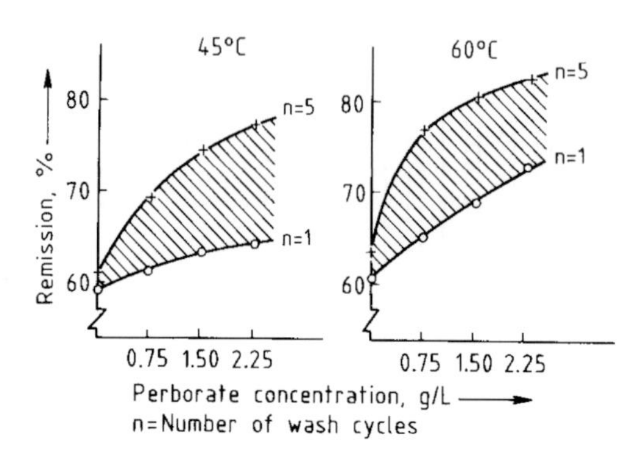

Figure 52. Bleach performance of sodium perborate vs. concentration (red wine soil) [89]

Monoperoxyphthalic acid
monomagnesium salt

Diperoxydodecanedioic acid
sodium salt

With these it is possible to obtain significant bleaching at a temperature as low as 30 °C. However, for several reasons (dye damage and cost effectiveness) a technical and commercial breakthrough with these peroxyacids has not been achieved so far.

Hypochlorite is used for bleaching in many global regions where laundry habits, such as cold water washing, cause sodium perborate to be less effective. In an alkaline medium, hypochlorite bleaches are converted to the hypochlorite anion:

$$HOCl + OH^- \rightleftharpoons ClO^- + H_2O$$

Hypochlorite bleach chemistry and applications have been reviewed [90] – [92].

Hypochlorite can be used in either the wash or the rinse cycle at concentrations between ca. 50 and 400 mg/L active chlorine (Fig. 53) and a wash liquor ratio (bath ratio) between 1 : 15 and 1 : 30.

Normally, an aqueous solution of sodium hypochlorite (NaOCl) is used as a source of active chlorine. Organic chlorine carriers (e.g., sodium dichloroisocyanurate), which are

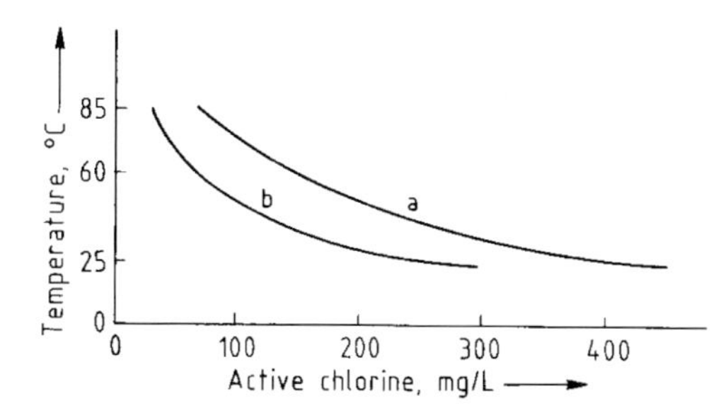

Figure 53. Recommended concentrations of active chlorine as a function of wash temperature [93]
a) Maximum concentration of application; b) Minimum concentration of application
"Active or available chlorine" is calculated as twice the actual weight percent of chlorine in the hypochlorite molecule

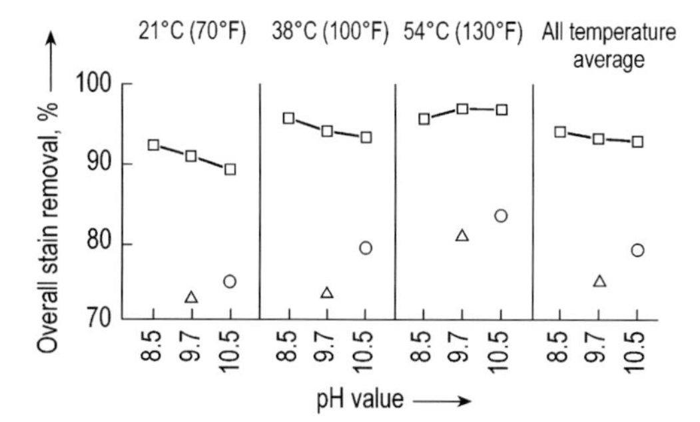

Figure 54. Stain removal comparisons
□ Hypochlorite (200 ppm active chlorine) + 0.15 % detergent; ○ Perborate bleach (28 ppm active oxygen) with protease enzymes + 0.15 % detergent; △ 0.15 % Detergent only[94]

hydrolyzed to hypochlorite in an alkaline medium, are only common in automatic dishwashing detergents. Figure 54 shows a comparison of the bleaching properties of sodium hypochlorite and sodium perborate as a function of temperature and pH value.

One of the major advantages of powdery sodium perborate over liquid sodium hypochlorite is the fact that whereas the latter must be added separately in either the wash or the rinse cycle, perborate can be incorporated directly into a powder laundry product. This results in a mild color-safe bleach. By contrast, successful addition of chlorine bleach solution, whether in a household or an institutional machine, is heavily dependent on experience of the user and on observing the manufacturer's recommendations. Incorrect dosage of sodium hypochlorite may easily occur, and this may cause significant damage to laundry and colors. A further advantage of sodium perborate is its long shelf life, whereas sodium hypochlorite solutions have limited storage stability, which depends very much on the content of impurities, particularly traces of heavy metal ions. On the other hand, hypochlorite bleaches can be used in both wash and rinse cycles independent of temperature. They provide effective bleaching and disinfection (hygiene) even at very low temperature. However, because of its great reactivity and extraordinarily high oxidation potential, sodium hypochlorite, in contrast to sodium perborate, may cause problems with textile dyes and most fluorescent whitening agents, both of which often show poor stability in the presence of chlorine. Studies of washing and bleaching habits show that peroxide bleach use dominates in Europe, chlorine bleach being used predominantly in Mediterranean countries. Hypochlorite bleach in either the wash or the rinse cycle is still the preferred bleaching agent in a large part of the world (cf. Table 17). The use of peroxide

Table 17. Washing habits in different regions

Washing conditions	United States	Japan	Europe
Washing machine	vertical axis/agitator type	vertical axis/pulsator type	horizontal axis drum type
Heating equipment	no	no	yes
Fabric load, kg	23 – 5	4 – 8	2 – 5
Amount of water, L	medium loads: ca. 50 – 60 large loads: ca. 65 – 75	low: 30 high: 65	low: 10 – 16 high: 20 – 25
Total water consumption, L (regular cycle)	120	120 – 150	55 – 90
Wash liquor ratio	ca. 1 : 25	ca. 1 : 10 – 1 : 15	ca: 1 : 4 – 1 : 10
Washing time, min	8 – 18	5 – 15	50 – 60 (90 °C) 20 – 30 (30 °C)
Washing, rinsing, and spinning time, min	20 – 35	15 – 35	100 – 120 (90 °C) 50 – 70 (30 °C)
Washing temperature, °C	hot: 55 (130 °F) warm: 30 – 45 (90 – 110 °F) cold: 10 – 25 (50 – 80 °F)	5 – 25	90 60 40 30
Water hardness (average), ppm $CaCO_3$	relatively low, 100	very low, 30	relatively high, 250
Detergent dispenser	mostly no	mostly no	yes
Recommended detergent dosage*, g/L	1.0 – 1.2	0.5 – 0.8	4.5 – 8.0
g/kg laundry	15 – 30	5 – 7	15 – 35
Peroxide bleach	mostly added separately	mostly dosed separately	mostly incorporated in detergent
Chlorine bleach	added separately	dosed separately	predominantly in mediterranean countries

* In the United States and Japan mostly without bleaching components.

("oxygen") bleach in laundry products has increased significantly in the USA, Japan, and other regions of the world during the 1980s and 1990s.

3.3.2. Bleach Activators

To achieve satisfactory bleaching with sodium perborate and sodium percarbonate at temperatures $\leq$ 60 °C, so-called bleach activators are commonly utilized. These are mainly acylating agents incorporated in laundry products. When present in a wash liquor of pH value 9 – 12, these activators preferentially react with hydrogen peroxide (perhydrolysis) to form organic peroxy acids in situ. As a result of their higher oxidation potentials relative to hydrogen peroxide, these intermediates demonstrate effective low-temperature bleaching properties. Due to the low concentration, resulting from the in situ generation, these peroxyacids are much less aggressive to fabric dyes and fluorescent whitening agents than sodium hypochlorite. However, activated peroxy bleach systems may adversely affect the color fastness of fibers treated with some specific dyes upon multiple washing [95]. Among the wide variety of bleach activators investigated [96], only the following compounds have been incorporated on a large scale in laundry products worldwide:

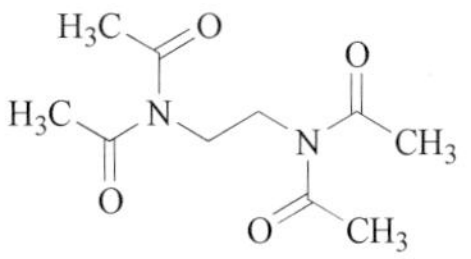

Tetraacetylethylenediamine (TAED) [97].

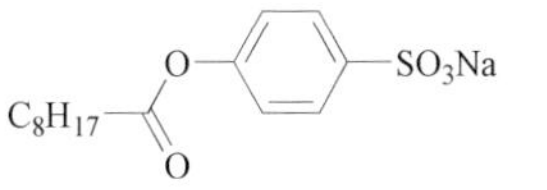

Sodium *p*-nonanoyloxybenzenesulfonate (NOBS) [104]

The first activator reacts with hydrogen peroxide according to the following scheme to produce peroxyacetic acid:

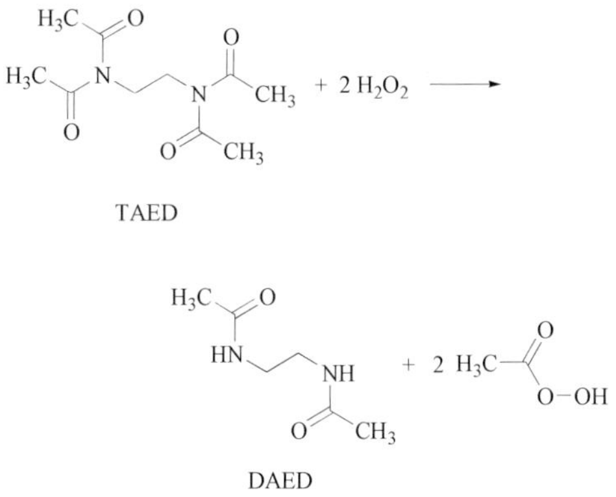

TAED $+ 2\ H_2O_2 \longrightarrow$ DAED $+ 2\ H_3C{-}C(=O){-}O{-}OH$

Under the conditions of the washing process the reaction activates only two acetyl groups. TAED has been used in laundry products in Europe since 1977 and has became state-of-the-art in many laundry detergents and bleaches. TAED has remained the bleach activator of choice in Europe, whereas NOBS has been predominantly used in the Americas and Asia. The broad spectrum of effectiveness of TAED has been reviewed [98] – [101]. Sodium *p*-nonanoyloxybenzenesulfonate (NOBS) reacts with hydrogen peroxide (is perhydrolyzed) to produce peroxynonanoic acid. Aside from their efficiency regarding improvement of low-temperature bleaching, these bleach activators markedly contribute to enhancing the hygiene of laundry due to the biocidal effect of the peracids generated in situ [102], [103]. The effectiveness of the system sodium

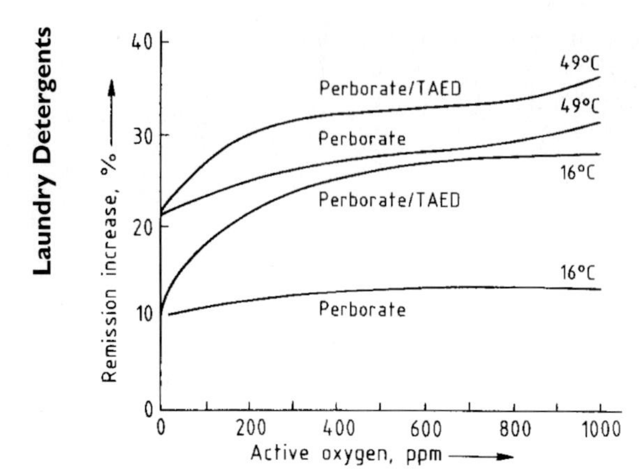

Figure 55. Bleaching effect as a function of bleach concentration [105]

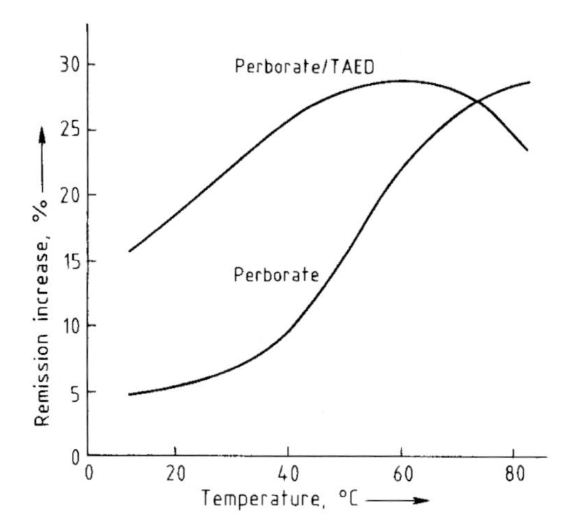

Figure 56. Bleaching effect as a function of temperature [105]

perborate – TAED on tea stains as a function of concentration and temperature is shown in Figures 55 and 56.

Perborate – TAED bleach systems are not sufficiently effective at cold washing temperatures used in large regions of the world. However, soaking of the laundry in a detergent containing activated perborate prior to washing can greatly enhance the removal of bleachable stains due to the prolonged time of soaking. The problem of graying of white textiles caused by dye transfer during soaking can be reduced by the addition of perborate – TAED to a detergent [106].

3.3.3. Bleach Catalysts

Numerous attempts have been made to introduce into sodium perborate and/or percarbonate based laundry products small amounts of catalysts that would increase their bleaching performance, especially at low temperature. In most cases, traces of heavy metal ions have been used. While addition of heavy metal ions alone can indeed cause decomposition of sodium perborate, it does not lead to a better bleach. In fact, bleaching action is usually diminished, and serious fiber and color damage may occur. The patent literature describes a number of examples in which the bleaching power of sodium perborate can be increased by heavy metal ion chelates. However, the effectiveness of such systems remains controversial, and catalysts have not been accepted as a part of the washing process. Excellent catalytic bleach enhancement by manganese complexes derived from 1,4,7-trimethyl-1,4,7-triazacyclononane and related ligands has been reported [107]. However, under certain conditions severe dye and fiber damage has been observed in connection with this catalyst system. For this reason detergents using manganese catalysts introduced on the European markets in 1994 were later withdrawn.

Photobleaching with aluminum or zinc phthalocyaninetetrasulfonate represents another form of bleach catalysis. With photobleaching, oxygen from the atmosphere is catalytically activated by the phthalocyanine compound, and the active oxygen generated in turn bleaches oxidizable stains. Suitable conditions for the few detergents that contain photobleaches include — apart from soaking in sunlight — slow drying of the laundry under conditions of high humidity and high light intensity. Therefore, photobleach catalysts are of interest and have been employed in laundry products only in regions with intense solar radiation.

3.3.4. Bleach Stabilizers

Traces of ions such as copper, manganese, and iron catalyze the release of oxygen from bleach systems. This reduces bleach effectiveness and at the same time causes damage to fabrics and dyes. The addition of small amounts of finely divided magnesium silicate largely suppresses this catalysis as a result of absorption of the heavy metal ions. Therefore, magnesium silicate was used in perborate based laundry detergents until the 1980s.

Another possibility for elimination of trace heavy metal ions is the addition of selective complexing agents. Sodium triphosphate has only a modest complexing effect with heavy metal ions and is incapable of causing sufficient bleach stabilization in their presence. By contrast, complexing agents such as sodium hydroxyethanediphosphonate (HEDP) or sodium diethylenetriamine pentakis(methylenephosphonate), as well as other phosphonic acids, exhibit a marked stabilizing effect. The use of these phosphonates is state of the art in many modern laundry products. Amounts of much less than 1 wt % of these strong, selective complexing agents suffice in detergent formulations.

3.4. Further Detergent Ingredients

Surfactants, builders, and bleaches are quantitatively the major components of modern detergents; the auxiliary agents — also called additives — discussed in this section are introduced only in small amounts, each to accomplish its own specific purpose. Today, their absence from current detergent formulations is difficult to imagine. They have greatly contributed to making laundry detergents more multifunctional during the last 30 years.

3.4.1. Enzymes

Stubborn proteinaceous stains derived from sources such as milk, cocoa, blood, egg yolk, and grass are resistant to removal from fibers by enzyme-free detergents, particularly when stains have dried-on. So are chocolate and starch-based food stains as well as greasy/fatty stains, particularly in low-temperature washing. Proteolytic, amylolytic, and lipolytic enzymes are usually capable of eliminating such soil without major problems during washing.

Use of enzymes in detergents was first described by Otto Röhm in a 1913 patent application. Enzyme-containing detergents failed to play a major role in the following decades because the only available proteolytic enzyme preparation, a pancreatic extract obtained from slaughtered animals, was too sensitive to the alkaline and oxidative components of detergents. Only in 1959 were Röhm's ideas realized by the use of proteolytic enzymes from fermentation with specific strains of bacteria (*Bacillus subtilis*, later on *Bacillus licheniformis*). These *protease* enzymes were highly resistant to alkali and showed adequate stability at temperatures as high as ca. 65 °C for the time period required by normal washing processes. The first successful detergents on the market containing these proteases were Bio 40 from Gebr. Schnyder in Switzerland in 1959, followed by Biotex produced by Kortman & Schulte in The Netherlands in 1963. Commercial production of detergent enzymes experienced rapid expansion in the years that followed. For example, by 1969 nearly 80 % of the detergents marketed in the Federal Republic of Germany contained proteases as enzyme additives. Industry reduced this proportion to $< 50\,\%$ as a result of public discussion of certain toxicological concerns. The reservations regarding detergent enzymes have since been addressed and overcome by technical modifications, particularly those involving the delivery form in which enzymes are introduced into detergents: enzymes have been manufactured since 1971 as dust-free granulates, prills, extrudates, or pellets. Today's detergent enzymes are perfectly safe. As a result, nearly all detergents produced in Europe, North America,

Table 18. Historical review of the preparation and use of detergent enzymes [108], [109]

Year	Enzyme	Enzyme-containing detergents
1913	Otto Röhm claims the use of tryptic enzymes for detergents	
1931		detergents containing pancreatic enzymes (Burnus, Germany)
1959	Bacterial protease	Bio 49 (Gebr. Schnyder, Switzerland)
1960	Alcalase and Maxatase, microbial proteases made available on a commercial scale by Novo Industri, Bagsvaerd, Denmark and Gist-Brocades, Delft, The Netherlands, respectively	
1963		prewash detergent Biotex (Kortman & Schulte, The Netherlands)
1966		first heavy-duty detergent with microbial proteases
1969		microbial proteases contained in 80 % of all detergents in Germany
1970		severe setback of production of microbial proteases due to public criticism (the "allergy debate")
1972	Additional microbial enzymes made commercially available for use in detergents (amylases)	enzymes in detergents declared to be safe by the German Federal Health Agency
1973	Amylase introduced in Germany	presoak and prewash detergent (Mustang, Henkel, Germany)
1975		market share of enzyme-containing detergents in Germany is 80 %
1975	Liquid proteases for HDL (Maxatase LS, Alcalase L)	
1980s	High-alkaline proteases become available (Savinase, Novo; Maxacal, IBIS; BLAP, Henkel; Purafect, Genencor)	improved efficacy and storage stability for enzymes in heavy-duty detergents
1986	Cellulase (Celluzyme, Novo)	
1987		heavy-duty detergent containing cellulase KAC (Attack, Kao, Japan)
1988	Lipase (Lipolase, Novo)	heavy-duty detergent containing lipase (Hi Top, Lion, Japan)
1988 – 1989	Bleach-resistant proteases (Maxapem, Gist-Brocades; Durazym, Novo)	
1990s	Enzymes spreading out in detergents worldwide (China, India, South America)	roll-out of multifunctional detergents with more than one enzyme in Europe, Japan, and the USA

Japan and many other countries worldwide contain enzymes today. Table 18 provides a brief historical review of the production and application of detergent enzymes.

The effectiveness of proteolytic, amylolytic, and lipolytic detergent enzymes is based on enzymatic hydrolysis of peptide, glucosidic, or ester linkages, respectively. To be well suited for use in detergents, enzymes must exhibit the following properties [108]:

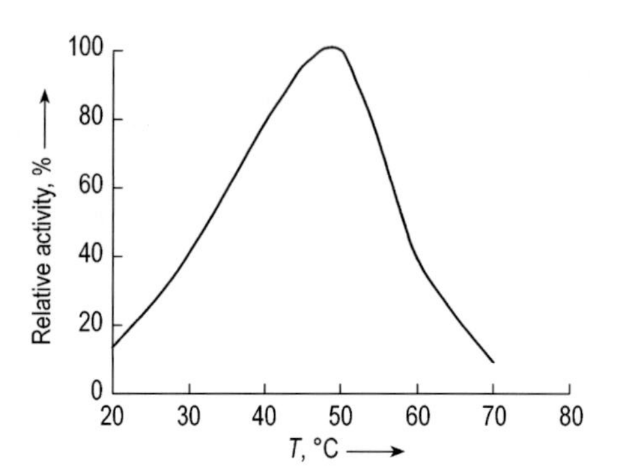

Figure 57. Activity of Savinase at different temperatures [110]

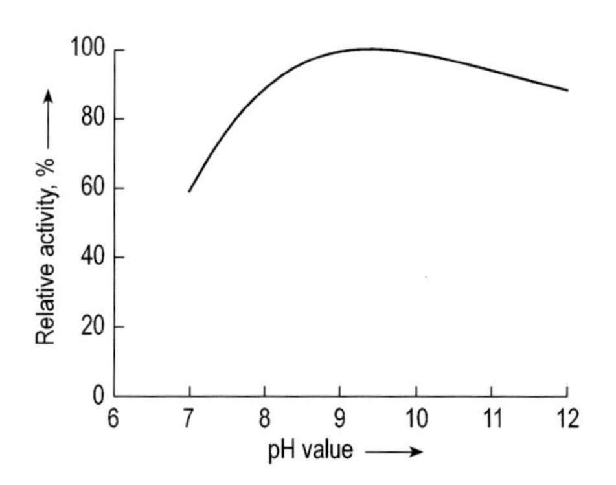

Figure 58. Activity of Savinase at different pH values [110]

Activity optimum at alkaline pH
Efficacy at low wash temperatures of 20 – 40 °C
Stability at wash temperatures up to 60 °C
Stability in the presence of other detergent ingredients, such as surfactants, builders, and activated bleach, both during storage and use
Low specificity to soils, i.e., a specificity broad enough to enable the degradation of a large variety of proteins, starches, and triglycerides

The activity of enzymes depends on temperature and pH value of the wash bath. Figures 57 and 58 show these respective dependencies of a typical commercially available protease (Savinase) [110].

Ever since detergent-grade enzymes were introduced in the 1960s there has been made considerable progress in terms of

Delivery form
Reduced dust content
Improved alkali resistance
Increased low-temperature activity
Increased activity per unit weight of finished product
Storage stability
Resistance to active oxygen

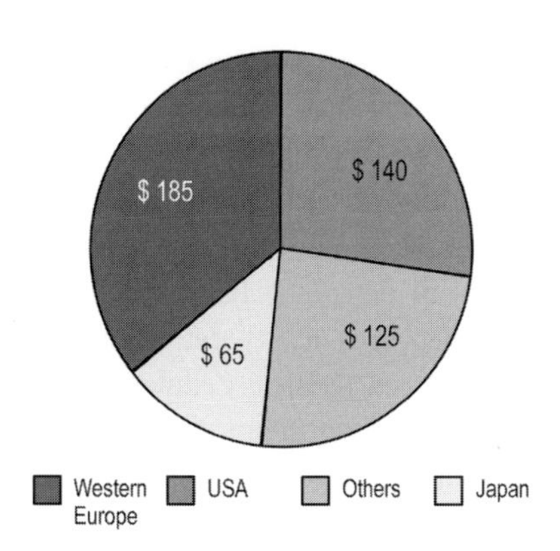

Figure 59. Regional detergent enzyme market values (1995, 10^6 US$)

Effectiveness, functionality
Genetic and enzyme engineering
Decreasing market prices

Enzymes in detergents have therefore meanwhile become state-of-the-art worldwide. Their use spread dramatically in Western Europe, North America, and Japan in the 1970s and 1980s. The 1990s saw also the opening up of the vast detergent markets of China, South America, and India for enzymes [109]. Figure 59 [111] shows the geographical use and value of enzymes in the world. Penetration of enzymes in the U.S. market is 75 %, compared with 95 % for both Europe and Japan. The mainstay of the market have been the protease types.

Only above a certain temperature of the wash bath (mostly 55 °C) does the activity of enzymes decrease, and then very rapidly. Other modes of decomposition present little problems. For example, intensive research has shown that detergent-grade enzymes currently available are not subject to significant damage during storage or in the wash liquor, provided detergents containing them are properly formulated. Stability problems with respect to anionic surfactants and active oxygen were once the subject of debate, but usually these are readily solved by the proper choice of enzyme type and the appropriate formulation.

Especially in the U.S. and European markets, *amylases* have been added to detergents along with proteases since 1973 to capitalize on the activity of the amylases toward starch-containing soils. Of the different amylases available, only α-amylases are used for detergents. They are able to catalyze the hydrolysis of the amylose and amylopectin fractions of starch, i.e., cleavage of the α-1,4-glycosidic bonds of the starch chain [112]. This eases the removal of starch-based stains by the detergent.

In this context, the *lipases* which, due to their substrate specificity, ease the removal of triglycerides-containing soils are also worth mentioning. The activity of lipases is highly dependent on temperature and concentration. Studies on model greasy soils have shown that removal of solid fats at very low temperatures (e.g., 20 °C) is largely due to the action of added lipases [113]. Lipases produced economically by applying genetic engineering techniques have been commercially available since 1988. The removal of triglyceride-based fatty stains is mostly evident after several washings and drying cycles. At present lipases are contained as further enzymes in premium branded detergents.

Cellulases are capable of degrading the structure of damaged (amorphous) cellulose fibrils, which exist mostly at the surface of cotton fibers after multicycle washing and using. Cotton textiles treated with a cellulase-containing detergent have a smooth surface. For these reasons cellulases are preferably applied in specialty detergents for delicate fabrics or in color heavy-duty detergents for colored textiles. Cellulases also help prevent redeposition of soil on cotton fibers. As such, garments treated with cellulase-containing detergents look new for a longer time, i.e., whites and coloreds do not look dull or dingy after multicycle washing. Today many detergents contain blends of two, three, or even four enzymes [114].

3.4.2. Soil Antiredeposition Agents, Soil Repellent/Soil Release Agents

(see also Section 2.4.3)

The principal characteristic expected of a detergent is that it will cause soil to be removed from textile fibers during the washing process. Removed soil is normally finely dispersed, and if a suboptimal detergent formulation is employed, or too little detergent is used, some may return to the fibers. This is termed a wash liquor showing "insufficient soil antiredeposition capability". The problem becomes especially apparent after multicycle washing as a distinct graying of the laundry, which then looks dull and dingy.

Redeposition of displaced soil can be largely prevented by carefully choosing the various detergent components (surfactants and builders), but addition of special antiredeposition agents is also helpful. Such agents act by becoming adsorbed irreversibly, i.e., in a way that prevents their removal by water, on both textile fibers and soil particles. Adherence of the soil to fibers is thereby hindered. This phenomenon is called soil repellent or soil release effect. Classical antiredeposition agents are carboxymethyl cellulose (CMC) derivatives bearing relatively few substituents. Analogous derivatives of carboxymethyl starch (CMS) have played a similar role. These substances are effective only with cellulose-containing fibers such as cotton and blends of cotton and synthetic fibers. With the increasing replacement of these natural fibers by synthetics, on which CMC has virtually no effect, the need arose to develop other effective antiredeposition agents and soil repellents. Some surfactants have been found to be well-suited to the purpose (Fig. 60), as are such nonionic cellulose ethers [115], [116] as the following:

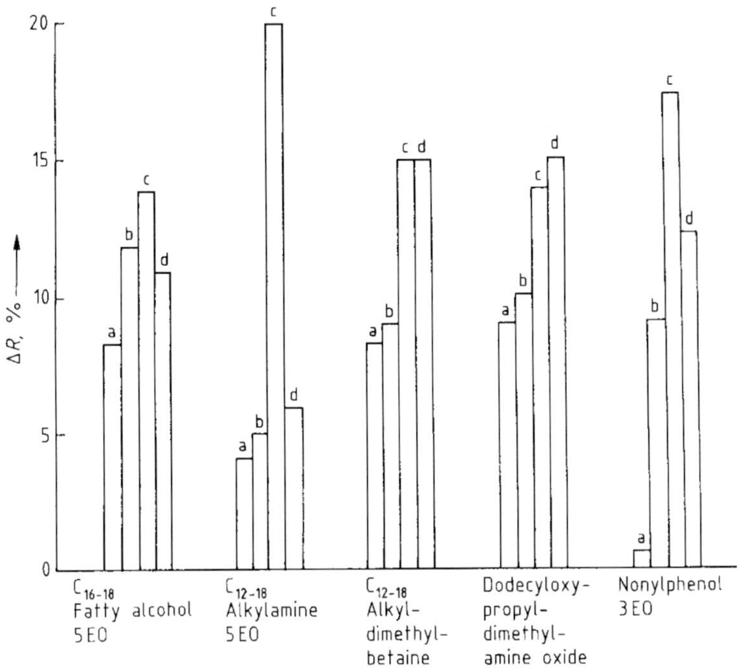

Figure 60. Influence of specific surfactants on the soil antiredeposition capability of detergents [40]
ΔR indicates the improvement in the soil antiredeposition capability conferred by a given additive
a) Cotton, 95 °C; b) Permanent-press cotton, 95 °C; c) Polyester, 60 °C; d) Polyester/cotton, 60 °C
5 g/L Detergent; 0.2 g/L surfactant; number of wash cycles: 3; time: 30 min

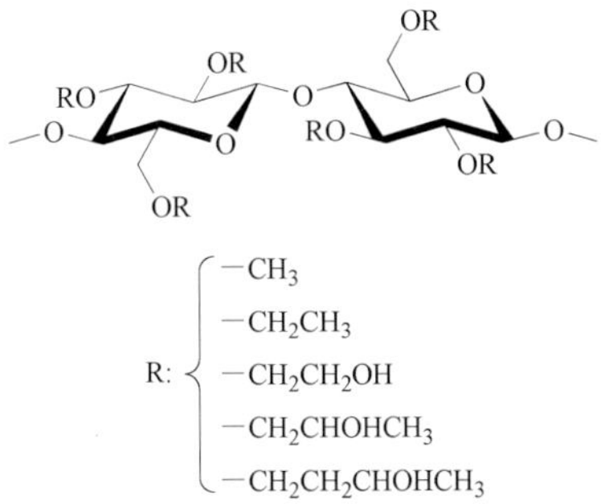

Also anionic derivatives of polymers from terephthalic acid and polyethylene glycol have proved to be very effective soil repellents, particularly on polyester fibers and polyester – cotton blends. They impart hydrophilic properties to these fibers and thus strongly repel oily/greasy soil [117] – [119].

Because of the multitude of fabrics currently on the market, manufacturers supplement most modern detergents with mixtures of anionic and nonionic polymers (e.g., carboxymethyl cellulose – methyl cellulose) and polymers from polyethylene glycol and terephthalic acid and their anionically modified derivatives.

Polymer cobuilders, such as sodium polycarboxylates, and cellulase enzymes have also proved to exert a very good soil antiredeposition effect with cotton fibers.

3.4.3. Foam Regulators

For soap detergents, which are very popular in regions with low per capita income worldwide, foam is understood as an important measure for washing performance. With detergents based on synthetic surfactants, however, soap has lost virtually all of its former significance in the industrialized countries. Nonetheless, most consumers — apart from those using horizontal axis drum-type washing machines — still expect their detergent to produce voluminous and dense foam. The reason for this seems to be largely psychological (i.e., foam provides evidence of detergent activity and it hides the soil). Consequently, detergents designed for use in vertical-axis washing machines (i.e., products mainly for the non-European market) are quite frequently given the desired foam characteristics by incorporating small amounts of foam boosters. Compound types suited to the purpose include

Fatty acid amides
Fatty acid alkanolamides
Betaines
Sulfobetaines
Amine oxides

In Europe, horizontal axis drum-type washing machines are very common; with such machines, only weakly or moderately foaming detergents are permissible. Thus, foam boosters have lost their former significance in the European market, with the exception of the role they retain in certain specialty detergents (e.g., detergents for hand washables). Especially at high temperature, heavy foaming may cause overfoaming in drum-type machines, often accompanied by considerable loss of active ingredients. Furthermore, large amounts of foam reduce the mechanical action to which laundry is subjected in horizontal axis machines and hence the washing performance. Moreover, excessive foam may result in poor rinsing, spinning, and draining of the washing machine. On the other hand, too little foam must be avoided because consumers have the perception that low foam levels are a sign of poor cleaning performance. A certain level of foam can be beneficial for delicate fabrics to reduce fabric damage. For these reasons, foam regulators — often somewhat incorrectly described as "foam inhibitors" — are commonly added to minimize detergent foaming tendencies.

Ensuring effective foam regulation requires that the regulating system be precisely matched to the other detergent components present. The most difficult cases are surfactants with exceptionally high foam stability, such as alcohol sulfates, alcohol ether sulfates, alkylpolyglycosides, and alkylglucamides. By contrast, an unstable foam,

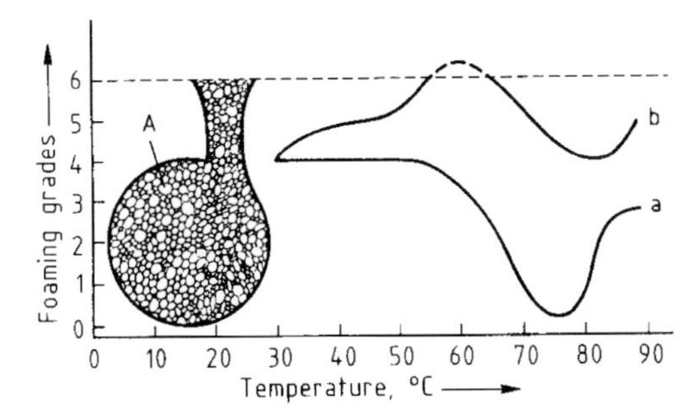

Figure 61. Foaming behavior in a drum-type washing machine [4]
A) Schematic drawing of the inner drum of a drum-type washing machine, including dispenser
a) 7.5 g/L Detergent based on 40 % sodium triphosphate; b) 7.5 g/L Detergent based on 40 % NTA
Foam grading: 0, no foam; 4, foam at the upper edge of the front door; 5, foam in the dispenser; 6, foam overflow

even if present in large amounts, like that produced by LAS or SAS, seldom causes a problem. The principal requirements for a substance to show foam regulating capability are extremely low water solubility and high surface spreading pressure. Typically, an antifoam is a dispersion of hydrophobic solid particles in a hydrophobic liquid. Foam regulators show a wide range of physicochemical properties, but their mechanism of action can usually be explained by assuming that they either force surfactant molecules away from interfaces or they penetrate interfaces that are already occupied by surfactants, thereby creating defects. These defects weaken the mechanical strength of the foam lamellae and cause their rupture.

A great many foam regulators are described in the patent literature, but relatively few have had any real impact. The detergents on the market that are based on LAS/alcohol ethoxylates can be effectively controlled by soaps with a broad chain length spectrum (C_{12-22}). Such soaps have limitations, however, for the following reasons:

1) The foam-depressing activity of soaps is largely due to calcium salts that form during the washing process, with the calcium originating either from hard water or from calcium-containing soil. In other words, satisfactory foam regulation occurs only when a sufficiently high calcium ion content is assured. It is absent when exceptionally soft water is used and when the laundry is only lightly soiled, and foaming problems may result.
2) The foam regulating power of soap is substantially lower with detergents based on anionic surfactants other than alkylbenzenesulfonates (e.g., α-olefinsulfonates, alcohol sulfates, α-sulfo fatty acid esters).
3) When soap is used as a foam regulator, the only complexing agents that can be added as builders are those with limited stability constants; i.e., even in the presence of builders, the calcium ion concentration must remain high enough to permit formation of sufficient lime soap able to act as a foam regulator. An advantage of sodium triphosphate, widely used as a complexing agent in detergents, is that stability constant values for calcium triphosphate complexes are quite low. Other complexing agents (e.g., nitrilotriacetic acid, NTA), that have higher stability constants, prevent the formation of foam-regulating lime soaps (Fig. 61).

Intensive investigations have shown that specific silica – silicone mixtures or paraffin oil systems are considerably more universal in their applicability and that their effectiveness is independent of both water hardness and the nature of the surfactant – builder system employed [120] – [123]. Therefore, most heavy-duty detergents in Europe have silicone oil and/or paraffins as foam depressors. Soap has almost lost its importance as foam regulator. Silica – silicone systems, frequently called silicone antifoams, are usually commercially available as concentrated powders. They are robust and effective over a wide range of washing conditions for a wide range of laundry products at quite low levels in the formulation (typically 0.1 – 0.4 %). The key silicone oils used for antifoams are dimethylpolysiloxanes [124]. Silicone antifoam powders are easily processed as postaddition components for detergent manufacturing. It is not recommended to process them in the detergent slurry during production, because the antifoam performance is largely destroyed during spray drying.

3.4.4. Corrosion Inhibitors

Washing machines currently on the market are constructed almost exclusively with drums and laundry tubs of corrosion-resistant stainless steel or with an enameled finish that is inert to alkaline wash liquors or plastic. Nevertheless, various machine components are made of less detergent resistant metals or alloys. To prevent corrosion of these parts, modern detergents contain corrosion inhibitors in the form of sodium silicate. The colloidal silicate that is present deposits as a thin, inert layer on metallic surfaces, thereby protecting them from attack by alkali.

3.4.5. Fluorescent Whitening Agents

Properly washed and bleached white laundry, even when clean, actually has a slight yellowish tinge. For this reason, as early as the middle of the 19th century, people began treating laundry with a trace of blue dye (blueing agents, e.g., ultramarine blue) so that the color was modified slightly and a more intense visual sensation of improved whiteness was produced. Blueing agents are still popular in some regions of the world. Modern detergents contain fluorescent whitening agents (FWA), also known as *optical brighteners,* to accomplish the same purpose. Fluorescent whitening agents are organic compounds that convert a portion of the invisible ultraviolet light into longer wavelength visible blue light. It is well known that the yellowish cast of freshly washed and bleached laundry is a result of partial absorption of the blue radiation reaching it, resulting in reflected light that is partially deficient in the blue region of its spectrum. The radiation emitted by optical brighteners makes up for this deficiency, so that the laundry becomes both brighter and whiter (Fig. 62).

This process of compensating for differences in yellow and blue radiation is fundamentally different from treatment with a blueing agent (washing blue). The latter

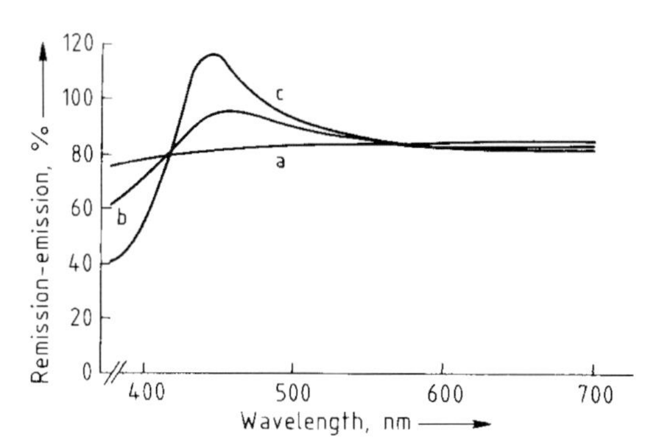

Figure 62. Remission – emission curve of a bleached fabric after application of a fluorescent whitening agent [36]
a) Bleached fabric; b) Fabric treated with small amounts of whitener; c) Fabric treated with large amounts of whitener

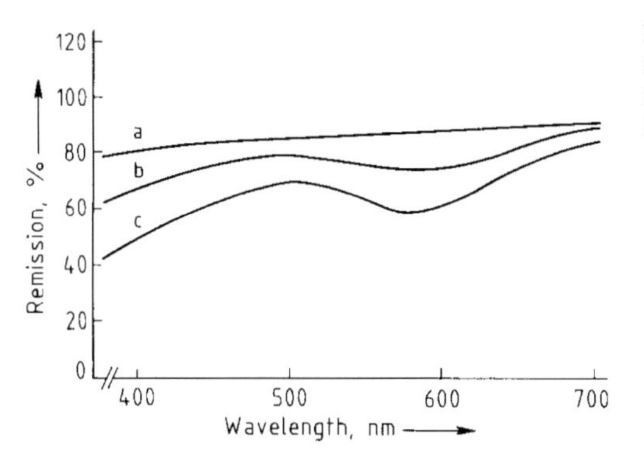

Figure 63. Remission curve of a bleached fabric after application of ultramarine blue [36]
a) Bleached fabric; b) Weakly blued fabric; c) Intensely blued fabric

entails subtractive absorption of yellow light, which results in an overall reduction in brightness (Fig. 63).

Limits exist for the extent of whitening that can be achieved by fluorescent whitening agents. This is due in part to the fact that these agents are themselves dyes that exhibit a certain extent of reflectance in the visible region in addition to their emission. Thus, any agglomeration or buildup of fluorescent whitening agent on the fibers eventually becomes apparent. Furthermore, the absorption edge gradually shifts into the visible region of the spectrum, leading to emission at higher wavelengths. These factors can work in combination with the innate color of the fabric, however, and eventually result in a visible shift of the hue (discoloration).

The major commercially available optical brighteners for laundry products are based on four basic structural frameworks: distyrylbiphenyl, stilbene, coumarin, and bis(benzoxazole). Table 19 provides an overview of the most important optical brighteners. Those mentioned here are estimated to cover more than 95 % of optical brighteners for laundry products on the market.

The application of FWA in the wash liquor is essentially a dyeing process (FWA are actually dyes). In the case of cotton and chlorine-resistant FWA, binding occurs through the formation of hydrogen bonds to the fibers. Permanent-press easy-care cotton is generally less susceptible to the effects of FWA. Whitening effects achieved with

Table 19. Summary of key fluorescent whitening agents for laundry products [125]

Structure	Chemical Name
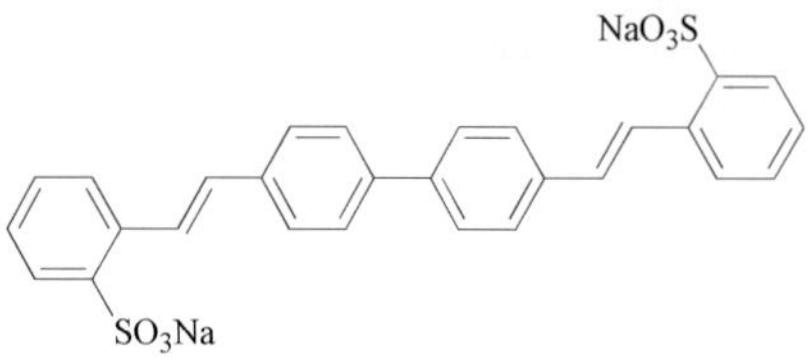 NaO_3S SO_3Na	4,4′-di(2-sulfostyryl)biphenyl disodium salt
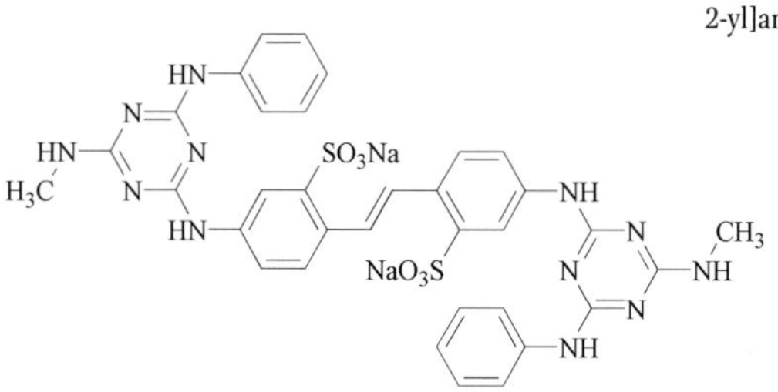 HN N N HN H_3C N HN SO_3Na NH NaO_3S N CH_3 N NH N NH	4,4′-Bis[[6-anilino-4-(methylamino)-1,3,5-triazin-2-yl]amino]stilbene-2,2′-disulphonic acid, disodium salt
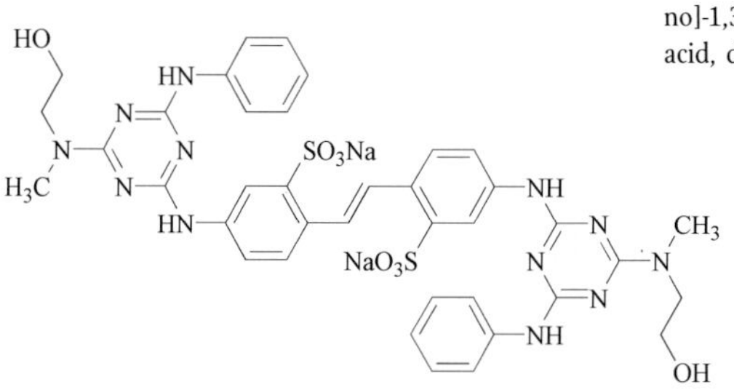 HO HN N N N H_3C N HN SO_3Na NH NaO_3S N CH_3 N N N NH OH	4,4′-bis[[6-anilino-4-[(2-hydroxyethyl)methylamino]-1,3,5-triazin-2-yl]amino]stilbene-2,2′-disulphonic acid, disodium salt
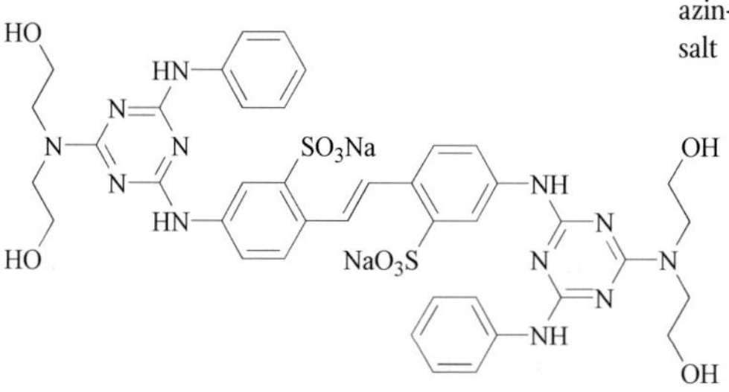 HO HN N N N N HN HO SO_3Na NH NaO_3S OH N N N N NH OH	4,4′-Bis[[2-anilino-4-[bis(2-hydroxyethyl)amino]-1,3,5-triazin-6-yl]amino]stilbene-2,2′-disulfonic acid, disodium salt

Table 19. (continued)

Structure	Chemical Name
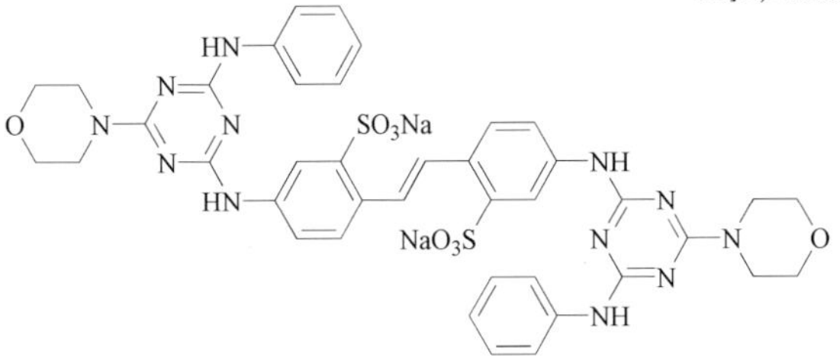	4,4′-Bis[(2-anilino-4-morpholino-1,3,5-triazin-6-yl)amino]-2,2′stilbenedisulfonic acid, disodium salt
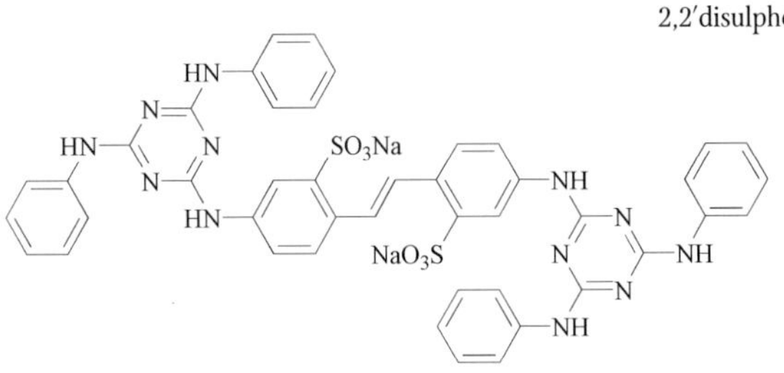	4,4′-bis[(4,6-dianilino-1,3,5-triazin-2-yl)amino]stilbene-2,2′disulphonic acid, disodium salt
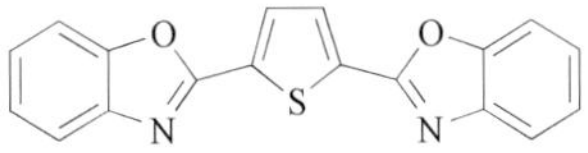	2,5-bis(benzoxazol-2-yl)thiophene
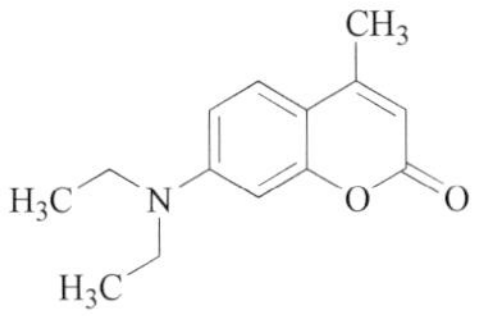	7-diethylamino-4-methylcoumarin

polyamide whiteners, and particularly polyester whiteners (which tend to be less adsorbed), are due largely to the diffusing power of whitening agent molecules present at the fiber surfaces.

Fluorescent whitening agents are evaluated not only on the basis of their affinity for the relevant fibers, but also on their stability and fastness. “Stability” means their resistance to chemical change during the washing process prior to their adsorption on fibers. “Fastness” refers to their chemical resistance after adsorption. Of primary concern is fastness against light and oxygen, as well as good oxygen stability and, in countries where chlorine bleach is used, fastness and stability to hypochlorite bleaches.

Most detergents currently on the market contain fluorescent whitening agents. A concern to detergent formulators is the difficulty to define optimal FWA type(s) and concentration in laundry detergents for sufficient whiteness maintenance because of the wide variety of washing habits and fiber types used for white textiles [126]. The use of FWA-containing detergents on certain pastel fabrics can cause unwanted color shifts; for this reason special FWA-free detergents are very popular [e.g., color detergents (see Section 4.1.5) and detergents for woolens)].

3.4.6. Dye Transfer Inhibitors

Frequently used dye transfer inhibitors are poly(*N*-vinylpyrrolidone) and poly(vinylpyridine *N*-oxide). These inhibitors form stable complexes with dye molecules that detach from colored textiles and diffuse into the wash bath. As such, the dye transfer inhibitors impede the re-adsorption or redeposition of the dyestuff on other textiles. The result is that white and colored garments look new longer after multicycle washing. A dull and dingy look of the textiles is avoided.

3.4.7. Fragrances

Fragrances were first added to detergents in the 1950s. Their presence is more than simply a fad or a matter of fashion. Thus, apart from their role in providing detergents with a pleasant odor, an important function of fragrances is to mask certain odors arising from the wash liquor during washing. This is particularly important because most washing machines are installed in the living areas of homes. Fragrances are also intended to confer a fresh, pleasant odor on the laundry itself. For this, long-lasting fragrances on dry laundry, resulting either from detergents or from fabric softeners have become a more and more important factor in the 1990s.

Detergent fragrances are generally present only in very low concentrations, usually $< 1\%$ in powder products. They are all complex mixtures of many individual ingredients (Table 20).

Several factors are involved in establishing the composition of a fragrance mixture apart from odor and the cost of the often expensive individual components. In particular, the detergent formulation must be taken into careful account, as must the properties of the fabrics to be washed. Chemical stability relative to other detergent ingredients is especially important, as is the limited volatility of the individual fragrances. Temperatures to which the detergent will be exposed during storage must also be considered.

The fragrance industry has at its disposal for the preparation of detergent fragrance oils not only a large selection of synthetic compounds, but also a wide range of natural fragrances. The overall selection is, nevertheless, restricted considerably due to the instability of many fragrances and the fact that some have a tendency to discolor

Table 20. Hypothetical formulation of a fragrance blend for detergent use

Ingredient	Quantity, g
Cyclohexyl salicylate	200.0
Iso E Super	100.0
Vertofix Coeur	80.0
Boisambrene Forte	50.0
Sandelice	40.0
Dihydromyrcenol	70.0
Benzyl acetone	35.0
Verdyl propionate	30.0
Verdyl acetate	20.0
Isoraldein 70	75.0
Floramat	25.0
O-tert-butylcyclohexyl acetate	20.0
Cironellol	20.0
Geraniol	20.0
Linalyl acetate	50.0
Hedione	50.0
Neroli phase oil	40.0
Linalool	20.0
Diphenyl ether	10.0
Anisaldehyde	5.0
Cyclovertal	3.0
Styrolyl acetate	5.0
Isobornyl acetate	15.0
Ambroxan	2.0
Herbavert	15.0

detergents to which they are added. There are also restrictions for toxicological and ecological reasons (cf. Chaps. 10 and 11).

3.4.8. Dyes

Until the 1950s, powder detergents were more or less white, consistent with the color of their components. Thereafter, products were commonly encountered in which colored granules were present along with the basically white powder: certain components had been deliberately dyed to make the finished detergent more distinctive, featuring speckles. Uniformly colored detergents have also appeared on the market, and the idea of introducing coloring agents has become quite common. The preferred colors for both powders and liquids are blue, green, and pink.

There are two important criteria for selecting a coloring agent:

1) Good storage stability with respect to other detergent components and to light and temperature
2) No significant tendency to adversely affect textile fibers through adsorption after multiple washing

3.4.9. Fillers and Formulation Aids

The usual *fillers* for powder detergents are inorganic salts, especially sodium sulfate. Their purpose is to confer the following properties on a detergent:

Flowability
Good flushing properties
High solubility
No caking of the powder even under highly humid conditions
No dusting

Compact detergents usually contain no fillers.

Formulation aids are materials required in the manufacturing of liquid detergents. The most important of these have the assignment of ensuring through their own hydrotropic characteristics that the other detergent components can be combined in a stable way in an aqueous environment. Above all, these ingredients must prevent phase separation occurring with time, and precipitation occurring as a result of shifts in temperature. Commonly used materials include short-chain alkylbenzenesulfonates (e.g., toluenesulfonate and cumenesulfonate) and, less frequently, urea, as well as low-molecular mass alcohols (ethanol, 2-propanol, glycerol) and polyglycol ethers [poly(ethylene glycols)]. These alcohols are frequently called solubilizers.

4. Household Laundry Products

Laundry products currently on the market in various parts of the world can be classified into the following groups from a product standpoint:

Heavy-duty detergents
Specialty detergents
Laundry aids

Quite large differences exist in the per capita consumption of detergents in various countries of the world. As depicted in Figure 64, it amounted to 2 – 3 kg/a in countries such as Brazil, China, and Russia and to more than 10 kg/a in Mexico and some European countries in 1997 [127], [128].

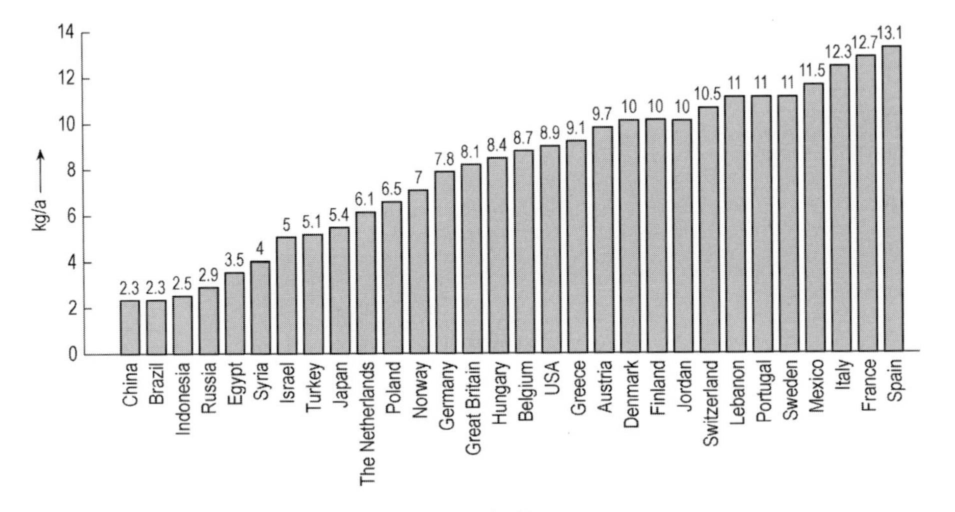

Figure 64. Per capita consumption of detergents in 1996 (kg/a)

4.1. Heavy-Duty Detergents

The category of heavy-duty detergents includes those detergent products suited to all types of laundry and to all wash temperatures. They are offered in the delivery forms of conventional powders, extrudates, tablets, bars, and liquids and pastes. Figure 65 indicates that worldwide powders have been the dominant delivery form of detergents over the years. They account for ca. 65 % of the world production of detergents. Compact powders worldwide accounted for 24 % of total detergent powder production in 1998. Depending on quality, product concept, manufacturing process, local standards, local regulations, and voluntary agreements, great differences are found from one formulation to another. For example, detergent manufacturers market laundry detergents with and without phosphate in various regions and countries for ecological and regulatory reasons. Figure 66 provides an overview of global production shares of individual heavy-duty detergent types in 1994 and 1998.

The world production of laundry detergents amounted to 21.5×10^6 t in 1998. Table 21 shows the proportions of the tonnages of individual delivery forms of detergents in 1994 and 1998. As can be seen, the total produced volume remained nearly constant in the 1990s worldwide.

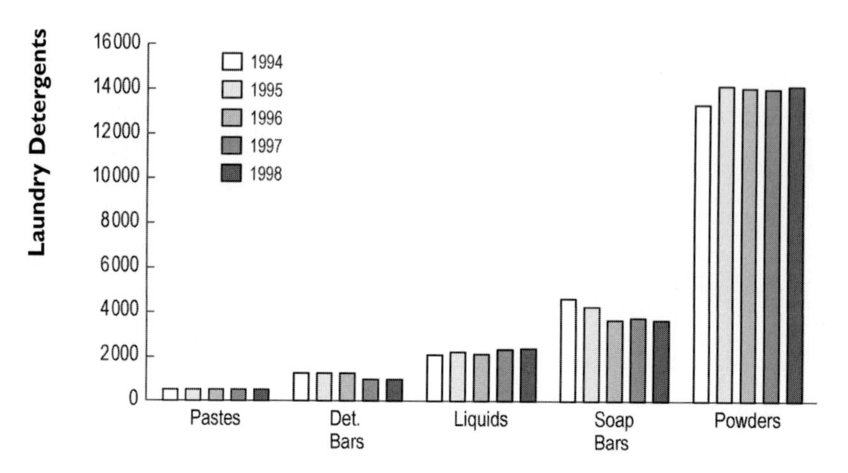

Figure 65. World production of individual heavy-duty detergents (in 10^3 t) (source: Ciba Speciality Chemicals Ltd., Switzerland)

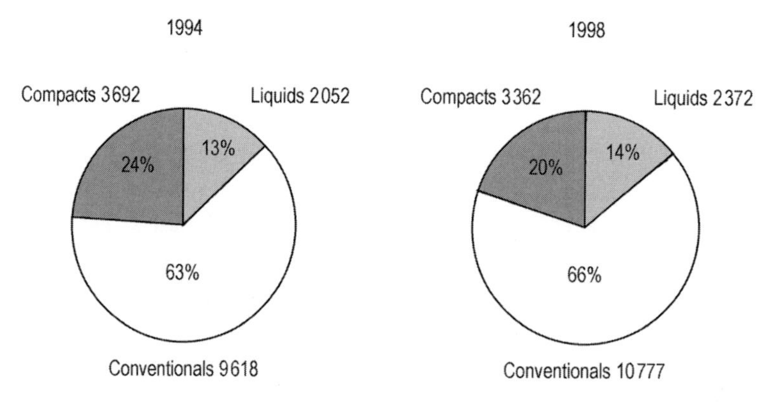

Figure 66. Detergent production, world total (10^3 t) (source: Ciba Speciality Chemicals Ltd., Switzerland)

4.1.1. Conventional Powder Heavy-Duty Detergents

Significant differences in composition exist among powder heavy-duty detergents around the world. An approximate breakdown of detergent formulations for various global regions is provided in Table 22.

Recommended detergent dosages in Europe generally lie between 3 and 8 g detergent per liter of wash water, whereas values of 0.5 – 1.5 g/L are usual in the Americas and Eastern and Southern Asia. There are a number of reasons for the latter being so much lower; for example, differences in washing habits (different washing machines, hand-

Table 21. Worldwide production Detergent Industry

Production, 10^3 t	1994	1998
Soap bars	4579	3651
Detergent bars	1251	968
Powders	13 310	14 137
Pastes	496	505
Liquids	2052	2372
Total	**21 688**	**21 633**
World Average per capita consumption, kg/a	4.5	3.9

(Source: Ciba Speciality Chemicals Ltd., Switzerland)

Table 22. Powder heavy-duty detergent formulations around the world in 1999[a]

Ingredients	Examples	Composition, %				
		USA	South America	Europe	China, India	Japan
Surfactants	alkylbenzenesulfonate, alcohol sulfate, alcohol ethoxylate	10 – 25	18 – 25	10 – 25	8 – 18	25 – 40
Builders	zeolite, sodium tiphosphate, Na-citrate, Na-silicate, Na-carbonate/Na-bicarbonate	40 – 65	40 – 55	30 – 55	30 – 50	40 – 55
Cobuilders	sodium polycarboxylate	0 – 5	–	3 – 8	–	0 – 5
Bleaching agents	sodium perborate, sodium percarbonate	0 – 10	–	8 – 15	–	0 – 6
Bleach activators	TAED, NOBS	0 – 3[b]	–	1 – 7[c]	–	0 – 3[b]
Antiredeposition agents	carboxymethyl cellulose, cellulose ethers	1 – 2	0.5 – 1	0 – 1	0.5 – 1	0.5 – 1
Stabilizers	phosphonates	–	–	0 – 1	–	–
Foam regulators	soap, silicone oil and/or paraffins	–	–	0.1 – 4	–	–
Enzymes	protease, cellulase, amylase, lipase	0.3 – 2	0.3 – 0.8	0.3 – 2	0.3 – 0.8	0.3 – 1.5
Optical brighteners	stilbene-, biphenyldistyryl derivatives	0.1 – 0.3	0.1 – 0.3	0.1 – 0.3	0.1 – 0.2	0.1 – 0.2
Soil repellents	poly(ethylene glycol terephthalate) derivatives	0 – 1	–	0 – 1.5	–	–
Fillers/processing aids	sodium sulfate	5 – 30	20 – 35	0 – 30	20 – 40	5 – 15
Minors	Fragrance	+/–	+	+/–	+	+
Water		5 – 15	5 – 15	5 – 15	5 – 15	5 – 15

[a] All figures expressed as 100 % active material, except for enzymes for which figures relate to % granulate.
[b] Nonanoyloxybenzenesulfonate (NOBS).
[c] Tetraacetylethylenediamine.

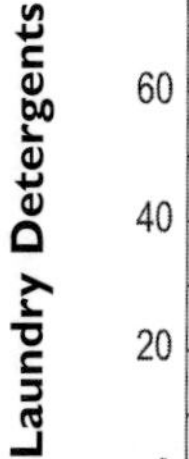

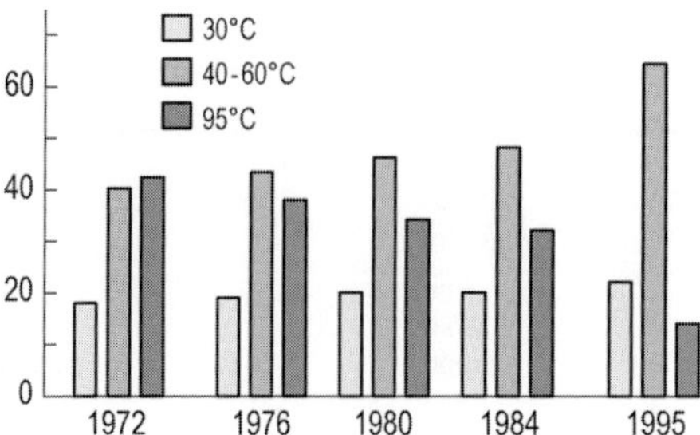

Figure 67. Percentage of wash loads versus washing temperature in Germany (source: Leading German Market Research Institute)

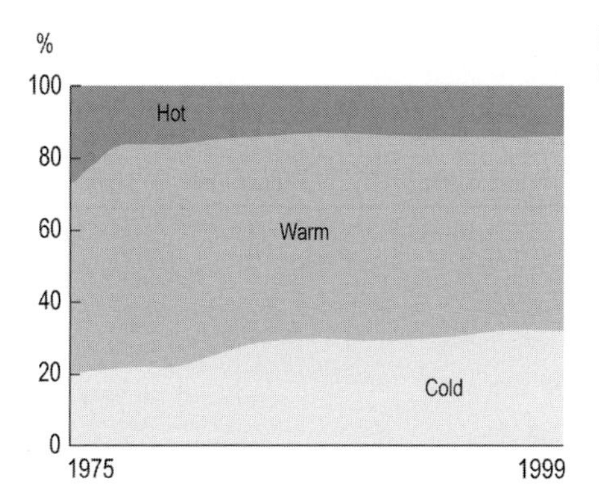

Figure 68. Home laundry washing temperature trends in the USA

washing), softer water, higher bath ratios, separate addition of bleach, e.g., hypochlorite solution or sodium perborate, and different detergent formulations. Detergents in North America, Japan, and Brazil are designed for agitator and impeller type washing machines (cf. Chap. 13). These detergents lack foam suppressors. They sometimes even contain foam boosters instead.

The rise in popularity of ever more colored and easy-care fabrics in the last 30 years, along with the trend toward energy conservation, has resulted in a large decline in the once common European practice of washing at 95 °C, while the 30 – 60 °C wash temperatures have significantly gained favor. The change is illustrated in Figure 67 for the time period 1972 to 1995 in Germany. A similar trend from hot ($\geq$ 50 °C) to warm (27 – 43 °C) and cold washing is apparent in the USA (Fig. 68). Detergent manufacturers have responded to these changes by increasing the content of surfactants that are effective in low-temperature washing, as well as by adding various enzymes, bleach activators, soil repellents, and phosphonates. Detergents have become more and more multifunctional over the last decades, particularly those for use at low washing temperatures.

Table 23. Composition of conventional versus compact heavy-duty detergents in Europe, % [127]

Ingredients	Conventional	Compact
Anionic and nonionic surfactants	10–15	10–25
Builders	25–50	25–40
Cobuilders	3–5	3–8
Bleaching agents	10–25	10–20
Bleach activators	1–3	3–8
Antiredeposition agents	0–1	0–1
Corrosion inhibitors	2–6	2–6
Stabilizers	0–1	0–1
Foam regulators	0.1–4.0	0.1–2.0
Enzymes	0.3–0.8	0.5–2.0
Optical brighteners	0.1–0.3	0.1–0.3
Soil repellents	+/–	+/–
Fillers/processing aids	5–30	–
Minors, water	balance	balance
Bulk density, g/L	500–650	600–900

4.1.2. Compact and Supercompact Heavy-Duty Detergents

Apart from a few minor and temporary attempts in Japan, Germany, and the USA in the 1970s and early 1980s [56], this heavy-duty detergent category was introduced on a large scale for the first time in Japan in 1987. Within a couple of years it almost completely replaced the conventional powders. The wave of compact detergents reached Europe in 1989. North America followed shortly afterwards [131]. The global introduction of compact and supercompact powder detergents in the late 1980s and early 1990s has had a major impact on detergent categories of large markets such as Japan, Europe, and the USA, as well as on composition and manufacturing processes of heavy-duty detergents [131]. Production processes of concentrated detergents are described in Chapter 6. A comparison of the composition of conventional versus compact heavy-duty detergents in Europe is depicted in Table 23 [127], [128]. For reasons of significantly reduced recommended dosage along with increased detergency performance compact detergents have markedly higher contents of surfactants, active oxygen for bleach, bleach activators, and enzymes than conventional detergents. On the other hand, the total level of alkaline builders has been reduced and the conventional filler/processing aid sodium sulfate has been omitted to a large extent (Fig. 69 [127], [128]). The apparent density of most compact detergents is above 0.75 kg/L. Most of the first-generation compact detergents were produced by the method of downstream compaction of tower powder.

Heavy-duty detergents featuring bulk densities of from 0.8 to 1.0 kg/L are called supercompact detergents or second-generation compact detergents. The prevailing manufacturing methods used are wet granulation/compounding and extrusion. Super-

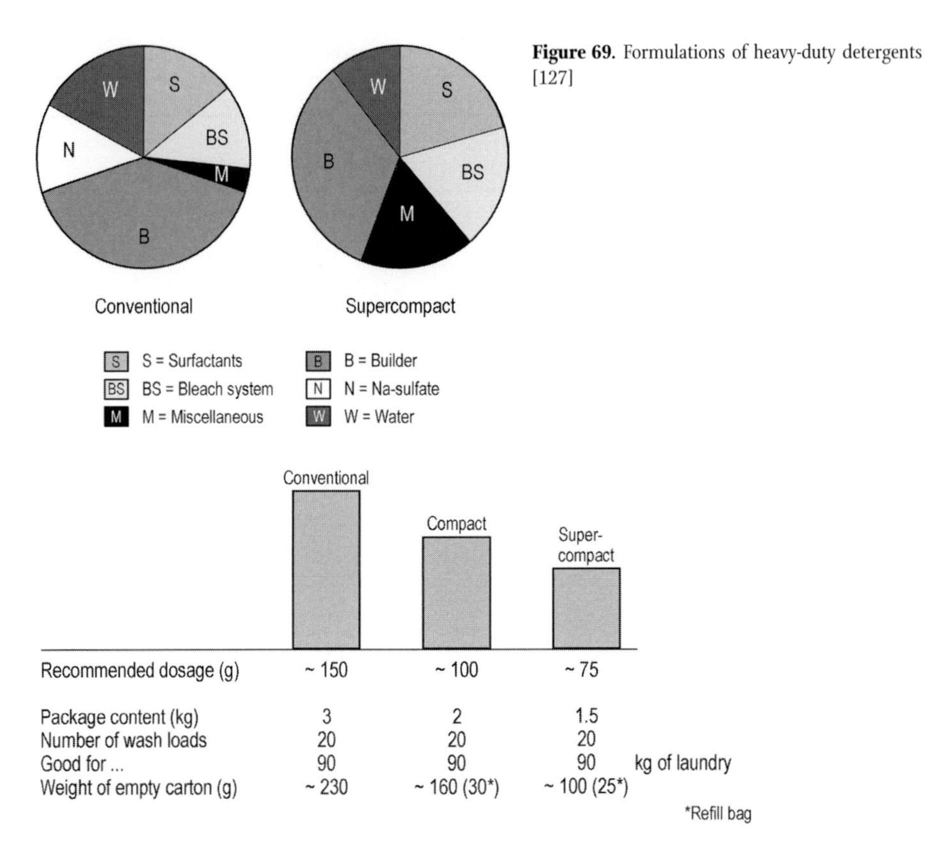

Figure 69. Formulations of heavy-duty detergents [127]

Figure 70. Consumer-relevant characteristics of heavy-duty detergents

compact detergents have made their way worldwide since 1992 and are meanwhile state of the art in many countries. Consumer relevant properties of (super)compact vs. conventional detergents are depicted in Figure 70 [127], [129]. Market shares of compact/supercompact powder heavy-duty detergents in Europe as compared with conventional powders and liquid detergents are indicated in Figure 71 [129] for 2000. The market shares significantly vary from country to country and from region to region. In the USA compact powders accounted for 80 % of total detergent powders, whereas their share in Japan was even 92 % in the total powder category in 1998.

Sodium triphosphate has been increasingly losing its importance as a key ingredient in detergents in Europe, the USA, and some East-Asian countries since the mid-1980s. 1998 market shares of nonphosphate powder detergents in most European countries are shown in Figure 72 [131]. Triphosphate has been replaced mainly by zeolite A, special silicates, sodium carbonate, specific polycarboxylates, and citrate (see Section 3.2). A key driving force for the increasing market shares of nonphosphate powder heavy-duty detergents has been the proliferation of supercompact detergents [127],

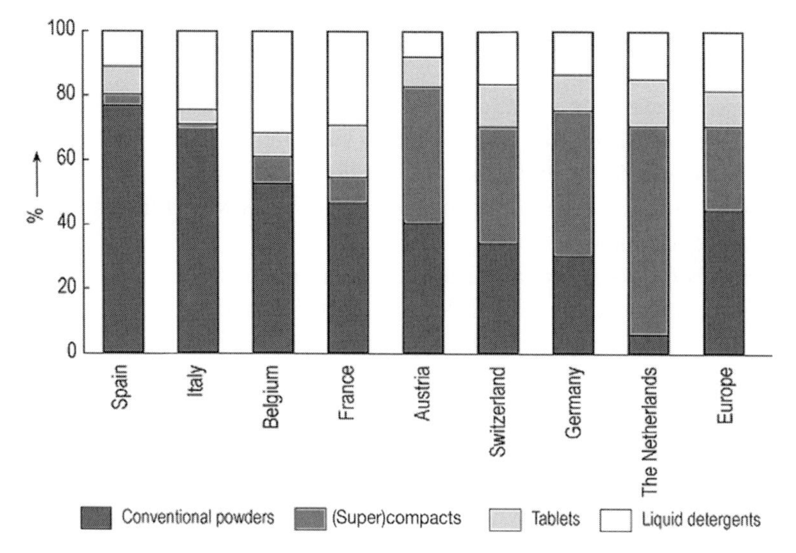

Figure 71. Individual heavy-duty detergents in Europe, market shares (value) 2000 in % [129]

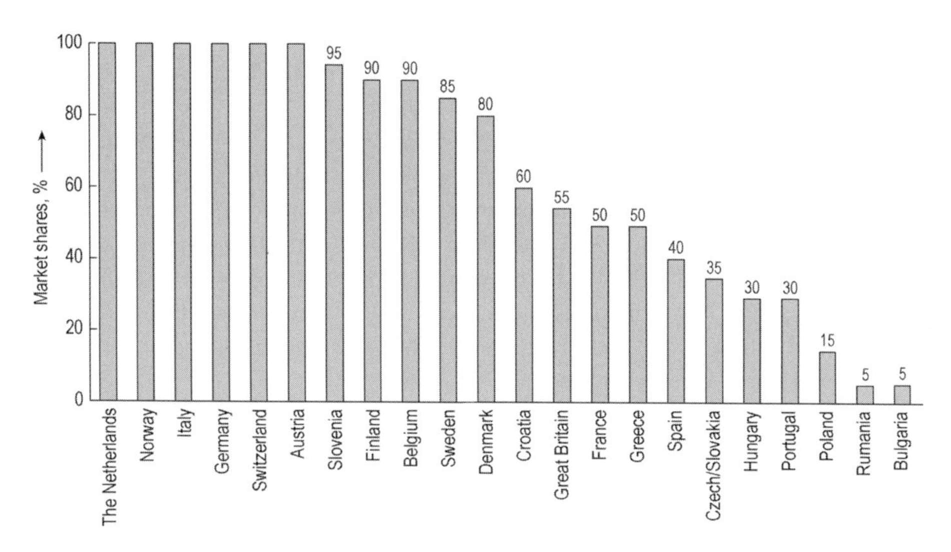

Figure 72. Market shares of nonphosphate HDD powders in Europe in 1998 [131]

[128]. Generally, the use of (super)compact detergents has contributed to decreasing their influx into wastewaters, much to the benefit of the environment [137].

4.1.3. Extruded Heavy Duty Detergents [136]

Extruded detergents have been an innovative and unconventional delivery form of second-generation compact detergents. The large-scale extrusion process for detergents was applied for the first time in 1992 and is described in Chapter 6 [138] – [141]. Extruded detergents have since then been marketed under the registered trademark of Megaperls® by Henkel throughout Europe.

Extruded detergents feature spherical particles of uniform particle size (for example 1.4 mm) whereas conventional spray-dried detergent powders and granulated supercompact detergents are characterized by a broad particle size distribution. The density of an extruded particle is approximately 1.4 kg/L. Extruded detergents have one of the highest densities achieved in detergent manufacturing.

Apart from being an extraordinary basis for manufacturing supercompact products, extruded detergents feature additional advantages such as complete absence of dust particles, very high homogeneity, no segregation of particles, and excellent free-flowing characteristics. Extruded detergents allow anionic surfactant contents of more than 20 % along with very high densities.

4.1.4. Heavy-Duty Detergent Tablets [142] – [145]

Busier consumer lifestyles have increased demand for laundry products that emphasize convenience and ease of use, as exemplified by detergent tablets. Apart from very few regional distributions in the past, detergent tablets were first introduced on a large scale in Europe in late 1997. Ever since they have spread out on the European market and held a share of 10 % in the heavy-duty detergent category by mid-2000. The Japanese, Canadian, and U.S. markets saw their introduction during the year 2000. In the wide field of laundry products, tablets meanwhile also cover the categories of bleach boosters and water softeners in Europe.

Consumer-relevant characteristics of tablets are ease of dispensing and convenience in handling, i.e., no dosing and dispensing aids are needed. Other advantages include precise dosing, smaller packages as compared with powder products due to the highly concentrated form, extra portability, and more accurate sense of how many washes remain in the detergent box.

Tablets are the most compact delivery form of nonliquid detergents. They have densities of 1 to 1.3 kg/L. This results in benefits in terms of lower packaging volume, ease of transportation and storage, and reduced shelf space. Detergent tablets therefore belong to the supercompact detergent category (see Section 4.1.2).

One of the key requirements of laundry detergent tablets is fast disintegration. Upon their first contact with water, be it in the washing machine dispenser of a European drum-type washer or in the basket of top-loading agitator (North America) or pulsator machines (East Asia), tablets must disintegrate within seconds or at least within a minute or so. This is a necessary prerequisite for the next step of dissolving, particularly with regard to the short washing times used in Japan, the USA and, recently, Europe. Special precautions must therefore be taken to avoid intermediate formation of viscous liquid-crystalline phases (gelling effect) of the surfactants in the tablet detergent during contact with water. To enable fast disintegration laundry detergent tablets generally contain special disintegrants. These can be classified into four groups:

1) Effervescents such as carbonate/hydrogencarbonate/citric acid
2) Swelling agents like cellulose, carboxymethyl cellulose, and cross-linked poly(*N*-vinylpyrrolidone)
3) Quickly dissolving materials such as Na (K) acetate or Na (K) citrate
4) Rapidly dissolving water-soluble rigid coating such as dicarboxylic acids

or combinations therof.

Another specific demand made on heavy-duty detergent tablets is sufficient hardness to enable handling during packaging, transportation, and in-home use. The conflicting desired properties of sufficient hardness and fast disintegration have to be well balanced. Usually, laundry detergent tablets have weights of ca. 35 to 45 g and diameters in the range of 40 to 45 mm. The most common recommended dosage in Europe is two tablets per wash cycle. One- and two-phase (colored) tablets are available on the market. Two-phase tablets allow separation of certain detergent ingredients that otherwise might adversely affect each other during storage, for example, enzymes and activated bleach.

In terms of their composition, heavy-duty detergent tablets differ from other supercompact detergents mostly with regard to the disintegrants. Table 24 lists the composition of European detergent tablets. All compact heavy-duty detergents in Europe have been nonphosphate zeolite-based thus far [127]. The reappearance of phosphate in the detergent tablet category in Europe (see Table 24) may therefore be considered a remarkable development [142].

4.1.5. Color Heavy-Duty Detergents

In 1991, a new category of heavy-duty detergents appeared on the West European market, for which the name color detergents was coined. So far, this type of compact detergent has mainly spread out on the market in Western Europe. 1999 market shares of color detergents in some countries on the European continent are shown in Figure 73. Color detergents differ from conventional European heavy-duty detergents in that they contain neither bleach nor optical brighteners and in that they feature specific dye transfer inhibitors such as poly(*N*-vinylpyrrolidone) or poly(vinylpyridine *N*-oxide). Color detergents are recommended and used for washing colored laundry, for which improved color care is claimed.

Table 24. Composition of heavy-duty detergent tablets on the European market [142]

Ingredients	Zeolite-based, %	Phosphate-based, %
Surfactants	13 – 18	15 – 18
Bleaching agents	13 – 15	12 – 16
TAED	3 – 7	4 – 7
Zeolite	11 – 25	
Sodium triphosphate		25 – 47
Layered silicate	0 – 9	–
Sodium polycarboxylate	2 – 3	2
Disintegrants	5 – 17	0 – 12
Enzymes	2 – 4	2 – 3

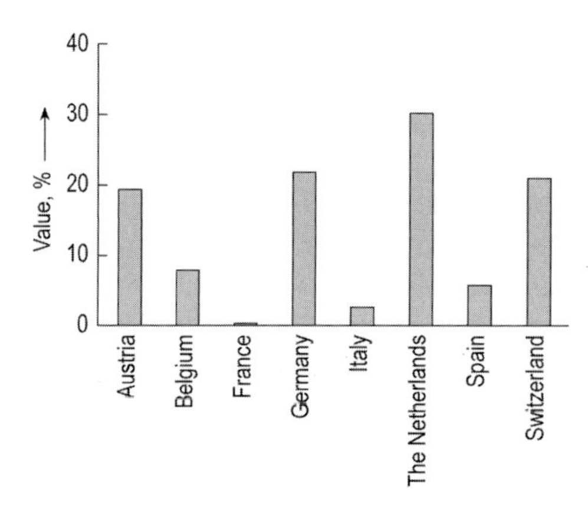

Figure 73. Market Shares (value) of Color Heavy-Duty Detergents in the Total HDD Category (source: Leading International Market Research Institute)

4.1.6. Liquid Heavy-Duty Detergents

Liquid detergents have been common in the USA since the 1970s for a variety of reasons. Today, the USA is the largest single market of liquid heavy-duty detergents (HDL) in the world. The market share of HDL in the USA was some 50 % in 1999. Until 1987, such products were insignificant in Europe, although liquid heavy-duty detergents had been available on the European market since 1981. In 1998, liquid heavy-duty detergents represented 12 % of the market in Europe, and their worldwide share was 14 % of total detergent production.

Liquid heavy-duty detergents are distinctive because of their usually high surfactant content (up to ca. 50 %). They rarely contain builders such as zeolite or triphosphate and are generally devoid of bleaching agents, the latter for reasons of loss of active oxygen and incompatibility with enzymes during storage. They are most effective in removing greasy and oily soil, especially at wash temperatures below 60 °C. A comparison of the detergency characteristics of European liquid and powder heavy-duty detergents is depicted in Table 25 [27]. From the characteristics of the two types of detergent it is obvious that both products have different but distinct fields of optimal

Table 25. Performance of powder versus liquid heavy-duty detergents

Product characteristics	Powder 60 – 90 °C	Liquid 30 – 60 °C
Removal of		
Fatty/oil soil	++	+++
Particulate soil	++	+
Bleachable stains	+++	++
Proteinaceous stains	++	++
Detergency performance on		
Cotton	++	+
Cotton/polyester blends	++	++
Synthetics	+	++
Wool	+	++
Antiredeposition/whitening		
30 to 60 °C	++	++
90 °C	++	+
Prevention of inorganic salt deposits	+	++
Fabric care	+	++
Solubility/dispersability	+	+++

+++ = excellent; ++ = very good; + = good

Table 26. Composition of liquid heavy-duty detergents in Europe (2000), %

Ingredients	Unstructured	Structured
Anionic surfactants	7 – 18	10 – 25
Nonionic surfactants	15 – 30	6 – 10
Soaps	10 – 22	4 – 6
Builders	0 – 8	15 – 30
Solubilizers	0 – 12	0 – 5
Alcohols	8 – 12	0 – 5
Enzymes	0 – 2.5	0 – 1.5
Optical brighteners	0.05 – 0.25	0.05 – 0.25
Stabilizers	+/–	+/–
Fragrance	+	+
Minors, water	30 – 50	30 – 50

performance. Depending on washing habits, HDL are alternatives to powders especially for washing in the lower temperature range [146].

HDL are marketed both with and without added builders. Formulations for liquid heavy-duty detergents with (structured) and without builders (unstructured) in Europe are given in Table 26 [127]. Liquid detergents with builders typically contain 15 – 30 % surfactants; the total amount of surfactants in products without builders can be as high as 50 %. The unstructured liquid detergents have continued to dominate the market. Besides having a high content of synthetic surfactants, they include a large proportion of soap with a defined carbon chain length as a water-softening component, as well as ingredients needed as formulation aids. The total content of washing active components can be as high as 60 %. The liquid products still contain water and are free of bleaching agents. Although many patents have been filed for liquid detergents that contain bleaching agents, none have yet been marketed successfully. In common with the increasing multifunctionality of powders, liquid heavy-duty detergents show a steady trend towards a higher enzyme content and incorporation of more types of enzymes.

Increasing concentrations of total active ingedients have been introduced in liquid detergents in the first half of the 1990s, although the degree of compaction has not been as drastic as with powdered products. Conventional liquid heavy-duty detergents still dominate the market in some European countries.

The physical appearance of liquid laundry detergents remained unchanged for a long time. They were marketed as low-viscosity (250 – 300 mPa · s) products with a Newtonian flow behavior. This changed with the launch of detergent gels in Europe. Gel-like pourable heavy-duty detergent concentrates that feature a specific rheological behavior represent a new delivery form in the liquid segment in Europe [147]. These gel detergents have much higher viscosities than conventional HDL (ca. 1500 – 3000 mPa · s). Gels were introduced in almost all European markets in 1997. Because of their high viscosity and rheological properties gels show improved detergency performance in pretreating stains as compared with conventional HDL [147]. Due to the continuous increase in the use of liquid concentrates and the reduction of the recommended dosage, it has been possible to reduce package sizes without decreasing the number of wash loads per package. This has resulted in large savings on packaging material and transportation costs.

4.2. Specialty Detergents

Specialty detergents play a relatively minor role outside Europe, but they are quite important in Europe.

Specialty detergents are products developed for washing specific types of laundry. Such detergents are generally used in washing machines but also for hand washing. They usually require the use of special washing programs (e.g., a wool cycle to prevent felting or a gentle cycle to prevent wrinkling). They can be characterized by the fact that fabric care ranks high and is a more important feature than with heavy-duty detergents, for which soil and stain removal is the key characteristic.

4.2.1. Powder Specialty Detergents

Distinction is made among products of the following types:

Detergents for delicate and colored laundry
Detergents for woolens
Detergents for curtains
Detergents for washing by hand

Many of these products are appropriate for either machine or handwashing. Most of them are marketed with an average bulk density of 0.4 – 0.5 kg/L.

Detergents specially designed for *delicate and colored laundry* contain neither bleach nor fluorescent whitening agents, which may adversely affect sensitive dyes used with some of these fabrics. These detergents are particularly useful for laundry colored with dyes sensitive to oxidation or for pastel fabrics that otherwise might experience color shifts if treated with fluorescent whitening agents. Most products today contain cellulases, which help fabrics look new longer and keep colors bright. Dye transfer inhibitors are added to some specialty detergents.

Detergents for *wool* are primarily intended for use in washing machines. Particular care must be exercised to prevent damage to the sensitive fibers that constitute natural wool, including applying low-temperature washing, short washing times, high bath ratios, and avoiding vigorous mechanical action.

High-foaming detergents for *washing by hand* are intended for washing small amounts of laundry in the sink or in a bowl. Due to the increased number of household washing machines, the use of specialty handwashing detergents has been gradually declining in Europe through the last decades. Their share was ca. 3 % in the entire detergent category in 1999.

Table 27 provides the compositions of various types of specialty detergents.

4.2.2. Liquid Specialty Detergents

Liquid specialty detergents have been on the market in Europe for a long time. Some liquid specialty detergents are intended for handwashing. However, most specialty liquid products are intended for washing machine application, e.g., detergents for woolens. The latter may be free of anionic surfactants, in which case they usually contain mixtures of cationic and nonionic surface active agents. The cationic agents act as fabric softeners to help keep wool soft and fluffy. Table 28 describes typical formulations for liquid specialty detergents.

Table 27. General formulations of West European powder specialty detergents

Ingredients	Composition, %			
	Detergents for delicate textiles	Detergents for woolens	Detergents for curtains	Handwashing detergents
Anionic surfactants (alkylbenzenesulfonates, fatty alcohol ether sulfates)	5 – 20	10 – 20	10 – 16	12 – 25
Nonionic surfactants (alcohol ethoxylates)	2 – 10	2 – 20	2 – 7	1 – 4
Soaps	1 – 5	0 – 5	2 – 6	0 – 5
Cationic surfactants				0 – 5
Zeolite A	20 – 40	5 – 40	25 – 40	7 – 30 *
Sodium perborate			0 – 15	
Sodium silicate	0 – 7	0 – 3	2 – 5	2 – 9
Sodium polycarboxylate	2 – 5	2 – 5	2 – 5	2 – 5
Antiredeposition agents	0.2 – 1.5	0 – 1.5	0.5 – 1.5	0.5 – 1.5
Enzymes	0 – 1			0 – 1
Optical brighteners	0 – 0.2		0.1 – 0.2	0 – 0.1
Dye Transfer Inhibitors	0 – 0.5			
Fragrances	+	+	+	+
Fillers and water	balance	balance	balance	balance

* Some hand washing detergents contain 15 – 30 % sodium triphosphate instead.

Table 28. Frame formulations of European liquid detergents for woolens

Ingredients	Composition, %	
	With incorporated softeners	Without incorporated softeners
Anionic surfactants (alkylbenzenesulfonates, fatty alcohol ether sulfates)	0 – 1	10 – 15
Nonionic surfactants (alcohol ethoxylates, fatty acid amides)	15 – 20	0 – 2
Cationic surfactants (esterquats)	2 – 4	–
Solubilizers (ethanol, propylene glycol)	0 – 5	0 – 5
Hydrotropes (toluenesulfonates, xylenesulfonates, cumenesulfonates)		0 – 3
Fragrances, dyes	+	+
Water	60 – 70	60 – 80

4.3. Laundry Aids

Laundry aids are products developed to meet the varying needs of often widely divergent washing habits and practices in different parts of the world. For example, most powdered heavy-duty detergents in Europe contain oxygen bleach. Most in the Americas, Africa, and Asia do not, and hence separate addition of bleach may be

Table 29. Frame formulations of soil and stain removers in Europe and the USA

Components	Composition, %				
	Europe		USA		
	Pastes	Sprays	Liquid/ sprays	Sticks	Gels
Anionic surfactants (alkylbenzenesulfonates, alcohol sulfates, soaps)	15–30		0–3	10–15	0–0.5
Nonionic surfactants (nonylphenol ethoxylates, alcohol ethoxylates, amine oxides)	3–10	15–40	1–15	25–35	8–45
Hydrotropes (sodium xylenesulfonate)			0–10	0–5	0–5
Builders (sodium citrate, sodium carbonate)			0–3.5	1–2	0–2
Oxidants (hydrogen peroxide)			0–3.5	–	–
Bleach activator			0–0.5	–	–
Solvents, solubilizers		30–80	0–12	25–35	0–35
Enzymes			0.2–1	2–3	0–2
Dyes, fragrance, stabilizers, water	balance	balance	balance	balance	balance

required. Depending on their use, laundry aids can be divided into the following categories:

Pretreatment aids (prespotters, water softeners)
Bleaches
Detergent boosters
Fabric softeners
Stiffeners
Laundry dryer aids
Refreshing products for dryer application
Odor removers

The global laundry aids market was worth $ 38.7 \times 10^9$ in 1998. Western Europe accounted for 30.4 %, while the USA ranked second with 20.8 % of global sales [148].

4.3.1. Pretreatment Aids

The most important pretreatment aids are soil and stain removers and water softeners. Special presoaking agents that were frequently applied in the past have gradually lost their market importance in the industrialized countries.

Soil and stain removers, also called prespotters or pretreatment aids, are frequently products with high surfactant contents. Their direct application helps removing greasy and/or bleachable stains (Table 29). Such products are applied to soiled areas prior to

Table 30. Frame formulation of water softeners in Europe

Ingredients	Composition, %
Zeolite A	45 – 60
Sodium polycarboxylate	10 – 15
Sodium silicate	10 – 15
Sodium sulfate, minors, water	balance

washing. They are supplied as pastes, sticks, or aerosols or in trigger spray bottles. The major targets of such products are fabrics that require low-temperature washing. The regular use of soil and stain removers can also help prevent the buildup of soil on problem areas such as shirt collars and cuffs after multicycle wear/washing.

Spray stain removers mainly consist of mixtures of solvents and surfactants. Solvents similar to those employed in dry cleaning loosen grease so that it can subsequently be emulsified by appropriate surfactant blends. Sprays generally work more rapidly than paste products. Proper application of such products prevents formation of undesirable rings around stains, a consequence that is nearly unavoidable when stains are subjected to local treatment with pure solvents instead.

Water softeners are used both in the USA and in Europe. European manufacturers recommend that these softeners be used in combination with detergents for applications involving water of medium to high hardness. However, the great majority of heavy-duty laundry detergents contain sufficient builders and cobuilders to prevent build-up of lime deposits upon multicycle washing even in hard water. European water softeners generally contain ion exchangers, usually zeolite A and polymeric carboxylic acids (Table 30).

4.3.2. Boosters

Boosters are products that can be added separately to detergents to exert specific influences on the washing process and thereby improve its effectiveness. The principal types of boosters offered on the market are bleaching agents and laundry boosters.

Bleaching agents are found worldwide on the market. They have been common in the Americas and Asia for a long time, where they are available in both powder and liquid form. Powder bleaching agents generally contain sodium perborate or sodium percarbonate (Table 31), whereas most liquid bleaching agents are dilute solutions of sodium hypochlorite or hydrogen peroxide. Bleaches are usually applied optionally when needed along with detergent. Sodium hypochlorite is to be used for whites only, whereas hydrogen peroxide solutions and powder and tablet peroxygen bleaches (perborate/percarbonate) are considered color-safe products.

Table 31. Frame formulations of all-fabric bleaches in the USA

Ingredients	Composition, %	
	Powders	Liquids
Builders (sodium carbonate, zeolite, sodium citrate, sodium silicate)	60–75	–
Anionic surfactants (alkylbenzenesulfonates, alcohol sulfates, alcohol ether sulfates)	5–15	2–10
Nonionic surfactants (alcohol ethoxylates, amine oxides)	0–trace	0.3–5
Bleach (sodium perborate, sodium percarbonate, hydrogen peroxide)	3–15	3–4*
Fillers (sodium sulfate, sodium chloride)	5–20	0–2
Fluorescent whitening agents	0.2–0.5	0.1–0.2
Enzymes	0–1	–
Dyes, fragrances, pH-regulators	+	+
Water	balance	balance

* Hydrogen peroxide.

Table 32. Frame formulations of laundry boosters in the USA

Ingredients	Composition, %	
	Powders	Liquids
Builders		
Sodium carbonate	30–40	–
Sodium borate	0–50	0–1
Sodium silicate	0–5	–
Anionic surfactants (alkylbenzenesulfonates, alcohol sulfates, alcohol ether sulfates)	0–10	0–5
Nonionic surfactants (alcohol ethoxylates, amine oxides)	–	0–1
Hydrotropes	–	+
Fluorescent whitening agents	0–1	0–0.5
Enzymes	0–2	0–1
Dyes, fragrances, neutralizing agents	+	+
Water	balance	balance

Laundry boosters are marketed largely in the USA; various formulations exist. They contain either sodium silicate, sodium citrate, sodium borate, or sodium carbonate, usually in combination with surfactants. Enzymes are often present as well (Table 32).

4.3.3. Aftertreatment Aids

After the washing process has been completed and soil removal has been accomplished, fabrics are sometimes subjected to some type of aftertreatment. The goal is to increase the usefulness of laundry by restoring textile characteristics that have suffered in the course of the wash. Needs in this respect can vary considerably, depending on the fabric involved. Thus, products in this category may be called on to provide elastic stiffness, improved fit and body (shirts and blouses); smoothness, sheen; good drape (curtains); fluffiness and softness (underwear, towels, bath gowns); or antistatic properties (garments, underwear, hosiery, and other easy-care articles made from synthetic fibers). To achieve such effects, the following product groups are marketed:

Fabric softeners
Stiffeners
Laundry dryer aids

4.3.3.1. Fabric Softeners

Textiles washed by machine are subjected to greater mechanical stress than those washed by hand. Indeed, machine-washed laundry may be so severely jammed that the pile of the fibers at the fabric surface is reduced to an extreme state of disarray, especially in the case of natural fibers such as cotton and wool. During subsequent drying in static air (e.g., when laundry is dried hanging indoors), this condition tends to become fixed in the fabrics, and the laundry acquires a harsh feel. Addition of a liquid fabric softener in the final rinse (rinse-cycle softener) results in fabrics that feel softer.

The conditions described above are frequently found in Europe, Japan, and other regions of the world. Fabric softeners play another role in the USA, where laundry is mostly dried in mechanical dryers. The tumbling of the laundry in the dryer accomplishes its own softening effect. The chief task of household fabric softeners in the USA is rather to impart antistatic properties and a pleasant odor to the laundry. Therefore, fabric softeners in the USA are frequently applied as sheets impregnated with active material which are added to the moist laundry at the beginning of the dryer cycle (see Section 4.3.3.3)

The principal active ingredients in commercially available rinse-cycle softeners are usually cationic surfactants of the quaternary ammonium type [60] – [63], [155], [156]. When applied in appropriate concentrations, cationic surfactants are adsorbed nearly quantitatively onto natural fibers, in contrast to their behavior with synthetic fibers (Table 33) [132] – [134].

To prevent undesired interactions between the anionic surfactants of a detergent and the cationic surfactants of a fabric softener, the latter must be introduced only in the last rinse cycle. Overuse of fabric softener must be avoided. Otherwise, the absorbency characteristics of the textiles decrease which in turn adversely affects the function of towels.

Conventional softeners, which contain on average 4 – 8 % active material, have been partially replaced in many countries by softener concentrates having some 12 – 30 % active material. This development is an answer to the increasing public environmental criticism centered on the bulky plastic packaging of conventional softeners. Bottles with

Table 33. Sorption of distearyldimethylammonium chloride* [58]

Textiles	Adsorbed amount**			
	mg/g	mol/g	mg/m^2	%
Wool	1.20	2.06	301	100
Resin-finished cotton	1.18	2.01	133	98
Cotton	1.17	2.00	169	96
Polyester/cotton	1.17	2.00	110	98
Polyamide	0.96	1.63	66	79
Polyacrylonitrile	0.90	1.53	68	74
Polyester	0.57	0.97	109	47

* Equilibrium conditions: time = 60 min, t = 23 °C, bath ratio = 1 : 10; initial concentration: 120 mg/L.
** Based on the mass and geometric surface of the textiles.

volumes of 4 – 6 L of conventional fabric softener need large amounts of plastic for their manufacture, and also present waste disposal problems.

Due to its poor biodegradability the formerly widespread distearyldimethylammonium chloride (DSDMAC) [60], the former active material of most softeners in Europe, the USA, and Japan, has been replaced by the readily biodegradable (see Section 10.5.1) esterquats in the 1980s and 1990s. Frame formulations of liquid fabric softeners are given in Table 34.

The formulation of products with higher amounts of active material requires a well-balanced system of selected emulsifiers to maintain good dispersion stability upon storage.

Fabric softeners have additional benefits. They impart good antistatic properties on fabrics. They prevent the build-up of electrostatic charges on synthetic fibers, which in turn eliminates such unpleasant phenomena as fabric cling during handling and wearing, crackling noises, and dust attraction. Also, fabric softeners make fabrics easier to iron and help reduce wrinkles in garments. In addition, they reduce drying times so that energy is saved when softened laundry is tumble-dried. Last but not least, they also impart a pleasant fragrance to the laundry. That is why branded softeners are frequently offered with different fragrances.

4.3.3.2. Stiffeners

If stiffness and body are desired rather than soft and fluffy laundry, stiffeners can be added as aftertreatment. Usual agents for this purpose include natural starch derived from rice, corn, or potato, which can be used to impart extreme stiffness to fabrics. Synthetic polymeric stiffeners are good alternatives. They impart a more modest degree of stiffness than natural starch does, which is more closely attuned to contemporary taste. Products of the latter type are generally liquid and are easier to apply than natural starch. Stiffeners are supplied as dispersions and contain, in addition to a small amount of starch, substances such as poly(vinyl acetate), which has been partially

Table 34. Formulations of fabric softeners in Europe, the USA, and Japan

Ingredients	Composition, %						
	Europe			USA			Japan
	Regular	Concentrate	Conditioner	Regular	Concentrate	Conditioner	
Quaternary ammonium compounds[a]	3–11	11–20	11–26	3–11	10–30	15–35	0–20
Aminoamide	0–2	0–2	0–2	0–2	0–2	0–2	0–5
Nonionic surfactants	0–0.8	0–3	0–3	0–0.8	0–3	0–3	0–5
Additional softening compounds[b]	0–1	0–4	0–4	0–1	0–4	0–4	0–1
Preservatives, antibacterial agents	0–0.5	0–0.3	0–0.1	0–0.5	0–0.3	0–0.1	0–2
Dye fixatives	0–0.2	0–0.2	0–0.2	0–0.2	0–0.2	0–0.2	
Ironing aids[c]	–	–	0–6	–	–	0–6	0–5
Dyes, fragrances	+	+	+	+	+	+	+
Ethanol/isopropanol solvents	0.2–2.5	0.5–3	0.5–4	0.2–2.5	0.5–3	0.5–4	0–3
Diol solvents	–	–	0–12	–	–	0–15	0–10
Viscosity regulators[d]	0–0.2	0–0.6	0–0.7	0–0.2	0–0.6	0–0.7	0–0.5
Water	balance	balance	balance	balance	balance	balance	balance

[a] Esterquat type: triethanol amine, diethanol amine, or epichlorohydrine esterified with tallow-based acids or oil-based acids are preferred in Europe. Diesters are preferred due to superior softening performance. Aside from esterquats, distearyl dimethyl ammonium chloride (DSDMAC) types, imidazolines, and amino ester salts are used in some countries.
[b] Fatty alcohols, fatty acids, triglycerides.
[c] Silicone oils, dispersed polyethylene.
[d] Magnesium chloride, calcium chloride, sodium chloride, sodium acetate.

Table 35. Formulation of liquid stiffeners in Europe

Ingredients	Composition, %
Alkyl/alkylaryl polyglycol ethers	0.1–2
Poly(vinyl acetates) (partially hydrolized)	10–30
Starch	0–10
Poly(ethylene glycols)	0.5–1.5
Fluorescent whitening agents	+/-
Dyes	+/-
Water	balance

hydrolyzed to poly(vinyl alcohol). Table 35 shows a frame formulation for typical European liquid stiffeners. Such products are offered not only as dispersions, but also as aerosol sprays. The latter allow local treatment of garments, for example, collars and cuffs.

Synthetic polymeric stiffening agents are called permanent stiffeners because, unlike starch, their effectiveness endures through several wash cycles. This property also has its negative aspects, however, since poly(vinyl acetate) film on a fabric can attract both soil and dyes, thereby leading to discoloration.

4.3.3.3. Laundry Dryer Aids

Laundry dryers are far more widespread in the USA than in Europe. They have held a significant place in the U.S. market since the early 1970s. Dryer aids are introduced into the dryer along with the damp, spin-dried laundry. During the drying process, they impart to the laundry both softness and a pleasant odor. Most importantly, however, they prevent static buildup on the fabric. The latter point is particularly significant in the USA, where synthetic fabrics are quite frequently used. Synthetic fabrics are less popular in Europe and elsewhere.

Laundry dryer aids are almost exclusively applied as sheets, which serve as carrier. The sheets are made from nonwoven material and impregnated with both fabric softeners and temperature-resistant fragrance oils. During the drying process, active ingredients from the dryer sheet are transferred to the laundry through frictional contact. The sheet materials are designed for single use and are discarded at the end of the drying cycle.

An additional reason for the popularity of dryer aids in the USA is the fact that washing machines in the USA often lack dispensers for automatic addition of liquid fabric softeners to the rinse cycle.

4.3.4. Other Laundry Aids

4.3.4.1. Refreshing Products for Dryer Application

Refreshing products form a new subsector in the laundry aids category. They were introduced in the USA and Europe in the late 1990s. These products are used for dry-clean and handwashable clothes. They do not replace professional dry cleaning, but they allow the consumers to refresh their garments any time at home. The custom products have two main features: a refreshing and a spot-cleaning action. The refreshing action is achieved inside the tumble dryer by using a moist impregnated sheet along with the item to be treated. Some systems use a plastic bag for producing a damp microclimate inside the dryer, whereas other systems work without a bag. In both cases the water vapor removes unpleasant odors from the garments. For spot cleaning either the moist sheet or a separate stain remover liquid is used in conjunction with an absorbent layer of cloth or paper placed underneath the stained garment to pick up dissolved soil.

4.3.4.2. Odor Removers for Washer Application

Laundry odor removers are liquid products. They were introduced on the U.S. market in the year 2000. Their formulation and technology allows the removal of (unpleasant) odors from laundry. The product is dispensed directly into the washer along with the detergent at the beginning of the wash cycle. Their main active component is based on cyclodextrins.

5. Industrial and Institutional Detergents

Even though the principles that determine the effectiveness of detergents for household and professional laundries are the same, detergents for large-scale institutional use generally differ insofar as they must be designed to meet the special circumstances associated with laundry on an industrial scale [135]. In contrary to home laundering, professional laundries have to deal with large volumes of textile garments and require therefore completely automatic processing with microprocessor-controlled machines and dosing units. They also need data retrieval systems for complete process control. The washing process is in general much shorter and is operated with soft water. For environmental and economic reasons energy and water recycling is essential. Since soil levels are commonly much higher in professional than in household laundry and the variety of textiles is very large, concentrated products, special bleaches, and stain-removal processes are often applied. In many cases disinfection of the laundry load is necessary. Specific packaging for laundry products in large quantities and safe storage systems are frequently required. On the whole, a professional laundry process is a thorough balance between economy, ecology, quality, and safety.

A large variety of products for professional laundries is available today. These products belong to one of the following three major categories:

Base detergents
Detergent additives
Specialty detergents

Table 36 shows the formulation of these products for institutional use.

Base Detergents. In principle, such products are completely built detergents and they are primarily responsible for a good basic detergency. Base detergents are usually highly alkaline and contain no bleach. The major ingredients are surfactants, builders (either phosphate or zeolite with polymer cobuilder), alkalies, fluorescent whitening agents, complexing agents, antiredeposition agents, and enzymes. Currently, they are used quite commonly in continuous batch washing machines with counterflow systems.

Table 36. Formulations of various types of detergents for institutional use

Components	Detergents		Partially built products		
	Base	Specialty	Surfactant boosters	Bleaching agents	Enzyme boosters
Surfactants	×	×	×		×
Sodium triphosphate or zeolite/polycarboxylate	×	×			
Alkalies (soda ash, metasilicate)	×	×	×		
Bleaching agents				×	
Fluorescent whitening agents	×		×	×	
Enzymes	×	×			
Complexing agents (phosphonates)	×			×	
Antiredeposition agents	×	×			×

Oxygen or chlorine bleach can be added depending on the degree of soiling. The addition of other detergent additives such as disinfecting agents or enzyme boosters is optional. Base detergents are available in form of powders, liquids, and pastes.

Detergent Additives. In order to improve the removal of heavy soilings and tough stains, which are quite common in professional laundries, detergent additives are frequently used. Special products like surfactant, alkaline, bleach, and enzyme boosters are commonly available. One of the most important additives is the disinfectant for hospital use. Such disinfectant additives are specifically designed to handle laundry originating from isolation wards and therefore require thermal and chemothermal sterilization. Common disinfectants for this purpose include peracetic acid – hydrogen peroxide mixture, sodium *N*-chloro-*p*-toluenesulfonamide (chloramine T), dichlorodimethylhydantoin, quaternary ammonium compounds, and other organic chlorine carriers which liberate active chlorine in the presence of water. Disinfecting procedures of this kind are mainly common in Europe. Laundry sterilization in the USA is accomplished either thermally or by active chlorine bleaches.

Specialty detergents are products formulated in such a way as to meet the demands of particular kinds of laundry or particular laundering processes. These include, for example, detergents without whitening agents for easy-care and colored fabrics, detergents with high contents of nonionics for heavily soiled working clothes, and enzyme-containing products for proteinaceous soils.

In many semiprofessional laundries like in small hotels and restaurants, detergents which are more or less comparable to those for the household sector are required. For example, perborate-containing detergents have been very popular especially in Europe. The formulation of a perborate containing detergent is given in Table 37.

Table 37. Formulation of a perborate-containing detergent for institutional use

Components	Examples	Composition, %
Surfactants	alkylbenzenesulfonates, soaps, alcohol ethoxylates	10 – 15
Sequestering agents	pentasodium triphosphate, zeolite, polymer cobuilder	20 – 30
Alkalies	sodium carbonate/silicates	10 – 20
Bleaching agents	sodium perborate, TAED	15 – 30
Fluorescent whitening agents	stilbene derivatives	0.1 – 0.3
Antiredeposition agents	carboxymethyl cellulose, polymers	0.5 – 2
Corrosion inhibitors	sodium silicate	3 – 5
Foam iregulators	silicone/paraffin	0.1 – 0.3
Stabilizers	phosphonates	0.2 – 2
Fragrance oils		0.1 – 0.2
Enzymes	protease, amylase	0.5 – 1.0

6. Production of Powder Detergents

Until 1987 the term powder detergent worldwide was synonymous to what today is called a conventional detergent. Powder detergents were manufactured mainly by hot spray drying and had densities of some 300 to 550 g/L. Triggered by the rapid market growth of compact detergents in Japan in the late 1980s, a significant change of the technology to manufacture high-density (> 600 g/L) heavy-duty detergents also took place in Europe and the USA, following the fast growing demands of the respective markets. In major parts of Europe and the vast majority of the USA, compact detergents became state of the art during the 1990s. In addition, detergent tablets have made their way in Europe since 1998, whereas their market introduction in North America began in 2000.

For compact powder detergents, the traditional manufacturing process comprising slurry drying of most detergent components in a spray tower and postaddition of temperature-sensitive components had to be extended by an additional densifying step.

In addition, quite a different development took place in the 1990s, namely, the use of nontower technology. This technology encompasses processes in which the drying step in the spray tower is replaced by a process which uses an increased number of dried raw materials and compounds (e.g., FAS on zeolite) than before. Initially, the bulk density of the tower powder was increased by dry mixing/milling the powder in fast-revolving, continuous, preferably high-speed mixers. In the 1990s, the European detergent industry developed new and more effective densifying processes which are based on tower powder, nontower granules, and dried raw materials or compounds. The achievable bulk densities depend on the detergent composition and the densifying process used (see Table 38).

Table 38. Densifying processes

Densifying process	Bulk density
Dry mixing/milling	650 – 720 g/L
Wet granulation/drying	750 – 900 g/L
Nonionic granulation	$\approx$ 800 g/L
Roller compaction/milling	700 – 900 g/L
Bar extrusion/milling	700 – 900 g/L
"Spaghetti" extrusion	700 – 900 g/L
Tableting	$>$ 1200 g/L

For each process many variants exist, e.g., wet granulation using

Water
Water-based binder, or
Pastes (surfactants)

in a

Dryer (fluidized-bed, linear, or round)
Mixer (Lödige, Fukae, Drais, Ballestra, Eirich, Schugi) [157], or
Combination of various mixers/dryers

A different technology has gained increasing importance as well, i.e., the process of mixing of compounds (see Section 6.2.4). In this process each raw material or part of it is separately agglomerated to dense particles. Therefore, anionic surfactant compounds, nonionic surfactants, builders, etc., can be easily varied in their content by dosing more or less of each compound. Compounds are manufactured by the detergent industry and — increasingly — by raw material suppliers, e.g., defoamer or tetraacetylethylenediamine granulates.

6.1. Technology Overview

Figure 74 depicts currently used detergent production technologies in a simplified way. Table 39 illustrates the main process alternatives, starting with drying/mixing, followed by a densification/shaping step. The processes are terminated with postaddition of thermally and chemically sensitive raw materials. Figure 75 additionally shows these alternatives as a chain.

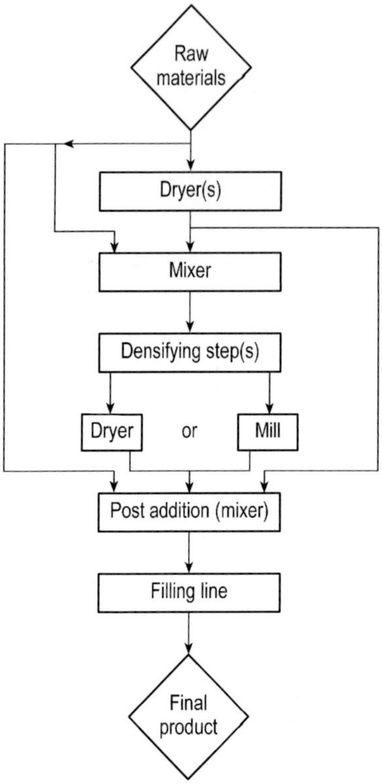

Raw materials
Dryer(s)
Mixer
Densifying step(s)
Dryer
or
Mill
Post addition (mixer)
Filling line
Final product

Figure 74. Manufacturing of compact powder detergents, schematic block flow diagram

Drying/mixing
Densification
Drying
Shaping
Post addition
Wet granulation
Tabletizing
Fluidized bed
- Dryer
- Cooler
Extrusion
- Spaghetti
- Bars
Mixer
Spheronizing
Roller compaction
Milling
Spray tower
Nio granulation
Tower
Nontower

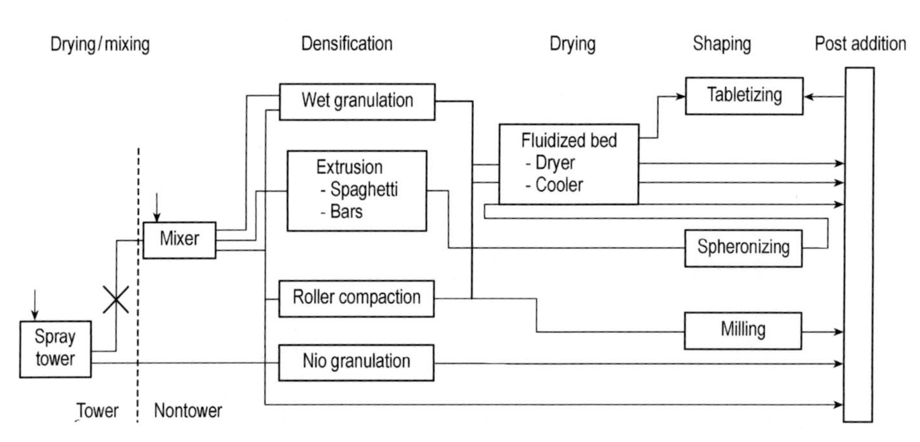

Figure 75. Chains of different processes

Table 39. Detergents manufacturing process alternatives

Process	Drying step	Densifying step
Tower process	Spray drying	Dry mixing/milling
Nontower (mixer(s))-process	– fluidized bed – chemical binding with e.g. zeolite, soda ash – neutralization with soda ash and chemical binding	– integrated in the high granulation step or – mixture is • extruded • tableted • roller compacted
Combination of tower/nontower "premix-process"	– one part is spray dried – another part is granulated with water and/or surfactants – third part is dry mixed	– without densifying – integrated in the high shear granulation step, or – mixture is • extruded • tableted • roller compacted
Compound process (mixer(s))	Compounds are – fluidized bed dried – roller compacted – granulated with surfactants	– integrated in the high shear granulation/extrusion/roller compaction step

6.2. Manufacturing Processes

6.2.1. Traditional Spray-Drying Process [158]

In the first step of the traditional spray-drying process (Fig. 76) a slurry of thermally stable and chemically compatible ingredients of the detergent is prepared. Solid and liquid raw materials are drawn off from silos or tanks and introduced batchwise into scales. Water is added as required to maintain a manageable viscosity. The liquid mixture and the solids are taken from the scales and mixed to form a slurry in a crutcher or by using some other type of forced mixer. The slurry is transferred to a stirred storage vessel from where the process runs continuously.

The slurry is transported by a low- or medium-pressure pump (up to 80 bar). Changes in pressure in the high-pressure portions of the system are compensated for by using an air vessel. The slurry is sprayed into the tower by two spraying levels connecting a series of nozzles. The number and type of nozzles must be designed such that overlapping of spray cones is avoided. Dried tower powder flows off from the tower at a temperature of 90 – 100 °C. To prevent lumping, an airlift is used for cooling.

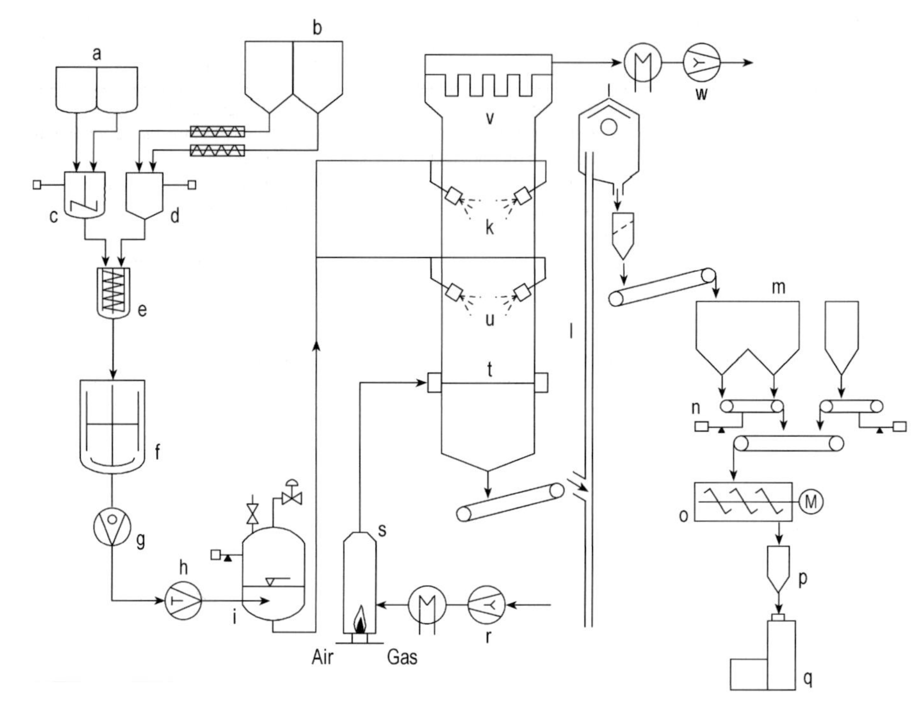

Figure 76. Traditional spray-drying process
a) Storage tanks for liquid raw materials; b) Storage silos for solid raw materials; c) Liquids weighing vessel; d) Solids weighing vessel; e) Mixing vessel; f) Intermediate tank; g) Booster pump; h) High-pressure pump; i) Air vessel; k) Nozzles; l) Airlift; m) Storage bunker; n) Belt conveyor scales; o) Powder mixer; p) Sieve; q) Packaging machine; r) Air inlet fan; s) Burner; t) Ring channel; u) Spray tower; v) Top filter; w) Exhaust

The spray-drying process can be run cocurrently or, most common, countercurrently [159]. If particles and air move cocurrently and the slurry enters the hottest zone of the tower, a rapid evaporation of water occurs, and blown-up, relatively light particles (beads) result. In the countercurrent process, drying starts in an area of high humidity and lower temperature, and beads are obtained that are characterized by thicker walls. The lower free-fall speed due to the countercurrent airstream results in a higher number of particles in the tower at any time. Agglomeration products resulting from a countercurrent system are heavier and coarser than those resulting from cocurrent spray drying. Figure 76 shows a countercurrent operation. The air is directly heated by exhaust gases from an oil- or gas-fueled burner and led into the tower at about 300 °C through a ring channel. Appropriate design of both the air inlet zone above the cone and the air outlet at the tower dome is important. Air entering the tower is “swirled”, i.e., it is accelerated tangentially and vertically. This causes homogeneous heat transfer between air and product. Water vapor, combustion gases, and drying air are withdrawn from the tower by means of a ventilating fan.

Fine particles are drawn off from the tower along with the exhaust gas and collected in a filter. In modern plants this filter is installed on top of the tower so that the separated fines and dust fall down into the agglomeration zone of the tower [160]. The resulting tower powder has a low content of fine particles. Heat exchangers are installed to use the exhaust gas energy for preheating the burner air. Another way is partially recycling the exhaust air (about 50 %) and mixing it with fresh air from the atmosphere [161]. In doing so, energy savings of about 10 – 20 % can be achieved. Plant control is possible by installation of an expert system [162].

6.2.2. Superheated Steam Drying

Environmental issues of spray drying using hot air are:

Emission of dust, normally < 10 mg/m^3
Emission of odor and volatile organic matter in very small amounts, normally < 50 mg/m^3
Potential dust-explosion and fire hazard
Loss of vaporization heat

These problems can be solved by switching over to drying with superheated steam instead of air [163]. The usage of superheated steam allows the heat of vaporization to be recovered [164]; the system is closed and no exhaust gas escapes from the tower [165]. Presently, this technology is being adapted in a pilot project [166].

The bulk density of steam-dried products is significantly higher and the water content is lower compared to hot-air spray drying.

6.2.3. Nontower Agglomeration Process

The nontower agglomeration process is carried out in a continuous mixer such as the Ballestra Kettemix Reactor or the Lödige CB-Mixer or in a combination of high- and low-shear mixers. Zeolite is applied in powdered form to bind surplus water, especially as zeolite P or X or as superdried zeolite A plus soda ash. In the first step the powder components are continuously added. Metering is carefully controlled at the entrance of the reactor. In this zone all liquids, such as nonionic surfactants, polymer solutions, fatty acids, sodium silicates, and linear alkylbenzenesulfonic acid, are separately dosed [167].

Due to the high energy input granules are formed and at the same time neutralization with sodium carbonate or sodium hydroxide takes place. The granules can take up limited amounts of nonionic surfactants. Optionally the granules are surface-treated with zeolite ("powdered") in an additional step to reduce product stickiness, which is mainly due to the nonionics. If the water binding capacity of zeolite and soda ash is not sufficient, e.g., in the case of neutralization with sodium hydroxide solution, added

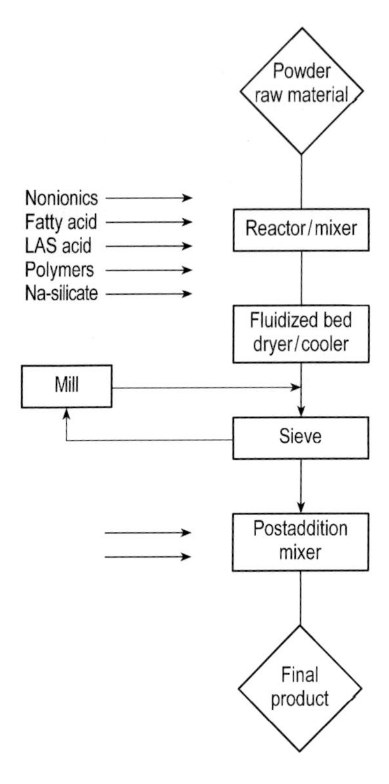

Figure 77. Nontower agglomeration process

water and reaction water have to be removed from the product in a further step. This is done in a fluidized-bed dryer, where the granules are dried with hot air (140 °C) in the entrance zone. In the outlet zone the granules are cooled down, normally using cold air (5 – 15 °C). The drying process is controlled by measuring the product water content and temperature. Depending on the formula the product temperature has to be maintained between 50 and 110 °C (preferably between 60 and 80 °C).

If the water balance is fitted by using small amounts of water and/or by neutralization with carbonates the granules are only cooled in the fluidized bed and not dried. The fluidization air is transported in a cycle with a cooling/heating step to ensure a constant water content of the air.

After sieving and milling, the granules are transported to the postaddition step.

Quite a few limitations with respect to the nontower agglomeration process exist. Firstly, the neutralization heat increases the granulation temperature. Therefore, the amount of acids in the feed material is strongly limited. Secondly, the water level must be lower than 14 %. The nonionic surfactant content depends on the detergent composition. At higher nonionics contents the granules become too soft and sticky. A typical nontower agglomeration process with one mixer and a fluidized-bed dryer is shown in Figure 77 as a block diagram. The more advanced variant with two mixers and a fluidized-bed cooler is illustrated in Figure 78 as a flow sheet diagram. The process uses a high-shear/low-shear mixer combination. Neutralization is effected by using 50 % caustic soda.

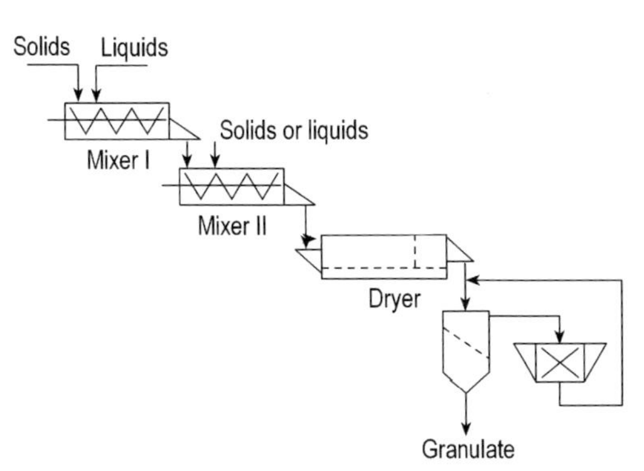

Figure 78. Flow sheet diagram of a typical two-mixer nontower process

6.2.4. Nontower Compound Technology

Another way of realizing a nontower process is application of compound technology. A finished laundry detergent is produced by blending the following components [168]:

Nonionic surfactant agglomerate	13.4 wt %
Anionic surfactant agglomerate	32.5 wt %
Layered silicate compacted granules	10.1 wt %
Granular percarbonate	22.7 wt %
Tetraacetylethylenediamine agglomerate	7.8 wt %
Suds suppressor agglomerate	6.5 wt %
Perfume encapsulate	0.1 wt %
Granular soil release polymer	0.4 wt %
Granular sodium citrate dihydrate	3.5 wt %
Enzymes	2.0 wt %

Each component is produced separately, in most cases by using the granulation technology. By applying different densification technologies an inhomogeneous product aspect results after blending of the compounds. Compound blending is presented in Figure 79.

6.3. Densification Processes

To increase the bulk density of the final product to values higher than about 650 g/L, one or more densifying steps must be used in detergent production and/or denser raw materials/compounds have to be applied. Usually, the particle size distribution of the densified product is adjusted by sieving/milling. In most cases the size distribution is similar to that of tower powder or standard products (0.2 – 0.8 mm). Larger and denser particles may retard dissolution of the finished product. Therefore, detergent manufacturers try to minimize the quantity of oversize particles.

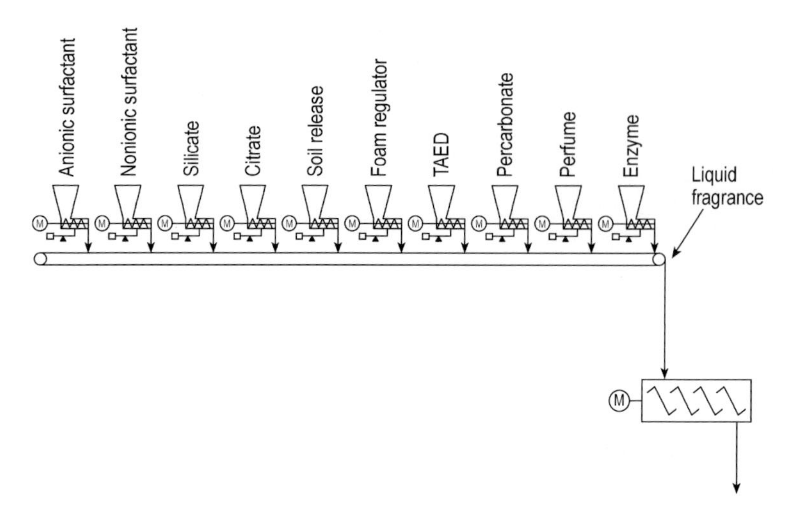

Figure 79. Dosing and mixing of different compounds

Many process alternatives for densification exist. Typical process technologies are described in the following.

6.3.1. Dry Densification in a Mixer

A very simple and cheap method for achieving a limited increase in detergent density is mixing/milling/spheronizing of tower powder in a continuous mixer such as the Lödige CB-mixer (Fig. 80). The densification yields an increase in bulk density of 50 – 100 g/L [169], [170]. Usually, a liquid (nonionic surfactant) is added to the CB-mixer to reduce stickiness, especially at the mixer walls and the tools.

6.3.2. Dry Densification in a Spheronizer

The bulk density of some products can be increased slightly by rounding off the particles in a spheronizer [158]. After entering the spheronizer the particles are accelerated by a high-speed rotating disk. They are driven against the vessel wall by centrifugal forces. There they roll against each other in a particle cloud. As a consequence, edges and surfaces of the particles are polished. In this way, the form of the particles is improved.

Figure 80. Lödige CB-mixer for dry densification of detergent tower powder (view of the internals)

6.3.3. Dry Densification in a Roller Press

Compacting of raw materials or compounds in a roller press is a very simple process that is followed by a relatively complex milling/sieving step (Fig. 81). Compacting in a roller press is accompanied by a dust problem, especially when using products with a high content of organic matter. The resulting product is very dense, but the particles are irregular in shape (splinters). The latest development in this field is a closed device in which rolls, mills, sieves, and recycling tubes are installed in one unit.

For "dry" densification an extruder can be applied as well. In this case, a binder/lubricant must be added to the powder mixture. The product is extruded to form bars. The final form can be achieved by milling/sieving, as shown in Figure 81.

6.3.4. Wet Granulation

Wet granulation is the most common process for densifying tower powders, raw materials, or compounds. Wet granulation is performed by adding water, polymer solutions, anionic surfactant pastes, or surfactant gels at high shear rates [171] – [174]. The wet granulation process consists of the following steps:

Step 1: grinding in a high-speed mixer or in a mill; addition of polymer solution (7 – 12 %) via spray nozzles to the powder in the mixer (pregranulation)
Step 2: granulation and conditioning in a high- or medium-shear-rate mixer
Step 3: evaporation of added water in a fluidized bed and/or cooling down
Step 4: sieving of coarse particles and milling; usually, fine particles remain in the product

Figure 82 illustrates one of several possible designs of a typical wet granulation process [175]. The use of two or more mixers of different types in a cascade increases process

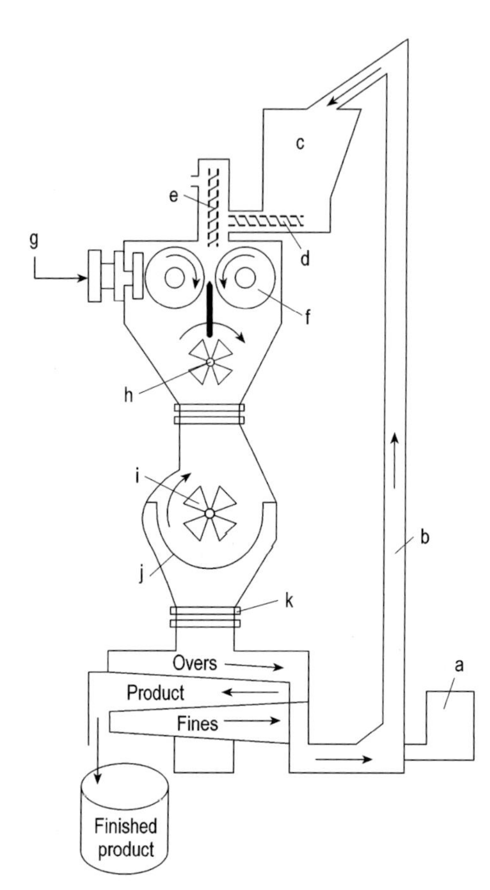

Figure 81. Typical compaction/granulation systems with screening and recycling of overs and fines
a) Original powder feeder (lower hopper), b) Recycle system, c) Upper feed hopper, d) Horizontal feed screw, e) Vertical screw, f) Compaction rolls, g) Pressure applied, h) Prebreak, i) Granulator, j) Screen, k) Screener

flexibility. Energy input, granulation temperature and time, dosage of liquids, powdering of the particles, and types of mixers represent the know-how of the detergent manufacturers.

A very promising new granulation technology is foam granulation [242], [176]. Usually, the granulation liquid is sprayed on the solids or dosed in a jet. The addition of a foam enables a more flexible granulation with less recycled material. A comparison of different granulation techniques is shown in Figure 83.

6.3.5. Spaghetti Extrusion

The spaghetti extrusion process represent a new approach to densifying detergents. Powders, raw materials, and compounds are mixed while a lubricant is added. The mixture is plastified in an extruder and pressed through many small holes. The resulting spaghetti-shaped detergent strands are cut in such a way that the length-to-diameter ratio of the resulting tiny cylinders is unity. The cylinders are rounded off in a

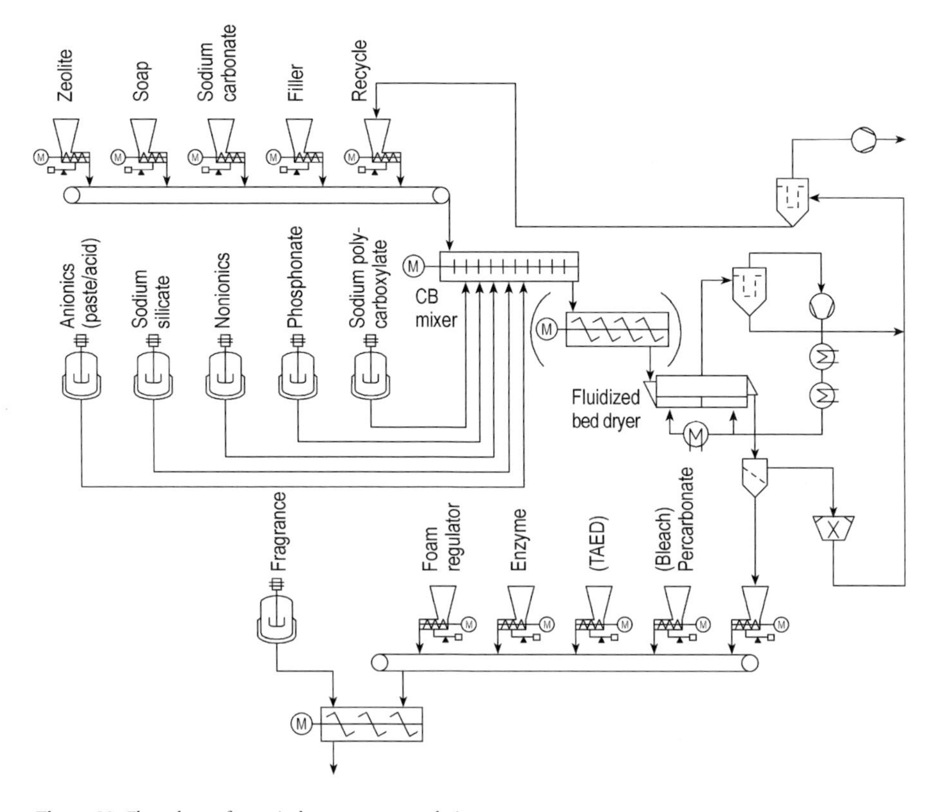

Figure 82. Flow sheet of a typical nontower granulation process

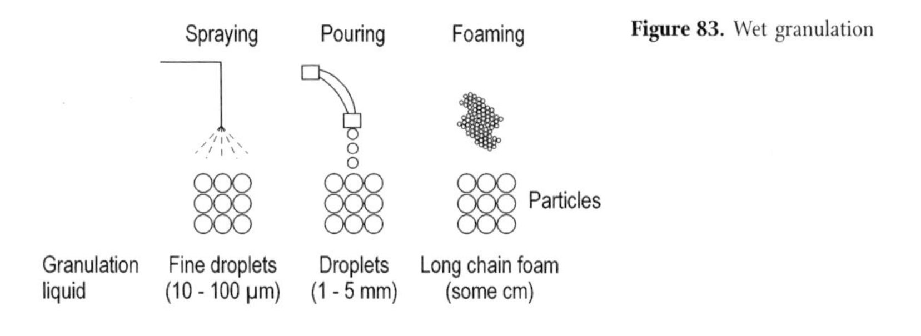

Figure 83. Wet granulation

spheronizer [138], [177]. To harden the product an additional cooling step can be necessary. This densification process has the advantage that a dust-free product is obtained with spherically shaped particles of uniform size, whereas all other processes deliver powders, granulates, splinters, or mixtures thereof, that have a more or less broad spectrum of particle sizes.

Densification by extrusion occurs before the postaddition process (Fig. 84). If the raw materials are used as they are, no further granulation step (e.g., tower drying or wet

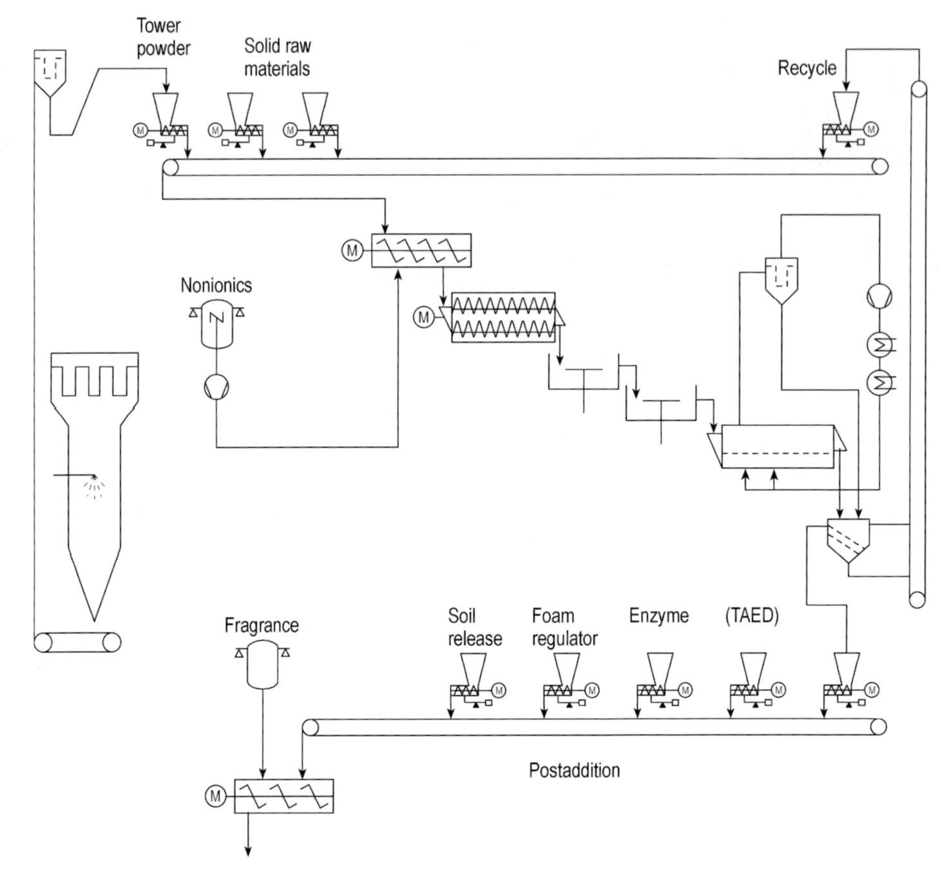

Figure 84. Flow sheet diagram of the extrusion (Megaperls®) process

granulation) is necessary. Both densification and particle shaping occurs inside the extruder by means of a specially constructed extruder head and knife.

The detergent mass is plastified in a cooled twin-screw extruder. In the extruder head a mass temperature of 60 to 80 °C prevails under a pressure of 70 to 110 bar. The mass is pressed through bore holes of 1.4 mm in diameter. There are about one hundred bore holes per plate and more than hundred plates per extrusion die.

The extruded particles differ significantly from other detergent particles, they are much larger, spherical, and nearly uniform (see Fig. 85).

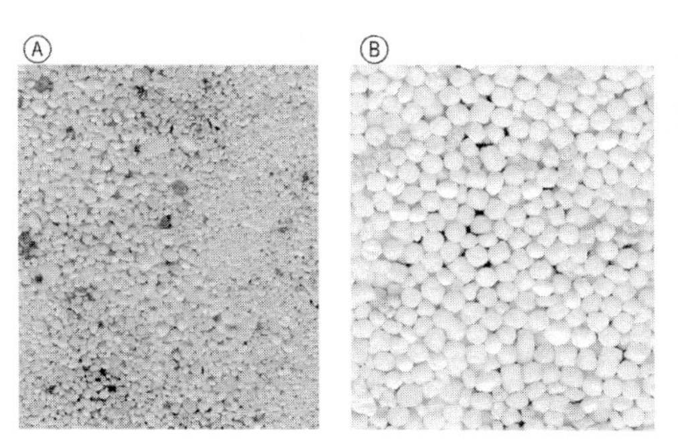

Figure 85. Appearance of market detergent products
A) Particles from granulation/compounding;
B) Particles from extrusion

6.3.6. Postaddition Process

The finishing process of granulated or extruded detergents includes postaddition of thermally sensitive raw materials and/or chemically incompatible ingredients such as bleaching components (sodium perborate, sodium percarbonate), bleach activators, enzymes, soil repellents, foam inhibitor concentrate, fragrances, and, occasionally, colored granules. After the postaddition step the finished product is ready for sieving and packaging.

Postadded granules or powders generally have to be adjusted in their particle size distribution to match the particles of the base powder. Smaller postadded particles, particularly when they have higher densities, tend to segregate in the package or in the process before.

6.3.7. Dry Densification in a Tablet Press

Tablets are a convenient application form for laundry products. Tablets may consist of one, two, or more layers. In Europe, usually two tablets are recommended to clean a 4.5 kg laundry load in the washing machine. To achieve maximum performance all ingredients need to be in a dissolved state from the very beginning of the wash cycle [178]–[182]. Two main problems must be solved:

1) The low surface-to-volume ratio of detergent tablets adversely affects the dissolution rate.
2) In the presence of surfactants, especially nonionics, gel phases may form upon first contact of the tablets with water. Gel formation impedes fast dissolution.

The different concepts that have been developed to formulate and to produce tablets that quickly disintegrate are presented in Section 4.1.4.

Independent of the product concept many physical parameters are identical for the finished tablet and its manufacturing (Table 40). All manufacturers use rotary die presses for production.

Table 40. Parameters affecting tablet quality and process efficiency

Product Assumptions	Process Conditions	Tableting Machine Arrangement
• Formula (formula splitting into two phases) • Content of liquids • Tableting aids • Particle size and distribution • Bulk density • Stickiness • Particle shape • Moisture content • Disintegrant in the tablet and/or coating	• Speed of tableting • Climatic conditions • Hardness of particles and tablets • Homogeneity of mixtures • Compressibility • Pressure of tableting • Temperature of the tablets	• Surface of the punches • Dosing systems • Tablet shape, weight, height and type • Control systems • One, two or three different phases • Tablets with an inscription

The critical points of tableting detergents are the strong reciprocal dependency of tablet hardness and dissolution or disintegration time. The tableting pressure depends on the mechanical stability of the tablet and on its disintegration time. Tablets produced under high pressure are very stable but have too long disintegration times. On the other hand, tablets produced under low pressure are friable. Only a narrow range of tableting pressures can be applied for producing stable, nonfriable tablets with a short disintegration time. The so-called operation area for the tableting press is not fixed, the values depend on formula, size and distribution of the particles as well as on the bulk density of the premixture.

As described in Section 6.3.6, the particle size distribution of all formula components has to be adjusted.

The ingredients of the tablet can be separated into two layers. In this way, for example, enzymes and bleach can be compacted together in one tablet but in different layers without interaction during storage. The manufacturing process for a single-layer tablet is given below. For the second and third layer the same procedure is applied in separate equipments to components with different composition and color:

Step 1: Granulation of the main components and drying
Step 2: Sieving of the granules in a range of 0.2 to 2 mm or a smaller range
Step 3: Postaddition/mixing/sieving (bleach, enzymes, salts, optical brighteners, polymer compounds, fragrances, suds and pH regulators, recycled materials, dyes)
Step 4: Tableting
Step 5: Flow wrapping of the tablets
Step 6: Packaging

The whole process must be temperature- and moisture-controlled. To prevent picking and sticking of granules the water content is limited to 6–9 %, depending on the formula. To ensure maximum action of the disintegrant, wrapping of the tablet in a

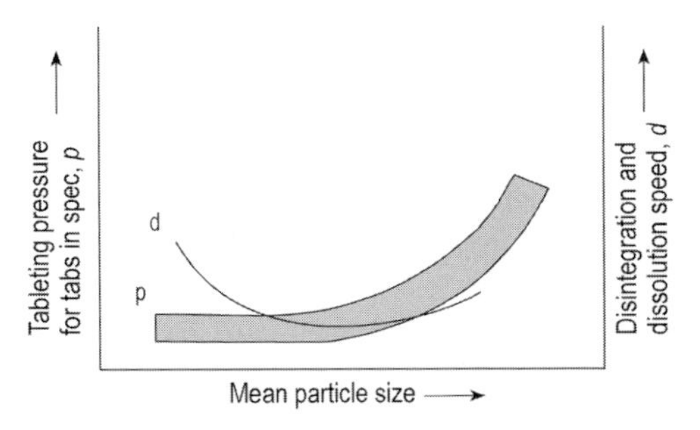

Figure 86. Dependence of the tableting pressure and dissolution rate on the particle size

Figure 87. Punches of a double-layer tableting machine [183]

so-called flow packer is necessary. Each flow pack usually contains two tablets for one wash cycle.

Figure 86 shows the dependency of the tableting pressure and dissolution rate on the particle size. Very fine particles need low pressures to form stable tablets. Since a disintegration agent with a small particle size develops only low forces, a minimum size of the disintegrant is needed, optimally adjusted to the particle size distribution of the granules. An interesting range of particle sizes may be, for example, 0.4 to 1.2 mm.

A contrary approach is the manufacturing of a friable, unstable detergent tablet with low tableting pressures. This kind of tablets already break by simply touching them. To stabilize the product it has to be coated with materials that form a very hard, easily melting layer or shell. A preferred technology may be using an enrober with single or double curtains. Suitable raw materials for enrobers are C_{12-14} dicarboxylic acids combined with a disintegrant.

Rotary die presses are used for tableting detergents [157]. Two-layer-tablet presses are equipped with an upper and lower punch and two filling stations. The residence time under maximum pressure and the press forces determine tablet stability and disintegration time. Figure 87 shows the punches.

6.4. Raw Materials

Each of the key detergent ingredients (see Chap. 3) consists of chemically and physically very different raw materials, also called systems, e.g., bleaching systems with sodium perborate monohydrate or percarbonate and TAED. The ingredients influence the choice of production technology.

6.4.1. Anionic Surfactants

For laundry detergents usually linear alkylbenzenesulfonates and alcohol sulfates, linear or branched, are used as anionic surfactants. As LAS do not crystallize, they are only available on the market as an aqueous paste or a dusty powder or powder compound. Alcohol sulfates crystallize more or less well, depending on their carbon chain distribution. Quite a few manufacturing processes exist for the production of AS compounds with up to 93 % active matter. Anionic surfactants and their manufacturing processes are amply described in the literature [184].

The starting materials for alcohol sulfates, alcohol sulfuric acid semiesters, are derived from sulfation of long-chain alcohols ($C_{12}-C_{18}$) with SO_3. The semiesters are not stable and have to be instantaneously neutralized with sodium hydroxide. One process is "low-water" neutralization with highly concentrated sodium hydroxide at elevated temperatures in a melt in a continuous neutralization loop [185]. Another technology applies a spray neutralization process, in which the alcohol sulfuric acid semiester and sodium hydroxide solution are mixed and sprayed into the tower using a special nozzle. The heat of neutralization evaporates the water, so that no further energy input is required for drying [186]. The resulting powder must be granulated under densifying conditions [187]. Another method embodies drying in a high-shear-rate dryer, in which the paste is dried in a thin film along the wall of the dryer [188].

A further process recommends applying vacuum neutralization in a special film reactor with a high-shear mixer on the top [167]. The reaction rate is controlled by continuously measuring the pH value of the product. Alcohol sulfates are most frequently used in a paste delivery form. The paste can be dried in a cocurrent spray-drying tower under inert atmosphere. The resulting powder is very light and fine and must be densified, e.g., by extrusion as noodles. The produced shape is shown in Figure 89.

The Ballestra company offers the Dryex process for the production of dry anionic surfactants, especially for AS. The neutralized and preheated (80 – 85 °C) pastelike surfactant is fed into the top of a wiped film evaporator. Rotating blades form a thin film over the evaporator surface and remove the dried product from the wall. Drying is supported by using a slight underpressure. The product can be flaked and milled inline. Similar processes are available from various manufacturers of thin-film evaporators.

In another more elegant and economic method the paste is dried in a fluidized-bed dryer/agglomerator. This process, shown in Figure 88, performs drying, densifying, and

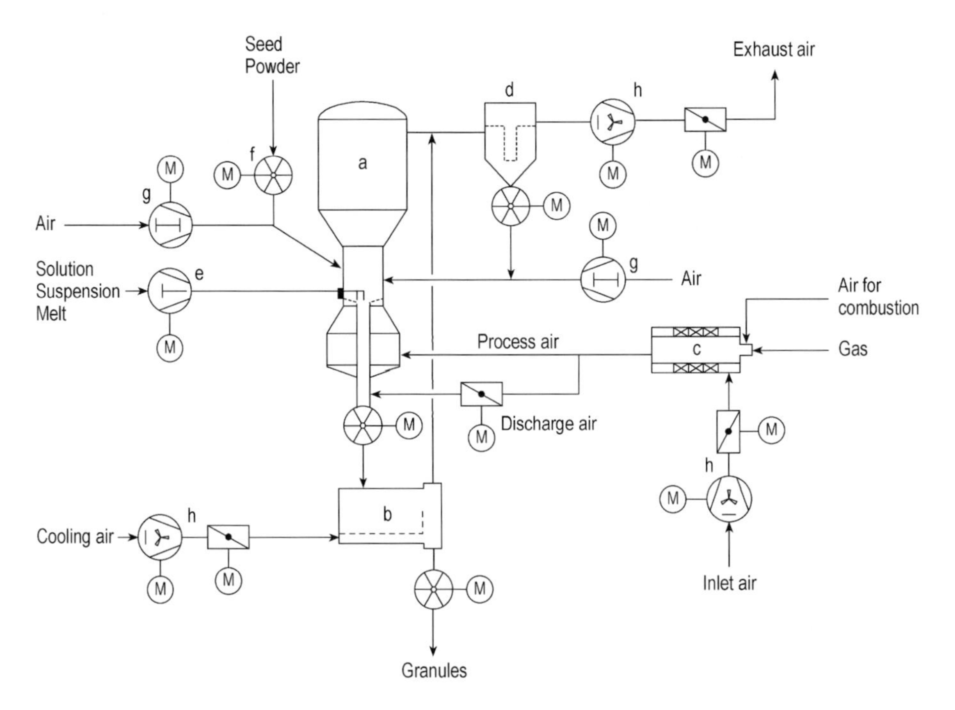

Figure 88. Flow sheet of the one-step agglomeration/drying process[190] a) Fluidized-bed dryer; b) Fluidized-bed cooler; c) Heater for process air; d) Deduster; e) Dosing pump; f) Metering rotary valve; g) Conveying air fan; h) Fan

sieving in one apparatus [189]. The paste is sprayed onto smaller particles, for example, product dust which is fluidized in the bed. The product is sprayed and dried shell by shell. Rotation and the method of drying provide products of high bulk density. The stickiness can be reduced by compounding with carrier materials such as zeolite, starch, and inorganic or organic salts.

The drying of AS paste is accompanied by emissions resulting from unavoidable small amounts of free fatty alcohols in the exhaust air. The problem may be partially resolved by reducing the inlet temperature.

Shape and particle size of the product depend strongly on the process type, as shown in Figure 89.

6.4.2. Nonionic Surfactants

Nonionic surfactants are derived either from natural renewable feedstocks or from petrochemicals. As *renewable raw materials* for surfactants, particularly tallow, coconut and palm kernel oil (laurics), and soybean oil are of paramount importance. The oils are purified in a first step with sulfuric acid in separators or decanters [191], [192]. The triglycerides are either split with steam at elevated temperatures and pressures to form

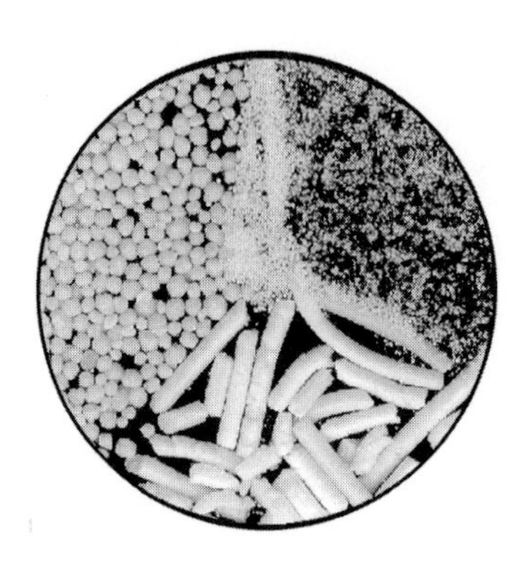

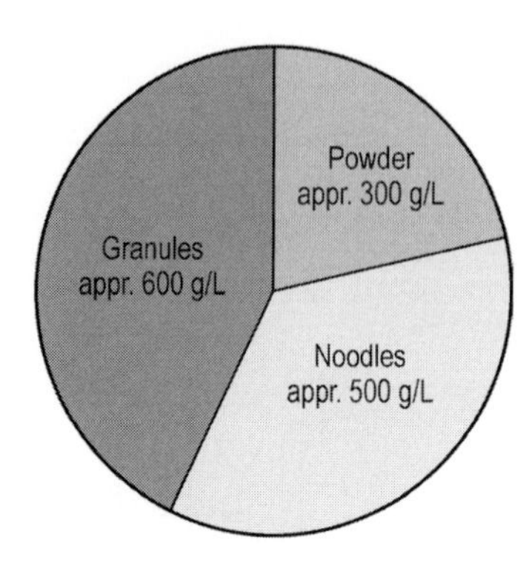

Figure 89. Particle shapes of dried anionic surfactant salts

fatty acids and glycerol, or transesterified with methanol to produce fatty acid methyl esters in a one-step reaction. The methyl ester are subsequently hydrogenated to form fatty alcohols.

For the production of fatty alcohols from *petrochemicals* (paraffins and ethylene) a range of processes exist [193]. The most well-known processes start with the catalyzed oligomerization of ethylene either according to the Shell Higher Olefins Process (SHOP) or the Ziegler trialkylaluminum process. The SHOP olefins are hydroformylated with carbon monoxide and hydrogen to form aldehydes and in a further step hydrogenated to low-branched alcohols. In the Ziegler process trialkylaluminum is oxygenated and in a further step hydrolyzed with sulfuric acid, which converts it to long-chain linear alcohols.

The manufacture of nonionic surfactants starts in most cases with chemical modification of fatty alcohols, fatty acids, or fatty acid methyl esters. Fatty alcohols are converted to alcohol ethoxylates and alkylpolyglycosides; fatty acids to fatty acid ethoxylates and alkanolamides; and the methyl esters to the corresponding ethoxylates and glucamides. Important nonionic surfactants are shown below:

Fatty alcohol ethoxylates $R{-}(OCH_2{-}CH_2)_nOH$

Fatty acid methylester ethoxylates $R^1{-}CO{-}O{-}(CH_2CH_2O)_n{-}CH_3$

Fatty acid ethoxylates $R^1{-}CO{-}O{-}(CH_2CH_2O)_n{-}H$

N-Methylglucamide $R^1{-}CO{-}N(CH_3){-}CH_2{-}CH(OH){-}CH(OH){-}CH(OH){-}CH(OH){-}CH_2OH$

Alkylpolyglycosides (APG)

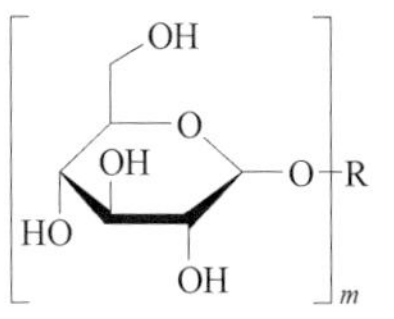

$R = C_8 - C_{18}$

$R_1 = C_7 - C_{17}$

$n = 2 - 10$

$m = 1 - 3$

Ethoxylated fatty alcohols are the most important nonionic surfactants. Also, ethoxylated fatty acids and fatty acid methyl esters [177] are valuable products for the detergent industry. Therefore, ethoxylation is a key technology for manufacturing nonionic surfactants. The ethoxylation process is conducted in batch reactors.

The ethoxylated substrates are chemically inhomogeneous because, firstly, the raw material has a chain length distribution (e.g., $C_{12} - C_{14}$ alcohols) and, secondly, the addition of ethylene oxide is carried out in a broad concentration range. The designation "fatty alcohol + 7 EO" indicates a mean value of seven added molecules of ethylene oxide per molecule of alcohol. In reality a mixture of alcohol ethoxylates with zero to eighteen EO moieties (broad-range ethoxylates with KOH catalyst) is obtained. When special catalysts are used, narrow-range ethoxylates are formed. These process parameters affect the product quality significantly and are part of the manufacturer's know-how.

6.4.3. Builders

Quite a few builders are applied in the detergent industry today. Worldwide, sodium triphosphate is a key builder used on a large scale. Since the early 1980s sodium aluminum silicates (zeolites) have become more and more important, particularly in Europe, the USA, and Japan. Today, the field of builders/cobuilders is more complex than ever. Apart from sodium triphosphate and zeolites, special layered silicates, amorphous silicates, and compounds made from sodium carbonate and amorphous silicate are applied [80], [81], [194]. Besides zeolite A, other zeolite types have gained importance, e.g., zeolite P and zeolite X. Production processes for zeolite A have been described [158].

Layered Silicates. Several methods have been established for manufacturing layered silicates. One method uses the crystallization of amorphous sodium silicates (SiO_2:Na_2O ratios from 1.9 to 3.5), evaporation of the water, and tempering at above 450 °C but below the melting point, to achieve complete crystallization. From this process α-, β-, γ-, and δ-silicate ($Na_2Si_2O_5$) modifications are available [195] – [196], [80]. Another process comprises spray drying of a water glass solution, milling of the product, and tempering in a rotary kiln at 400 to 800 °C for 1 h to crystallize the sodium disilicate [197]. Subsequent densification can be carried out by roller compaction, if necessary.

Amorphous Silicates. A simpler process is the manufacture of amorphous silicates. A spray-dried silicate solution having an $SiO_2 : Na_2O$ ratio of between 2 and 3.5, which is derived from the hydrothermal or water glass process, is compacted and partially loaded with a selected liquid, e.g., nonionics [198]. A variation of the densifying process is the granulation of milled silicate particles [185], [199].

In the 1990s, some raw material suppliers developed manufacturing processes for detergent-grade zeolite types, such as zeolite P and zeolite X, which feature increased loading capacity for liquids [167], [200] – [202]. The manufacturing processes differ from the zeolite A manufacturing technologies in the crystallization conditions.

6.4.4. Peroxygen Bleaches

The detergent industry has used sodium perborate tetrahydrate or monohydrate as a source of active oxygen since 1907. In addition, sodium percarbonate has gained increasing importance in the last decades.

Sodium Perborate. As raw materials for the sodium perborate tetrahydrate process boron minerals, e.g., kernite or tincal with B_2O_3 contents > 47 %, are utilized. In the first step the boron compounds are dissolved in an alkaline solution at about 90 °C to form sodium metaborate:

$$Na_2B_4O_7 + 2\,NaOH \rightarrow 4\,NaBO_2 + H_2O$$

After filtering and diluting the solution to 75 g/L, hydrogen peroxide is added with cooling (< 30 °C) in a crystallizer:

$$NaBO_2 + H_2O_2 + 3\,H_2O \rightarrow NaBO_3 \cdot 4\,H_2O$$

The sodium perborate tetrahydrate crystals are separated from the liquid phase in a centrifuge. They contain 3 to 10 % of surface-adsorbed water, which is removed in a fluidized-bed dryer with product temperatures below 60 °C. Above this temperature (65.5 °C) the product melts in its water of crystallization. Sodium perborate monohydrate is produced from the tetrahydrate by dehydration in a fluidized bed, conditioning, and drying with a steam – air mixture at temperatures not exceeding 100 °C. In its crystalline state sodium perborate forms a stabilizing anion (see Section 3.3.1) [85], [203]:

$$2\,Na^+ \left[\begin{matrix} HO & & O{-}O & & OH \\ & B & & B & \\ HO & & O{-}O & & OH \end{matrix} \right]^{2-}$$

$$+2\,H_2O \rightleftharpoons 2\,H_2O_2 + 2\,NaH_2BO_3 \qquad +4\,H_2O \rightleftharpoons 2\,H_2O_2 + 2\,NaB(OH)_4$$

Sodium percarbonate ($Na_2CO_3 \cdot 1.5\ H_2O_2$) is easily manufactured starting from a crystallization process or from spraying a 35 % aqueous hydrogen peroxide solution onto sodium carbonate in a mixer or, preferably, in a fluidized bed. Percarbonate stability depends on the degree of purity of the sodium carbonate used. Especially heavy metal ions adversely affect the stability by catalyzing the decomposition of hydrogen peroxide.

To enable storage of sodium percarbonate in silos it has to be protected by a coating layer. It is common to use inorganic materials as a coating layer, e.g., sodium carbonate, sodium perborate monohydrate, or sodium sulfate. Coating is performed in batchwise or continuously operating fluidized beds.

Stability. All bleaching agents mentioned above are storage stable up to 60 °C. The perborates are stable at this temperature even when they are loaded with nonionic surfactants. Dust and particles of percarbonate in direct contact with even small amounts of organic material may result in spontaneous ignition of the organic material. Caution measures are to be taken in case of a damaged coating layer or with broken particles.

6.4.5. Enzymes

Detergents may contain up to four different types of enzymes: proteases, amylases, cellulases, and lipases. Enzymes are effective at low temperatures. Even small amounts of these biocatalysts guarantee major effects.

Enzymes are produced in fermenters by microorganisms such as bacteria, yeasts, and fungi. For production of proteases and amylases *Bacillus* species are used, which today are usually genetically engineered. A flow sheet of production of enzymes for detergents is show in Figure 90.

In the fermenter, the extracellularly produced enzymes are dissolved and highly diluted in the fermentation broth. The first step of downstream processing is the separation of the biomass and other solids originating from the nutrient medium (Fig. 90).

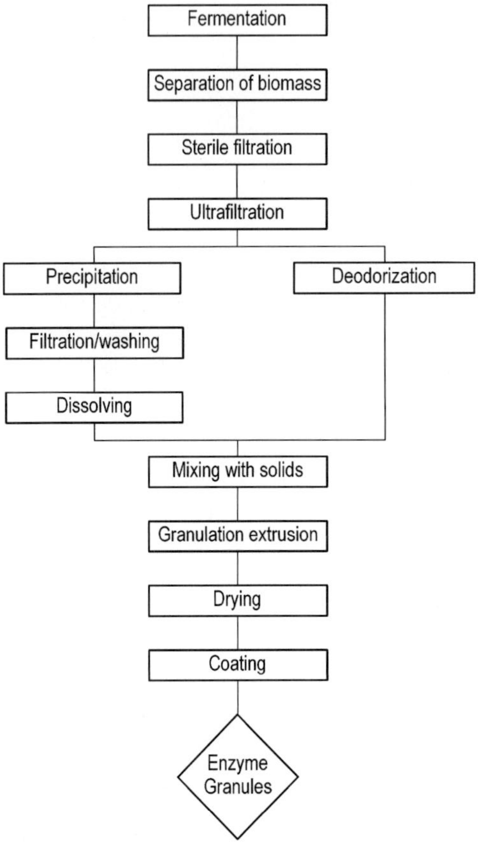

Fermentation
Separation of biomass
Sterile filtration
Ultrafiltration
Precipitation
Deodorization
Filtration/washing
Dissolving
Mixing with solids
Granulation extrusion
Drying
Coating
Enzyme Granules

Figure 90. Downstream processing in enzyme production

Three alternatives are in use:

Centrifuges with extreme centrifugal forces (separators)
Precipitation with salts
Microfiltration

This separation step has proved difficult in practice because the bacterial cells that produce the enzymes are very small (0.2 to 0.5 μm) and their density is close to that of the fermenter broth. Moreover, the fermenter broth has a relatively high viscosity, frequently combined with a non-Newtonian flow behavior. A further limitation is imposed by the poor temperature stability of the enzymes, so that, in many cases, the working temperature may not exceed 30 °C [204]. In addition, slimes can impede work-up.

To guarantee a microorganism-free product, a sterile filtration in a microfiltration plant follows. In the next step the highly diluted solution is ultrafiltered to achieve 15- to 25-fold concentration of the large protein molecules. For the next step two alternatives exist:

Figure 91. Extrusion process [205]

1) Precipitation of the enzymes, filtering the suspension, washing and dissolution of the cake.
2) Deodorization of the concentrated enzyme solution to remove unpleasant odors, generated during fermentation and sterilization. This is conducted in vacuum with superheated steam.

The concentrated enzyme solution is finally sprayed on solid carriers, granulated or extruded (Fig. 91), dried, and coated. Coating is performed in a fluidized bed by spraying a polymer solution on the particles under drying conditions. The coating layer must be perfect to prevent human contact with the particles and to make them dust-free.

7. Analysis of the Composition

The fact that detergents can consist of a large number of individual components whose structures vary greatly is apparent from Chapters 3 and 4. The dramatic industrial advances of the past 50 years together with ecological and economic issues have resulted in steady modifications of detergent formulas. These modifications have required adjustments of analytical methods, with routine procedures frequently becoming outdated.

The multitude of products and a constantly growing variety of raw materials, as well as the complexity of their constitution, makes a summary of the methods used for detergent analysis difficult, especially because the situation is complicated even further by the wealth of analytical techniques available.

The most important factor in choosing an analytical method is the nature of the question posed. In addition, the value of an analysis is highly dependent on the use of proper sampling techniques. Often several methods of analysis are employed in the search for a single answer.

The following procedure can be regarded as a typical example of how a general analysis of a powder detergent might be conducted:

First, the product is separated by extraction into ethanol-soluble and ethanol-insoluble fractions. The ethanol extract can usually be regarded as containing all of the surfactants. The various surfactants are then determined: anionic surfactants by two-phase titration, nonionic surfactants by passing the ethanol extract through both anionic and cationic ion-exchange resins, and the soaps either by titration of the alkalinity of the ethanol extract or by a second two-phase titration in alkaline medium. After suitable sample preparation, such major inorganic components as carbonates, silicates, phosphates, borates, and sulfates are determined either instrumentally or by classical wet methods. All of the more specialized auxiliary ingredients (e.g., soil antiredeposition agents, fluorescent whitening agents, enzymes, chelating agents, bleach activators, etc.) are present in relatively small quantities, and their qualitative and quantitative analysis requires considerable knowledge and experience. Thus, they would fall outside the bounds of a general investigation.

Following is a summary of the literature on detergent analysis. This is intended only as a useful overview. The information has been divided into five categories:

1) Detergent ingredients
2) Purposes of detergent analysis
3) Importance of sample preparation
4) Analytical methods
5) Sources of information

7.1. Detergent Ingredients: [149] – [154], [206], [207] – [210]

Inorganic components: [211] – [218]
Organic components: [219] – [222], [223], [224]

7.2. Purposes of Detergent Analysis: [225]

Raw material analysis (triphosphate and alkylbenzenesulfonate): [226] – [228]
Production control (EO products): [229]
Product quality control (detergent analysis): [230], [231]
Market analysis (detergents): [222], [223, pp. 991 – 1045], [232] – [237], [238, pp. 25 – 43 (1977)]
Trace analysis (vinyl chloride in PVC): [239] – [241]
Analysis of formulated products: [242]
Analysis of detergent formulation: [243]
Detergent analysis: [244]

7.3. Sample Preparation: [222, pp. 21 – 26]

Liquids, solutions, suspensions, emulsions, and pastes: [227, p. 6], [245, H – I 4 (65)]
Detergent bars[245, G – I 1 – 3 (50)], [246, Da 1 – 45]
Powders: [245, H – I 5], [246, Dc 1 – 59, Dd 1 – 59], [247, ISO 607 (1980)], [248], [249, DIN 53 911]

7.4. Analytical Methods

7.4.1. Qualitative Analysis: [222, pp. 27 – 49], [234, pp. 166 – 178]

General Analysis:
Solubility, pH: [245, H – III 1 (65)]
Elemental analysis: [245, H – III 5 (65)]
Analysis for Inorganic Ingredients: [250], [251]
Active oxygen: [250, p. 171], [251, pp. 492 – 493]
Peroxide accumulation in detergents: [252]
Active chlorine: [250, p. 159], [253, Part 2, vol. VII, p. 35]
Ammonia: [250, p. 241], [254, E 5]
Boron: [250, p. 230], [251, p. 776],
Water-insoluble inorganic builders: [154], [255] – [259]
Analysis for Organic Ingredients: [260] – [262]
Alkanolamines after hydrolysis: [222, p. 32],[224], [263]
Sequestrants (NTA, EDTA): [264], [265]
Urea: [266], [267, p. 235], [268]

Fluorescent whitening agents: [222, p. 414], [269] – [271]
Anionic surfactants: [272] – [275]
Alkylene oxide adducts (EO, PO, aromatics): [222, p. 46], [260, p. 138], [274], [276], [277]
Hydrotropic substances: [266]
Enzymes: [278]
Carboxymethyl cellulose: [279], [280], [281] – [283]
Amphoteric surfactants: [284]
Alkyl polyglycosides: [285], [286]
Polycarboxylates: [287]

7.4.2. Sample Preparation

Extraction, Perforation, and Sublation
Ethanol solubles: [222, p. 148], [245, H – III 4 (65)]
Soap fatty acids: [245, G – III 6 b (57) and G – III 8 (61)], [288, p. 1352]
Monosulfonate in alkanesulfonate: [227, p. 39],[245,H – IV 2 b (76)], [247, ISO 3206 (1975)]
Acetone solubles: [222, pp. 253 and 362]
Nonionic surfactants in wastewater: [289]
Nonionic surfactants in sludge: [290]
Polyglycol in ethylene oxide adducts: [291]
Distillation: [292]
Ion Exchange: [222, pp. 50 – 58], [293]
Nonionic surfactants, anionic surfactants, cationic surfactants, soap, fatty acids, and urea: [294] – [301]
Antimicrobicides: [302]
Phosphate types: [222, p. 467]
Carbon Methods
Sulfate: [245,H – III 8 b (76)]
Phosphate: [303]
Decomposition Methods: [304]
Residue on ignition, with hydrofluoric acid digestion: [305, p. 8]
Sulfate ash: [222, p. 335], [245, H – III 11], [306]
Borate: [222, pp. 474 – 476]
Borate decomposition for X-ray fluorescence analysis: [228], [307], [308]
Potassium disulfate melt for Al and Fe: [305, p. 12]
Soda – potash decomposition for water glass: [305, p. 9]
Wet Decomposition
Sulfuric acid – nitric acid, perchloric acid, nitric acid – perchloric acid: [309, pp. 188 – 209]

Measurement of zeolite, silicate, and phosphate: [259]
Chemical Transformations: [310]
Sodium hydrogen sulfite – acid hydrolysis: [221, p. 152], [245, H – IV 7 a], [311], [288, pp. 1474 and 1353], [312], [313]
Cleavage of ether sulfates: [222, pp. 279 and 323], [314] – [316] [317] – [319]
Esterification of fatty acids: [222, p. 572], [320] – [323]
Desulfonation (alkylbenzenesulfonates, alkanesulfonates): [222, p. 197], [315], [316], [324] – [329]

Acetylation (fatty alcohols): [330, p. 543]
Determination of industrial grade carboxymethyl celluloses: [281]
Alkaline Hydrolysis: [222, p. 150]
UV photolysis: [331]

7.4.3. Quantitative Analysis

Gravimetric Analysis
Inorganic Ingredients
Dry residue: [222, p. 434], [332]
Carbon dioxide: [253, Part 3, vol. IV aα pp. 225 – 229]
Silicon dioxide: [319, D 501 – 558], [333], [222, p. 447], [253, Part 3, vol. IV aα, p. 456]
Phosphorus pentoxide: [222, p. 452], [247, ISO/ TC 91 444F], [253, Part 3, vol. V aβ, p. 118], [334]
Sulfate: [245, H – III 8 b (76)], [335]
Magnesium: [222, p. 492], [253, Part 3, vol. II a, pp. 141 – 149]
Organic Ingredients: [288]
Ethanol solubles: [222, p. 148], [245, H – III 4 (65)]
Loss on drying: [222, p. 548], [332]
Unsulfated fraction (US): [222, p. 555], [247, ISO – 893 (78) and ISO – 894 (77)]
Nonsaponifiable fraction: [222, p. 555], [245, C – III 1 a and 1 b (77)]
Soap fatty acids: [222, p. 550], [245, G – III 6 b (57) and G – III 8 (61)], [288, p. 1352]

Nonionic surfactants: [222, p. 167], [245, H – IV 9 b]
Volumetric Analysis: [222, p. 215], [336], [337]
Acidimetry
Alkalinity: [222, pp. 334 and 472]
Sodium oxide content in soaps: [222, p. 554], [245, H – III 7 a (65)], [338]
Borate: [222, p. 474], [253, Part 3, vol. III, pp. 11 – 60]
Potentiometry: [339] – [341]
Phosphate: [222, pp. 455 and 480]

Borate: [222, p. 480], [342]
Chloride: [222, p. 446], [253,Part 3, vol. VII aβ, pp. 107 – 109]
Fluoride: [343], [344]
Nonionic surfactants in wastewater: [289]
Nonionic surfactants in detergents:[345]
Ionic surfactants: [346] – [353]
Ethylenediaminetetraacetate: [354]
Complexometry: [355], [356]
Ethylenediaminetetraacetate: [222, pp. 403 and 487], [247, ISO TC 91/490]
Nitrilotriacetate: [222, p. 403], [247, ISO TC 91/513 E]
Citrate: [357]
Calcium: [254, E 3], [358, p. 25]
Magnesium: [355, pp. 136 – 137], [358, p. 28], [254, E 4]
Nonionic surfactants: [359]
Conductometry: [360]
Sulfate: [361]
Two-Phase Systems: [222, pp. 215 and 218], [232, p. 427], [227, p. 16], [362, p. 221]
Anionic surfactants: [222, p. 234], [245, H – III 10], [363], [313], [364]
Cationic surfactants: [247, ISO 2871 (73)], [365]
Turbidity titration: [366]
Miscellaneous Titrimetric Analyses:
Sulfate: [222, p. 442], [227, H – III 8 a (76)], [245, H – III 8 a (76), p. 24; H – III 8 d], [367], [368]
Chloride: [222, p. 445], [253, Part 3, vol. VII a, pp. 98, 99]
Water by Karl Fischer method: [222, p. 435], [245, H – III 3 a (65)], [247, ISO 4317 (77)]
Active oxygen: [222, p. 483], [247, R 607], [253, Part 3, vol. VI aα, pp. 193 – 207]
Active chlorine: [222, p. 487], [253, Part 3, vol. VII aβ II, pp. 74 – 76]
Calcium and magnesium binding capacity: [369], [370]
Gluconate: [371]
Bleach activator: [372]
Spectrophotometric Determinations: [309]
Perborate: [373, A 124]
Phosphate: [222, p. 457], [334], [374] – [376]
Enzymes: [377], [378], [379]
Enzymatic analysis: [380], [381]
Alkylpolyglycosides: [382]
Polycarboxylates: [383]
Trace elements: [309]
Anionic surfactants in wastewater: [335], [384] – [386]
Cationic surfactants in wastewater: [387]
Size Exclusion Chromatography:
Polycarboxylates: [388]

Flame Photometry: [389]
Sodium and potassium: [389]
Atomic Absorption Spectrometry: [390] – [393]
Traces of Cu, Fe, Zn, and Mn: [394], [395]
Traces of As: [396]
Traces of Co [397]
Atomic Emission Spectrometry: [256], [398]
X-Ray Fluorescence: [228], [399] – [402]
X-Ray Diffraction: [403]
Phosphate phases I and II: [404]
Zeolite types: [405]
Radiometric Methods: [406], [407], [408]

7.4.4. Separation Methods: [409], [410, p. 21]

Capillary Zone Electrophoresis:
Anionic surfactants: [411], [412]
Polycarboxylates: [413]
Paper Chromatography: [414]
Phosphate types: [222, p. 460], [226], [415]
Alkanolamines: [416]
Alkanolamides: [222, p. 391]
Urea: [409, pp. 360 and 402], [414, p. 210], [417]
Thin-Layer Chromatography: [222, p. 45], [262], [418], [419]
NTA, EDTA: [420]
Akanolamines: [224]
Ether sulfates: [421]
Surfactant mixtures: [274], [275], [362, p. 47], [422]
Free fatty alcohols in ethylene oxide adducts: [423], [424]
Bactericides: [302], [425]
Amphoteric surfactants: [222, p. 371], [426]
Anionic surfactants: [427] – [429]
Optical brighteners: [430]
Nonionic surfactants in wastewater: [431], [432]
Column Chromatography: [222, pp. 51 – 63], [410], [433], [434]
Glycerides [288, Part I, p. 832], [435]
Fatty acid polyglycol esters: [436]
Petroleum sulfonates: [222, p. 61], [437]
Nonionic surfactants: [222, p. 167], [438]
Gas Chromatography: [222, p. 80], [439] – [442]
Fatty acid carbon-chain distribution: [323], [443] – [445], [446]

Alkylbenzenesulfonate chain distribution: [447]
Fatty alcohol carbon-chain distribution: [448]
Solvents from cleansers: [449], [450]
Propellants in sprays: [451] – [453]
Dioxane in ethylene oxide adducts: [454]
Antimicrobial agents: [425]
Alkoxy content of cellulose ethers: [280]
Nitro musk compounds: [455]
Pentaacetylglucose: [456]
Fragrances by head space analysis: [457]
Supercritical Chromatography:
Nonionic surfactants: [458]
Anionic surfactants: [458]
Liquid Chromatography: [441], [459] – [461]
Poly(ethylene glycol) in nonionic surfactants: [462] – [464]
Unreacted alkylbenzenes and fatty alcohols in their sulfonation or esterification products: [465]
Bleach activators: [466], [467]
Fatty acid amides: [468]
Separation of fluorescent whitening agents: [430], [469] – [471]
Bacteriostatic substances in soaps: [472]
Nonionic surfactants: [441], [473] – [476], [478]
Anionic surfactants: [474], [475], [477], [479] – [481]
Separation of sulfonic acids: [482], [483]
Separation of lower alcohols: [484]
Separation of sulfur-containing surfactants: [485]
Phosphonates, polyphosphates: [486]
Ion Chromatography: [487] – [489]
Phosphates: [490] – [492]
Phosphonates: [490], [493], [494]
Capillary Electrophoresis: [495] – [497]

7.4.5. Structure Determination

UV Spectrometry: [222, p. 74], [498] – [501]
Alkylbenzenesulfonates: [246, Dd 3 – 60], [502] – [504]
Toluenesulfonate, xylenesulfonate, cumenesulfonate: [222, p. 205], [505], [506]
Structural information of surfactants: [437], [507]
Toluenesulfonamides: [508]
IR Spectrometry: [222, p. 69], [234], [509] – [515], [516]
Petroleum sulfonates: [437]

Degree of branching in alkylbenzenesulfonates: [517], [518]
Surfactant sulfonates: [519], [520]
Soil antiredeposition agents, cellulose derivatives: [521]
NMR Spectrometry: [222, p. 86], [522], [523]
Ethylene oxide – propylene oxide adducts: [222, p. 301], [441], [524], [525]
LAS: [526], [527]
Mass Spectrometry: [222, p. 76], [528] – [530], [531]
Fragrances (GC – MS): [532], [533]
Bactericides (TLC – MS): [222, p. 426], [538]
Nonionic surfactants (LC/FAB – MS): [534], [535]
Nonionic surfactants (MS): [536]
Nonionic surfactants (GC/CI – MS): [537]

7.4.6. Determination of Characteristic Values

Elemental Analysis: [539]
Nitrogen: [222, pp. 124 and 444], [539, p. 178], [540]
Halogen: [222, p. 128], [541], [542]
Sulfur: [222, pp. 126 and 439], [539, pp. 252 and 307], [542]
Phosphorus: [222, p.129], [309, pp. 899 and 902], [539, p. 360]
Fat Analysis Data: [288], [543]
Acid number: [222, p. 182], [245, C – V 2 (77)]
Saponification number: [222, p. 183], [245, C – V 3 (57)]
Iodine number: [222, p. 183], [245, C – V 11 a – d (53)]
Hydrogen iodide number: [245, C – V 12 (53)]
Hydroxyl number: [222, pp. 186 and 303], [245, C – V 17 a + b (53)]

7.4.7. Analysis Automation: [544]

Phosphorus pentoxide: [545] – [547]
Enzymes: [548], [549]

7.5. Sources of Information: [238, p. 403 (1981)]

Analytical Commissions: [550], [551]
Germany
GAT (Gemeinschaftsausschuß für die Analytik von Tensiden) [552]; member organizations of GAT:

a) DGF (Deutsche Gesellschaft für Fettwissenschaften) Varrentrappstraße 40–42, 60486 Frankfurt, Germany

b) TEGEWA (Verband der Textilhilfsmittel-, Lederhilfsmittel-, Gerbstoff- und Waschrohstoff-Industrie e.V., Karlstraße 21, 60329 Frankfurt, Germany, and the IKW (Industrieverband Körperpflege und Waschmittel e.V., Karlstraße 21, 60329 Frankfurt, Germany.

c) NMP (Normenausschuß Materialprüfung im DIN) Burggrafenstraße 4–10, Postfach 11 04, 10787 Berlin, Germany

International Organizations

CESIO/AISE (Comité Européen d'Agents de Surface et Intermédiaires Organiques/ Association Internationale de la Savonnerie, de la Détergence et des Produits d'Entretien) Working Group on Analysis

ISO (International Organization for Standardization)

IUPAC (International Union of Pure and Applied Chemistry) Commission on Oils, Fats, and Derivatives in the Applied Chemistry Division

Collections of Methods, and Sources from Which They May Be Obtained:

DGF-Einheitsmethoden.

Single pages: Deutsche Gesellschaft für Fettwissenschaften e.V., Varrentrappstraße 40–42, 60486 Frankfurt, Germany.

DIN Standards, ASTM Standards, ISO Standards.

Beuth Verlag GmbH, Burggrafenstraße 6, 10787 Berlin, Germany.

AOCS Official and Tentative Methods.

American Oil Chemists Society, 508 South Sixth Street, Champaign, Illinois 61820, USA.

ISO International Standards

ISO Central Secretariat, 1 Rue de Varembé, 1211 Geneva 20, Switzerland.

8. Test Methods for Laundry Detergents

All detergent test methods have the establishment of product quality as their goal. “Quality” is taken to include all those subjectively judgeable and objectively measurable properties of a product that play a significant role in its application.

Many possible approaches are available for measuring detergent quality, but these can be subdivided into three major groups:

Laboratory methods
Practical evaluation
Consumer tests

8.1. Laboratory Methods

Practical evaluations require careful statistical analysis of increasing numbers of test series, most of which are time-consuming and associated with high labor and material costs. For this reason, development of preliminary and screening test procedures has become necessary, as well as suitable laboratory methods for obtaining useful and relevant information, especially for development work. Such tests are designed to approximate service conditions to which the products will actually be subjected. The tests generally produce valuable clues to product quality, but their results should not be overrated, since the circumstances under which they are obtained are not totally identical to those that apply in the field. Normally these tests are conducted with laboratory equipment specially designed for small-scale tests (e.g., Launder-ometer [553], Linitest equipment [554], Terg-o-tometer [555], and various foam testing equipment [556], [557]).

In place of naturally soiled domestic laundry, use of artificially soiled and stained textiles and test swatches is standard in laboratory tests. These swatches are prepared by using various standardized fabrics and knitware and various soils and stains. The swatches are often commercially available [558]. In addition, individual stains are prepared in the laboratory for specific purposes, either by hand or using special equipment. Artificially soiled fibers or fabrics must be prepared so that they respond selectively to different detergent components and different sets of washing conditions if adequate differentiation is to be assured. The amount of a certain soil on every swatch is identical and it is evenly distributed.

Criteria for testing detergents are subdivided as follows:

- Single wash cycle performance (soil and stain removal and bleaching)
- Multiple wash cycle performance, e.g., after 25 or 50 washes (soil antiredeposition properties, degree of whiteness, buildup of undesirable deposits, fiber damage, stiffness, color change, fluorescent whitening)
- Special characteristics (powder characteristics such as density, free flowability, dispensing in a washing machine, homogeneity, dusting properties, solubility, foaming, rinse behavior, and such storage characteristics as chemical and physical stability, hygroscopicity, color, odor, and tendency to form lumps)

The literature describes numerous methods for testing according to the above criteria, some of which are standardized. Standardization is a concern not only of national bodies (e.g., ANSI, the American National Standards Institute; JISC, the Japanese Industrial Standards Committee; DIN, Deutsches Institut für Normung [559]; AFNOR, Association Française de Normalisation; BSI, British Standards Institution), but also of international groups (e.g., ISO, International Organization for Standardization).

The above national organizations are all members of ISO and can, therefore, exercise influence on questions of international standardization [560]. Another particularly important organization concerned with international standardization of test methods is the CID (Comité International des Derivés Tensio-Actifs) with its subcommittee, the CIE (Commission Internationale d'Essai). This organization was disbanded in 1978, but in the meantime, its activities have been taken over and carried forward by the Working Group TMS (Test Methods for Surfactants) of the CESIO (Comité Européen d'Agents de Surface et Intermédiaires Organiques).

8.2. Practical Evaluation [561]–[565]

Practical wash evaluation is conducted with the exclusive use of commercial washing machines. In contrast to pure laboratory investigations and screening procedures, this work is carried out with laundry whose soil has been acquired naturally. The purpose of such experiments is to obtain results that can be regarded as far more realistic. Nonetheless, it is impossible to encompass the entire spectrum of conditions that might be encountered in practice. In many cases, so-called anonymous domestic laundry is tested: laundry of household origin, but not otherwise identified. Far better, however, even though more difficult to arrange, are wash-and-wear tests conducted by using specific new fabric items, e.g., terry towels, dishtowels, shirts, pillow cases, etc., that have been distributed for regular use during a certain period (usually a week) in a sufficient number of representative households. After each use cycle, the samples are collected in the laboratory and washed with the detergent being evaluated. The same item must always be washed with precisely the same detergent and under identical washing conditions. The number of necessary wash-and-wear cycles depends on the nature of the questions being investigated. Nonetheless, usually ca. 15–25 wash-and-wear cycles are necessary, occasionally even more. The detergent is generally used according to the manufacturer's directions, although effects of under- and overdosing may also be taken into account. Analytical precision is improved if instrumental methods are used to evaluate test results, but such measurements are sometimes difficult both to interpret and to conduct. Thus, an alternative evaluation is based primarily on visual evaluation, with physical measurements used only in a supporting role. Recommendations for comparative testing of detergent use characteristics have been the subject of a tentative DIN standard [566], as well as an ISO standard [567].

In the interest of both manufacturers and especially consumers, comparative use tests must be conducted under the most realistic conditions. This is particularly true of tests designed for consumer education, such as those carried out by consumer organizations; otherwise the results are of little value. Therefore, good communication and close contacts between the consumer organizations and the product manufacturers regarding test methods and criteria to be applied prior to conducting the tests is of paramount importance. Oversimplified tests, e.g., those using artificially soiled swatches only, are incapable of making the sorts of distinctions that consumers consider relevant, and are a disservice to all parties concerned. Ultimate product judgements must be based not only on wash results, but on all of the relevant factors affecting use value.

8.3. Consumer Tests

To obtain a final degree of certainty with respect to a detergent's use characteristics and acceptance, consumer tests are conducted in addition to laboratory tests as a regular control to determine consumer acceptance. In such tests, consumers subject a product to their own sets of conditions; i.e., they use their own washing machines on their own laundry and soil, and it is assumed that they introduce the detergent in the manner to which they are accustomed. The precision of such in-home use tests is of course limited, because consumers judge, they do not measure. Therefore, it is necessary that a large number of consumers (usually 200 – 300 households) participate if statistically meaningful conclusions are to be drawn. Experience has shown that such tests are of high value in addition to laboratory results, both for determining product weaknesses and for identifying particular strengths. By their very nature, consumer tests subject a product to a great many different sets of circumstances, and they lead to very meaningful results. Thus, consumer tests are very frequently conducted prior to marketing a new product.

9. Economic Aspects

The worldwide economic role of detergents represents a factor of considerable importance. Nonetheless, laundry detergent consumption per capita varies tremendously from country to country (Fig. 64 [127]). Changing patterns of use for laundry detergents and increased standards of living over the last 40 years reveal a remarkable growth in worldwide consumption, with the absolute quantity rising from approximately 10×10^6 t in the late 1950s to 21.5×10^6 t worldwide in 1998.

Total consumption of laundry, dishwashing, and cleaning detergents in Europe was 6.4×10^6 t in 1991. Eight years later it was 6.8×10^6 t, corresponding to a volume growth rate of nearly 1 % per year (Table 41 [568]).

Only a handful of large companies manufacture most of the world's laundry detergents and cleansing agents. The leaders are Procter & Gamble (Cincinnati, Ohio, USA) and Unilever (London, UK and Vlaardingen, The Netherlands). Further internationally operating major manufacturers are Colgate Palmolive (New York, USA), Henkel (Düsseldorf, Germany), Reckitt Benckiser (Windsor, UK), Kao (Tokyo, Japan), and Lion (Tokyo, Japan). In the U.S. market, four firms (Procter & Gamble, Lever, Dial, and Colgate) dominate. The remainder is split among many smaller local or regional producers. In Europe, for example, the Association Internationale de la Savonnerie, de la Détergence et des Produits d'Entretien (AISE), a trade association made up of

Table 41. World total surfactant consumption by end use, 1992 and 1998 (10^3 t) [569]

End use	1992	1998
Household		
North America	866	1115
Western Europe	1061	1119
Asia	1208	1599
Other regions	998	1235
Total	**4133**	**5068**
Personal care		
North America	204	210
Western Europe	143	163
Asia	119	194
Other regions	86	155
Total	**552**	**722**
Industrial and institutional cleaners		
North America	263	311
Western Europe	155	192
Asia	60	75
Other regions	42	53
Total	**520**	**631**
Industrial process aids		
North America	1191	1316
Western Europe	644	676
Asia	858	1085
Other regions	355	364
Total *	**3048**	**3441**
Grand total	**8253**	**9862**

* Does not include soap.

European national detergent industry groups, estimates that its members represent some 1200 individual firms.

9.1. Detergent Components

Production capacity and demand for the major detergent raw materials by volume are briefly reviewed in the following sections.

9.1.1. Surfactants [569]

World surfactant consumption was estimated at 9.9×10^6 t in 1998, corresponding to an increase of 20 % since 1992 (Table 41). The two largest end uses, household products and industrial processing aids, together account for for 86 % of the total. Household products alone accounted for nearly 5.1×10^6 t, or half of the world total. Household

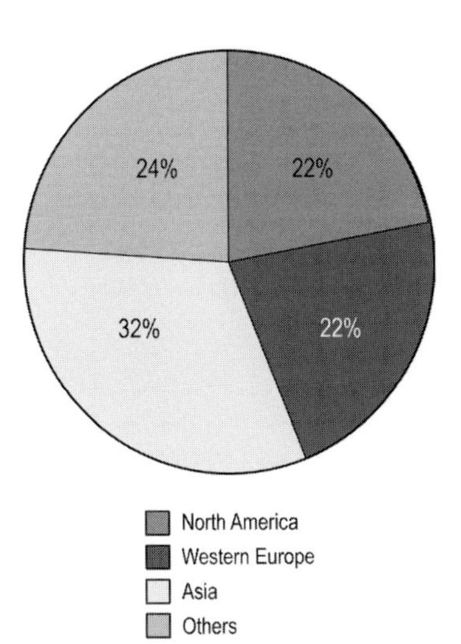

Figure 92. Surfactant demand for household use in 1998 [569]

products include heavy-duty powder and liquid detergents, light-duty liquids, and fabric softeners. The regional distribution of surfactants consumption for houshold use in 1998 is shown in Figure 92.

Industrial processing aids cover five major end uses: plastics and elastomers, textiles, agricultural chemicals, leather and paper chemicals, and other miscellaneous uses. The other major categories featured include personal care products and industrial and institutional (I&I) cleaners, which together consumed about 1.4×10^6 t of surfactants in 1998. I&I products include laundry products, hard-surface cleaners, and dishwashing detergents. From data shown in Figure 93 it is obvious that the two major sectors of the total world surfactant market are the household and I&I areas. Surfactant consumption grew most rapidly in Asia in the 1990s (> 30 %), whereas Europe showed only a moderate growth (< 10 %) in that period.

Breakdown of total surfactant consumption by type is depicted in Table 42. Apart from soap, which is the world's largest surfactant by volume (worldwide consumption 8.8×10^6 t in 1999 [570]), surfactant consumption is headed by anionic surfactants, viz. linear alkylbenzenesulfonates, alcohol sulfates, and alcohol ethoxy sulfates. Second in consumption are alcohol ethoxylates and alkylphenol ethoxylates. LAS has remained the most widely used synthetic surfactants. They have weathered the environmental concerns and experienced a comeback in Europe. They continue to dominate in the Americas and Asia.

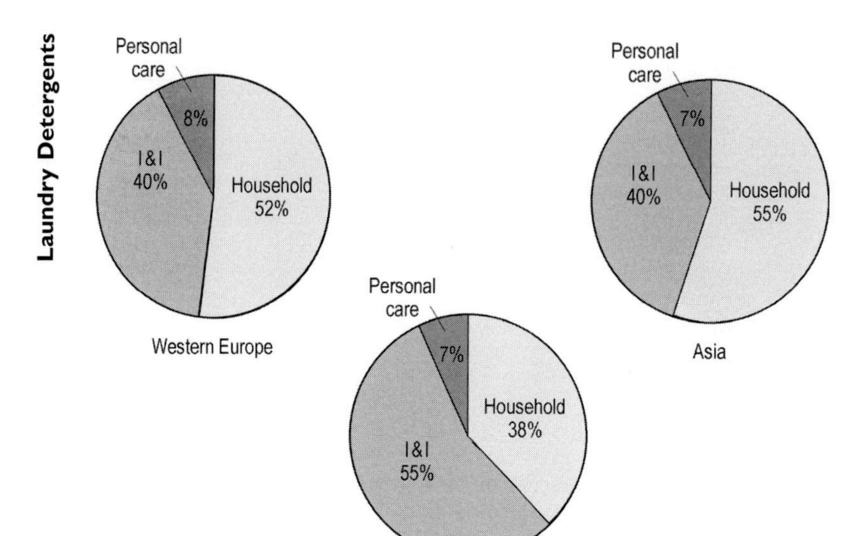

Figure 93. Total surfactant end user markets in 1998 [569]

Table 42. World major surfactant consumption by type, 1992 and 1998 (10^3 t) [569]

Surfactant type	1992	1998
Linear alkylbenzenesulfonates	2385	3027
Branched alkylbenzenesulfonates	411	198
Alcohol sulfates	466	479
Alcohol ethoxysulfates	511	911
Alcohol ethoxylates	742	849
Alkylphenol ethoxylates	652	701
Quaternaries	312	434
Others *	2774	3263
Total	8253	9862

*Does not include soap.

9.1.2. Builders

Sodium triphosphate has been the most important detergent builder in powder detergents since the 1950s worldwide. Its use has become a subject of intense public discussion in many countries in the last 30 years as a result of phosphate-induced regional overfertilization of surface waters. In countries where overfertilization has been particularly severe, this has resulted in both voluntary agreements and legal restrictions, which have caused a sharp decrease in the use of sodium triphosphate for washing and cleansing purposes, mainly in the USA, Europe, and Japan. Within ten years, the use of sodium triphosphate in detergents dropped by more than 50 % in these

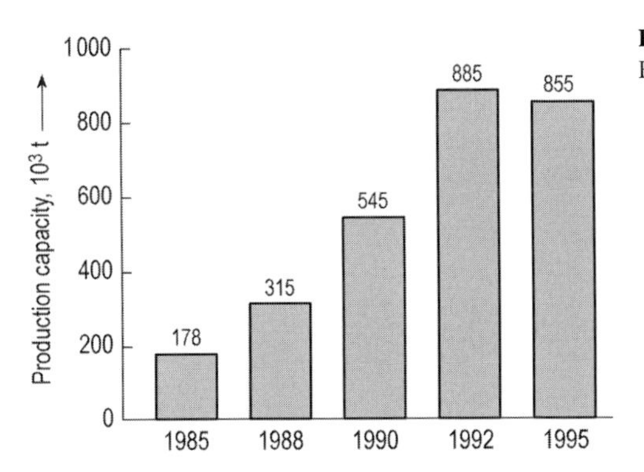

Figure 94. Production capacities of zeolite A in Europe (10^3 t) [80]

regions [571], [572]. Worldwide production capacity has risen dramatically for the most important phosphate substitute zeolite A. Figure 94 shows the increase in production capacities for zeolite A in Western Europe in the 1980s and 1990s [80]. Additional major manufacturing capacities are located in the USA, Japan, and Taiwan. Meanwhile, production capacities for detergent-grade zeolites have largely surpassed demand.

Quantitatively, detergent builder zeolites represent the largest application field for zeolites [572]. Almost 90 % or 1 050 000 t/a of zeolites produced worldwide in 1994 were used for detergents. Of this, ca. 950 000 t/a were zeolite A and some 100 000 t/a were of the zeolite P type.

In Japan, for example, since the mid-1980s more than 95 % of all detergents produced and sold are phosphate-free, containing zeolite instead. Also in the USA meanwhile nearly all laundry detergents are nonphosphate products. A large portion of them contains zeolite A as the main builder. The situation in Europe is illustrated in Figure 72 [131]. Here, too, zeolite is the main builder for phosphate-free laundry products. The replacement of phosphate by zeolite A has gained additional momentum as a result of changes in the prices for raw materials.

The economic situation with respect to other detergent ingredients is highly variable from continent to continent. Moreover, these materials are used in smaller quantities. Hence, their economic impact is less significant and will not be dealt with here.

9.2. Laundry Detergents

In 1998, total world production of soap bars and laundry detergents was 21.5×10^6 t. Again, the tonnage varies among the global regions and individual countries (Table 43).

Worldwide, powder detergents continue to constitute the largest section of the total textile washing and cleaning products market, both in volume and value terms. The powder detergents category is prominent due to the well-established nature of the standard powder delivery form, and latterly the market performance of the tablet laundry detergents, introduced into several markets in the late 1990s and 2000.

Table 43. World production of soap bars and powder, liquid and paste laundry detergents in 1998, 10^3 t

Country	Soap bars	Powder detergents	Liquid detergents	Paste detergents	Total laundry detergents
Austria	-	61	7	-	68
Belarus	-	-	31	-	31
Belgium	-	226	46	-	272
Bosnia-Herzegowina	-	2	-	-	2
Bulgaria	-	7	-	-	7
Croatia	-	31	2	-	33
Cyprus	-	4	-	-	4
Czechia	10	306	10	-	326
Denmark	-	29	5	-	34
Finland	-	26	-	-	26
France	-	467	111	-	578
Germany	-	578	46	-	624
Greece	11	48	5	3	64
Hungary	-	65	10	-	78
Ireland	5	41	1	-	47
Italy	20	485	190	1	695
Macedonia	1	20	1	-	23
Norway	-	6	-	-	6
Poland	31	196	10	-	237
Portugal	27	91	10	-	128
Rumania	-	42	1	5	43
Russia	70	340	6	-	421
Serbia	-	61	-	-	61
Slovakia	2	20	-	1	22
Slovenia	-	21	1	-	23
Spain	6	406	34	-	446
Sweden	-	9	4	10	13
Switzerland	-	41	-	1	51
The Netherlands	-	31	5	-	37
United Kingdom	-	337	80	-	417
Total Europe	*183*	*3997*	*616*	*21*	*4817*
Total North America	-	*1493*	*1493*	-	*2986*
Argentina	62	143	18	-	223
Brazil	550	640	10	-	1200
Mexico	241	846	41	4	1132
Others	544	584	13	10	1151
Total Latin America	*1397*	*2213*	*82*	*14*	*3706*
Congo	57	81	-	-	138
Egypt	220	350	-	1	571
Kenya	122	82	10	31	245
Nigeria	151	82	-	-	233
South Africa	63	163	9	2	237
Others	120	269	5	1	395
Total Africa	*733*	*1027*	*24*	*35*	*1819*
China	274	1508	13	107	1902
India	965	1500	9	-	2474
Indonesia	42	180	1	260	483
Iran	5	265	10	-	280
Japan	23	522	50	-	595
Philippines	244	61	-	-	305
Saudi-Arabia	12	122	-	4	138
South Korea	35	109	2	-	146
Turkey	82	230	2	-	314

Table 43. (continued)

Country	Soap bars	Powder detergents	Liquid detergents	Paste detergents	Total laundry detergents
Others	574	719	56	64	1413
Total Asia	*2256*	*5216*	*143*	*435*	*8050*
Total Australia	*1*	*115*	*44*	-	*160*
World Grand Total	*4570*	*14061*	*2402*	*505*	*21538*

(Source: Ciba Speciality Chemicals Ltd., Switzerland)

Exceptions are the USA, where liquid laundry products took over the lead in the 1990s; Indonesia, whose market is dominated by detergent pastes; and the Philippines, where the vast majority of consumers use soap bars. India is the world's largest single market by volume for soap bars, but here, too, powders dominate and continue to grow at the expense of bars.

In terms of value, total textile washing products amounted to sales of $ 17.7×10^9 in the seven major single markets (USA, Japan, Germany, UK, France, Italy, and Spain) in 1999 [130]. These markets are best characterized by having a high degree of maturity, along with fierce price competition and high advertising expenditures. Four of these markets registered positive gross value growth over the period 1995 through 1999. This translates into a positive real value growth of the U.S. market of almost 13 %, whereas Germany, Italy, and Japan recorded real value declines of 6.5 %, 11.7 %, and 7.6 %, respectively. Due to their high levels of saturation and maturity, continuous product innovations play an important role in stimulating these major markets.

Within the powder detergents sector, conventional products claim most of the sales in France, Italy, Spain, and the UK, whilst the (super)concentrated powders are more popular in Germany, Japan, and the USA for various reasons, such as space-constraints in the home (Japan), reduced packaging costs, space-saving benefits, and consumers' greater environmental awareness (Germany), as well as consumer perceived greater convenience and product benefits vs. retail price (USA).

In the liquid detergent sector, concentrated products form the larger subsector in Germany, Japan, the UK, and the USA. Conventional liquid detergents lead in France, Italy, and Spain.

Within the category of other detergents, fine-fabric specialty detergents lead in most of the seven major markets, ahead of handwashing detergents and detergent bars. Changing lifestyles and increased saturation of the washing-machine market have contributed to considerably reduced consumption of products of the last two subsectors.

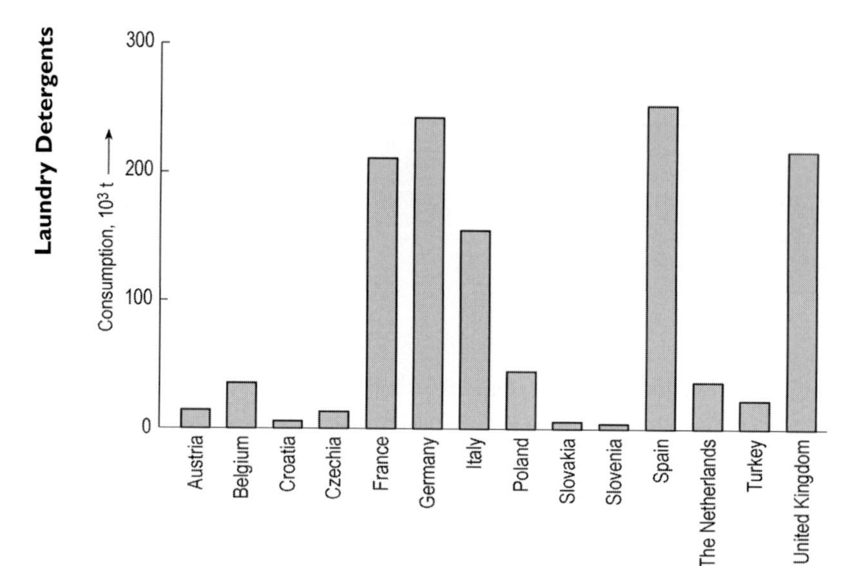

Figure 95. Consumption of fabric softeners in Europe in 1999 (source: Leading International Market Research Institute)

9.3. Fabric Softeners

Europe is the major consumer of rinse-cycle fabric softeners. A number of reasons explain this, including the fact that European tap water on average is relatively hard and the fact that the vast majority of drum-type washing machines are equipped with separate dispensers for laundry aids in Europe. Apart from the USA and Japan, products in this class continue to be relatively insignificant in volume in the rest of the world. Figure 95 provides information about European fabric softener consumption for the year 1999. These consumption figures represent both conventional fabric softeners, which contain 4 to 8 % active ingredients, and concentrated softeners, which usually have 12 to 30 % active cationics. After a period of decreasing consumption in the 1980s, fabric softeners enjoyed growth in both value and volume in the 1990s.

Concentrated fabric softeners have dramatically increased their market share at the expense of diluted conventional softeners. The shares of concentrated softeners in Europe in 1998 varied from some 20 % in Italy to 98 % in Germany. Use of concentrated products saves packaging materials and transportation costs and reduces the amount of waste. To further decrease waste, part of the concentrated fabric softeners are sold as refill packages in many European countries. Their share of total concentrates varies from 10 % in Belgium to some 65 % in France. Fabric softener sheets, which are very popular in the USA as laundry dryer aids, find only very limited use in Europe and Japan. The share of softener sheets ranges from 0 to 5 % of the fabric softener market in European countries.

Total consumption of fabric softeners in Japan was some 250 000 t in 1998, about 40 % of which were concentrated products.

9.4. Other Laundry Aids

In the seven major detergent markets, laundry aids comprise a wide number of subcategories, with spot and stain removers and laundry bleach claiming the leading positions in most key markets. Although chlorine bleaches have been used for textile washing for many years, the introduction of color-safe, non-chlorine bleaches led to a considerable consumption of this subcategory in the 1990s. Laundry bleach accounts for the largest share of sales in the USA, Spain, and Italy.

10. Ecology

10.1. Laundry, Wastewater, and the Environment

The washing process represents a complex interaction between soiled laundry, water, mechanical and thermal energy, and detergents. Laundry is repeatedly cycled through the system, and clean laundry can be regarded as the product. Wastewater is the troublesome byproduct which has the potential for causing a number of undesirable effects in sewage treatment plants and in the environment. Virtually all of the clean water brought into the process is later released to the sewage system in the form of contaminated wastewater containing additional energy (heat), soil from the laundry, lint, dyes, finishing agents, and detergents. Detergent components are released to the wastewater either in essentially unchanged form or as the products of reaction with other materials present.

Laundry wastewaters vary considerably in concentration and composition. These differences arise partly due to variations in laundry soil levels, although washing technology and the composition and amount of added detergent are also significant factors. The differences are particularly great between the wastewaters generated by household laundry and those from I&I laundries. The latter generally contain lower portions of contaminants originating from detergents, largely as a result of more economical use of typical I&I laundry products and the use of water-softening equipment. Little doubt exists that laundry wastewater must be generally regarded as a heavily contaminated medium; it should not be returned to receiving waters in untreated form.

Fortunately, the wastewater emanating from the hundreds of millions of household washing machines and from washing by hand worldwide is widely distributed, and in a pattern consistent with that of general water consumption. As a result of dilution in the public sewerage system and in sewage treatment plants, both the temperature and the relatively high pH value of wastewater are considerably decreased, and the effect of peak loads is also somewhat compensated for. Only for this reason is biological treatment of laundry wastewater in a normal sewage treatment plant feasible. If laundry wastewater were treated separately, major problems would be encountered in dealing with the load of organic pollutants introduced by household and commercial laundry operations. Dilution of laundry wastewater within a typical public sewage system normally exceeds a factor of ten.

10.2. Contribution of Laundry to the Sewage Load

Table 44 presents calculated estimates of average theoretical concentrations of the detergent and cleanser components in public wastewater. These estimates are derived from statistical data collected in Germany and depend on both product consumption and the average water consumption in households. These calculated average wastewater concentrations are "worst-case" assumptions, since the average per capita wastewater production rate (combined household and commercial) is around 200 L/d in the EU member states.

In addition, the calculated concentrations of some of the chemicals are only partly due to the use in laundry detergents since other applications (e.g., dish washing and cleansing products) may contribute to a considerable extent. Furthermore, actual values for concentrations and loads show considerable variation. Time of day is a factor, as is the day of week (with Saturdays and Mondays producing the highest values). Even seasonal variations are apparent, but all of these fluctuations are reasonable when examined in the light of known household laundry habits [574].

A significant portion of domestic water is used in the household laundering operation. According to [575], the per capita amount of laundry wastewater generated daily is around 17 L, which translates into 13 % of the domestic water consumption or about 8 % of the total hydraulic load reaching a sewage treatment plant. Not only in terms of the hydraulic load but also with respect to the chemical load, the laundry washing process is a major factor influencing municipal wastewater. As shown in Table 45, the detergents contribute about 10 % to the COD, the biodegradable organic matter (usually reported as the BOD_5), and the dry solid matter content of the municipal wastewater load, while the contribution to the nitrogen and phosphorus load is low. In addition, of course, the soil contained in soiled laundry represents another important contributor to the load of wastewaters. About half of the biodegradable contaminants load of mixed wastewater derived from laundering is due to laundry soil and stains. The amount of organic contamination originating from soil is doubled or tripled in institutional laundry wastewaters, compared to laundry wastewater from households.

Table 44. Consumption figures and average wastewater concentrations of ingredients of household detergents and cleansers (Germany, 1999)

Product ingredient	Consumption, t/a [573]	Per capita consumption, g/d[a]	Calculated average wastewater concentration, mg/L[b]
Total ingredients	688 000	23.0	177
Anionic surfactants	104 700	3.50	26.9
Nonionic surfactants	53 000	1.77	13.6
Cationic surfactants	30 400	1.02	7.8
Zeolites	139 000	4.64	35.7
Polycarboxylates	15 700	0.52	4.0
Na carbonate (soda ash)	98 800	3.30	25.4
Phosphates	23 600	0.79	6.1
Na citrate	15 600	0.52	4.0
Na perborate tetrahydrate	49 500	1.65	12.7
Na percarbonate	27 200	0.91	7.0
TAED	13 100	0.44	3.4
Phosphonates	2 900	0.10	0.7
NTA	400	0.01	0.1
Carboxymethyl cellulose	2 200	0.07	0.6
Dye transfer inhibitors	400	0.01	0.1
Silicates	22 300	0.74	5.7
Enzymes	6 000	0.20	1.5
Optical brighteners	500	0.02	0.1
Paraffins	1 200	0.04	0.3
Soil repellents	600	0.02	0.2
Perfumes	5 700	0.19	1.5
Dyes	100	0.003	0.03
Na sulfate	74 600	2.49	19.2

[a] Population in Germany (1998): 82×10^6 [573].
[b] Average per capita water consumption in households in Germany: 130 L/d (1998) [573].

10.3. Detergent Laws

Detergents represent a typical and important group of environmental chemicals, i.e., substances entering the environment after their use. The per capita consumption of detergents is nowadays at a level considerably below the figures of the 1980s (1980: 33.3 g per capita and day in Western Germany; 1997: 21.1 g per capita and day in Germany). Nevertheless, the current annual consumption is in the range of 660 000 t/a (Table 44). These large amounts of chemicals go down the drain and enter receiving waters directly or after sewage treatment. Since the wastewater treatment situation was relatively poor in most European countries a few decades ago, it is not surprising that certain detergent ingredients, i.e., surfactants and phosphates, exhibited an apparent specific impact on the aquatic environment and became subject of specific laws and regulations.

Table 45. Contribution of laundry washing (without laundry soil) to the chemical load of municipal wastewater

Chemical load parameter	Concentration in laundry waste water [576], mg/L	Per capita load of laundry wastewater *, g/d	Per capita load of municipal wastewater [576], g/d	Contribution of laundry washing to the chemical load in wastewater, %
COD	600	10.2	120	8.5
BOD_5	350	5.95	60	10.0
Dry solid matter	450	7.65	72	10.6
Total nitrogen	7	0.12	12	1.0
Total phosphorus	2	0.03	2.4	1.0

* Derived from a water consumption of 17 L/d for laundry washing [576].

10.3.1. Development of the European Detergent Legislation

One of the earliest pieces of detergent legislation was the German Detergent Law of 1961 [2]. With its specific requirement for a minimum level of biodegradability, this law subjected a fundamental product parameter to legal control for the first time. The trigger for setting minimum biodegradability standards for synthetic detergents was the fact that the first generation of synthetic anionic surfactants was represented by the poorly biodegradable tetrapropylenebenzenesulfonate. The highly branched chemical structure of the anionic surfactant prevented its microbial degradation in sewage plants and rivers and caused formation of large mountains of foam in sewage treatment plants, weirs, locks and rivers in Germany, particularly during the droughtlike summers of 1959 and 1960. The law and the statute that followed (1962) [3] placed a strict requirement of at least 80 % biodegradability on all anionic surfactants. This legal move successfully displaced TPS from detergent formulations by the readily biodegradable linear alkylbenzenesulfonates. Thus, by 1965 foaming problems in German sewage treatment plants and surface waters had been eliminated.

In the late 1960s most of the European countries had limited the use of TPS in household detergents. Based on an agreement of the Council of Europe [577] requiring all types of surfactants to be biodegradable, the EEC released two directives on the biodegradability of surfactants in 1973: The EEC Directive 73/404/EEC stipulated an average biodegradability of at least 90 % for all types of surfactants used in detergents, i.e., anionic, nonionic, cationic and amphoteric surfactants [578]. The Directive 73/405/EEC described the biodegradability test methods to be applied for anionic surfactants and specified a minimum of 80 % degradation in a single test [579].

On the national level these European Directives were translated into detergent laws by the member states, e.g., in Italy in 1974, France in 1977, The Netherlands in 1977, and the UK in 1978 [580]. In Germany the detergent law of 1975 [581] and its statutory orders of 1977 [582] and 1980 [583] regulated the "environmental compatibility of laundry detergents and cleansers". According to § 1 (1) of the German Detergent Law placement of detergents and cleansers on the market is permitted only if the absence of all avoidable deterioration in the quality of surface water can be ensured. Particular attention is paid to the potential role of surface water in the drinking water supply and to problems related to the operation of sewage treatment plants. The general requirements and authorizations for implementation apply directly to the following areas:

1) Unambiguous product labeling, including qualitative declaration of all significant components present
2) Deposition of frame formulations for all products at the Federal German Environmental Agency
3) Utilization only of biodegradable organic components, particularly surfactants and other organic compounds as specified in a statutory order
4) Provision of directions for dosage of detergents on all packages, including variations applicable to water of differing hardness, where hardness is defined in terms of four ranges
5) Obligation on the part of local water supply authorities to publish the hardness characteristics of their water
6) Limitations on permitted phosphate levels, including some cases of a total phosphate ban, provided that ecologically sound substitutes are available (see Section 10.3.4)

Products covered by the Detergent Law include: household laundry detergents; household cleansers; dishwashing agents; rinsing and other laundry aids; detergents for I&I laundries; industrial cleansers; and cleansing agents employed in the leather, tanning, paper, and textile finishing industries that might be released into wastewater.

Thus, the definition of detergent products according to the German Detergent Law is more comprehensive than the EU definition, that only covers surface active substances which are main components in detergents and designed for cleaning purposes [580].

10.3.2. Regulatory Limitations on Anionic and Nonionic Surfactants

The Directive 73/405/EEC [579] and its updated version Directive 82/243/EEC [584] specify the minimum biodegradability of anionic surfactants contained in detergents. In addition, the procedures are described to be applied for measurement of biodegradability. In close analogy the Directive 82/242/EEC [585] specifies an 80 % biodegradation pass limit for nonionic surfactants used in detergents, and the test methods to be used. A corresponding national regulation has existed in the Federal Republic of

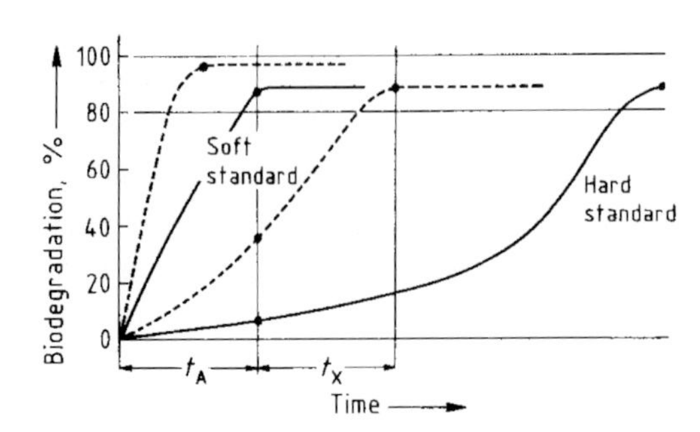

Figure 96. Biodegradability evaluation in the OECD Screening Test
t_A: maximum 14 d; $t_A + t_X$: maximum 19 d
--- examples for two surfactants; one is biodegrated more easily (t_A) than the second one ($t_A + t_X$)

Germany already since 1977 [582] ("Tensidverordnung"). The regulation requires a minimum of 80 % biodegradability for the anionic and nonionic surfactants present in a packaged detergent. Anionic surfactants are determined as "methylene blue active substances" (MBAS), i.e., materials forming a chloroform soluble complex with the cationic dye methylene blue. Nonionic surfactants are defined as "bismuth active substances" (BiAS), i.e., materials forming an insoluble complex with the bismuth-containing Dragendorff reagent [586].

Although not all anionic and nonionic surfactants can be analytically measured by the MBAS and BiAS method [587], these parameters are used for the evaluation of the biodegradability according to the legal requirements. Essentially, the percentage of MBAS or BiAS loss describes the disappearance of a specific analytical reaction of the parent surfactant. This loss of reactivity is due to microbial transformation of the chemical structure of the surfactant, resulting in the loss of typical surfactant properties such as surface activity and foaming power. Therefore, it is called "primary or functional biodegradation" (see Section 10.4.2.1).

10.3.3. Primary Biodegradation Test Procedures

According to the Directives 82/243/EEC and 82/242/EEC and corresponding national legislations two types of test methods are applicable for determining the (primary) biodegradability of anionic and nonionic surfactants, respectively. The first type is a discontinuous shake flask test operating with a low bacterial inoculum, that is incubated with 5 mg/L of MBAS or BiAS as the sole carbon source. In this *OECD Screening Test* [588] the loss of MBAS or BiAS is determined periodically up to 19 d. The results are compared with the degradation behavior of two control substances: the poorly degradable TPS (< 35 % MBAS decrease) and the readily biodegradable LAS (about 92 % decrease). The biodegradability determination in this test is illustrated in Figure 96. The second test type is represented by the *OECD Confirmatory Test* [588] simulating the biodegradation process occurring in an continuous activated sludge plant (Fig. 97). In this procedure, MBAS (20 mg /L) or BiAS (10 mg /L) is dissolved in synthetic sewage (containing peptone, meat extract, urea, and mineral salts) and fed

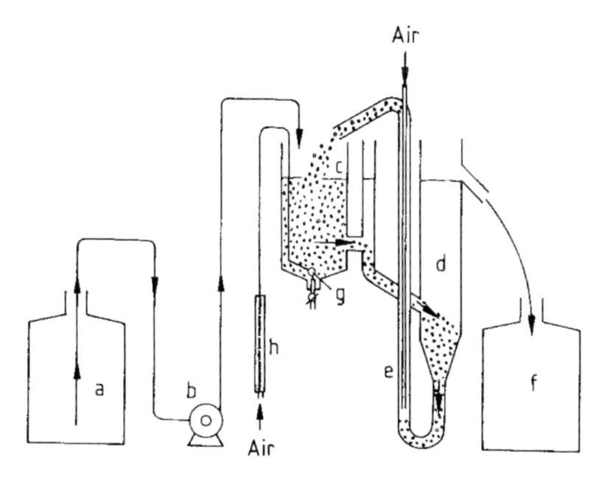

Figure 97. Experimental arrangement for the OECD Confirmatory Test as specified by the German Detergent Law
a) Sample container; b) Dosage pump; c) Activated sludge vessel (capacity 3 L); d) Settling vessel; e) Air lift; f) Collection vessel; g) Fritted disk; h) Air flow meter

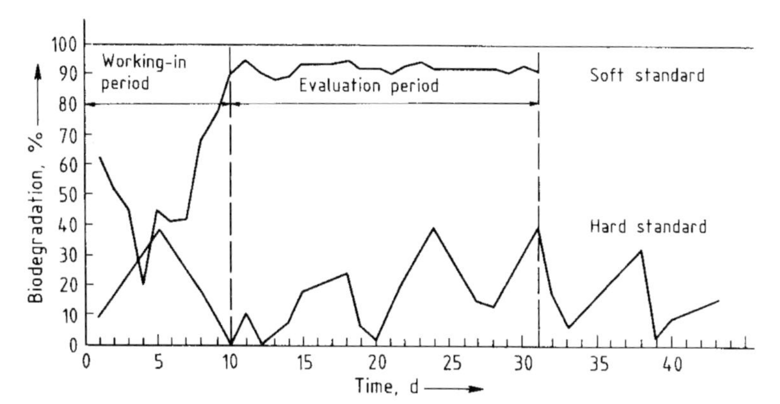

Figure 98. Determination of a rate of biodegradation according to the method specified in the OECD Confirmatory Test

continuously into a model sewage treatment plant with a hydraulic retention time of 3 h. After inoculation of the test system with the effluent of a predominantly domestic sewage treatment plant and after growth of the activated sludge, the effluent is analyzed periodically. This running-in period lasts six weeks at maximum and is followed by a 21-d evaluation period after the degradation rate has become regular. The average degree of biodegradation during the evaluation phase is calculated as a mean value of at least 14 separate results based on samples from 24-h collection periods (Fig. 98).

The discussed test types represent a hierarchical system of the biodegradability evaluation. Usually, surfactants are evaluated on the basis of die-away screening tests such as the OECD Screening Test or the French AFNOR Test. In addition, the British Porous Pot Test, which is a continuous sewage-works simulation test, may be used at this stage of data generation [584], [585]. However, if the degradation proves MBAS or BiAS loss lower than 80 %, or if doubt remains, a subsequent confirmatory test is required, and the outcome of this OECD Confirmatory Test is regarded as definitive.

10.3.4. Regulation of Maximum Phosphate Content in Detergents

The detergent regulations at the European Union level address exclusively the surfactants and their biodegradability. However, at national levels, regulations or voluntary agreements on the phosphate content of detergents exist in many countries, e.g., Austria, Germany, Italy, The Netherlands, Norway, Sweden, Switzerland, USA, Canada, and Japan.

In itself, phosphate is not harmful but is a natural and essential macronutrient for all living organisms. Normally, the phosphate concentration of surface waters is so low that it is a limiting factor for growth of algae and higher plants. Consequently, when excess phosphate is released into the aquatic environment, the resulting overfertilization leads to increased growth of algae. Due to the concomitant secondary processes (organic load of waters, oxygen depletion after the organic bio-mass is microbially degraded) the overall water quality may be considerably reduced. Detergent phosphates released with laundry wastewater are quickly converted into orthophosphate. Thus, the use of sodium triphosphate in detergents came under critical scrutiny although many other phosphate sources exist that contribute to eutrophication of surface waters.

According to a profound study on phosphates in Germany in 1975, some 60 % of the phosphate contained in municipal sewage originated from detergents and cleansers [589]. As a consequence of the partial removal of phosphate in sewage treatment plants and the input of phosphates by other sources (human excretion, food industry, agricultural fertilizers), the share of detergent phosphates in surface waters was estimated to be about 40 %. This balance showed already that the reduction of phosphates in detergents is an important but not the sole factor in solving the eutrophication problem of surface waters.

The most comprehensive approach involves chemical elimination of phosphates in the sewage treatment plant (tertiary treatment) removing the total phosphorus content of the wastewater. This approach is realized in some countries to a greater or lesser extent. Nevertheless, phosphate reduction in detergents, i.e., at its source, provides immediate relief to receiving waters and, ultimately, also to the coastal areas of the seas, which are increasingly confronted with eutrophication problems.The phosphate regulation of 1980 [583] implementing the pertinent authorization of the German Detergent Law of 1975 (see Section 10.3.1) had a considerable impact on the development of the phosphate loads in German rivers (Fig. 99). The phosphate reduction in detergents was enforced in a two-step decree starting with an overall relative decrease of 25 % of the detergents' phosphate in 1981 and around 50 % in late 1983. The second step required the availability of suitable phosphate substitutes for detergents (see Section 10.5.2). Their good performance in phosphate-reduced and -free detergents and the acceptance of nonphosphate detergents by the consumer has meanwhile led to a complete substitution of phosphate-containing detergents by phosphate-free ones in many countries, for example in Europe in Germany, the Scandinavian countries, Italy, Austria, The

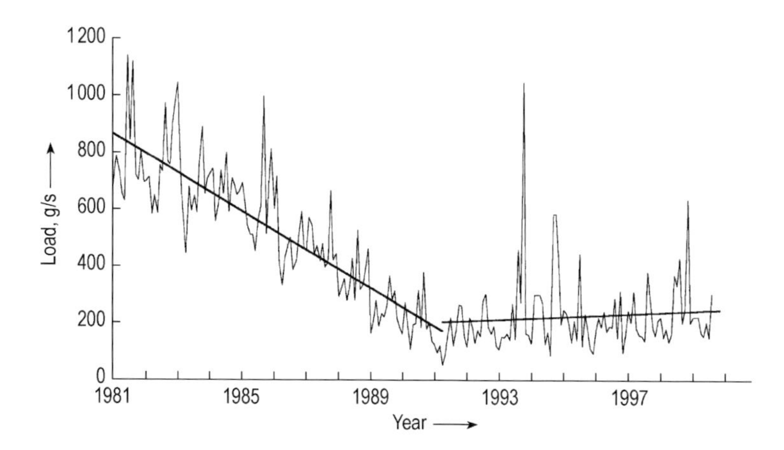

Figure 99. Variations of the orthophosphate load in the river Rhine from 1981 to 1999 (sampling site: Düsseldorf-Himmelgeist, Germany)

Netherlands, and Switzerland (see Fig. 72). Thus, the legal requirements on reduction or banning phosphates in detergents have been superseded in many cases by the industrial supply of efficient phosphate-free detergents and their acceptance by the consumer.

The legislative pressure towards phosphate reduction in detergents accompanied by the availability of suitable substitutes has resulted in a noticeable quality improvement of a number of surface waters today [590]. Phosphate balances of the river Rhine in 1979 and 1989 showed that the measured phosphorus load reduction of about 28 400 t/a within this period corresponded very well with the expected phosphorus reduction due to the use of phosphate-reduced or -free detergents (26 700 t/a) [591].

10.4. General Criteria for the Ecological Evaluation of Detergent Chemicals

The history and the development of the detergent legislation in Europe impressingly reflects the environmental relevance of this group of anthropogenic chemicals (xenobiotics) encountered already in times when the terms "environmental safety" or "environmental risk assessment" were not yet used. The lesson of the so-called detergent problem of the late 1950s was learned both by the legislators and the detergent industry long before general criteria were developed and applied for the ecological evaluation of chemicals. Now, such broadly applicable instruments and standards have been available for several years, e.g., the risk assessment directives for new and existing chemicals of the European Union [592], [593] and the accompanying Technical Guidance Documents [594] that provide an evaluation system applicable by authorities and industry as well. This chapter deals with the concept and the methods of the environmental compatibility assessment that form the basis for the evaluation of chemicals in general.

10.4.1. Concept

Production, use, and disposal of chemical products has a more or less significant impact on the environmental compartments water, soil, and air. To secure that such emissions will not negatively influence the biota of these compartments it is necessary that a possible ecological risk is already recognized before the product enters the environment. The environmental safety of chemical products is essentially based on the safety of their individual ingredients. Thus, the product safety assessment must concentrate on the identification of possible risks of the raw materials. Two main issues exist that are relevant to safety or risk assessment:

1) The environmental fate of detergent ingredients and the resulting concentrations in the affected environmental compartments. The final outcome of such an exposure assessment is the predicted environmental concentration (PEC).
2) The environmental effects of a chemical towards the living organisms of the compartments concerned. The final result of such an effects assessment is the predicted no-effect concentration (PNEC).

The comparison of these two assessment parameters, PEC versus PNEC, forms the basis of the internationally used concept for the environmental risk assessment of chemicals: a chemical is considered safe if the PEC is not higher than the PNEC. The environmental fate and effects of a chemical and the risk assessment based thereon are governed essentially by the inherent properties of the individual chemical such as biodegradation behavior and ecotoxicological properties. The connection of these intrinsic substance properties with the concrete conditions of use of the substance — e.g., the use pattern and the usage figures of the chemical, the sewage treatment situation, the application of safety factors, etc. — provides the data necessary for the calculation/measurement of PEC and PNEC (Table 46).

10.4.2. Environmental Exposure Assessment

The chemical compounds contained in detergents are typical representatives of xenobiotics intended to go down the drain after their use and thus to enter receiving river waters directly or after passing a sewage treatment plant. Depending on their physicochemical properties (for example solubility, adsorption behavior, volatility), these substances can be distributed in environmental compartments such as the water phase, sediment, sludges, soil, etc. The knowledge of the fate and the resulting environ-

Table 46. Elements of environmental safety assessment of chemicals

	Environmental fate	Environmental effects
Substance-inherent properties (hazard identification)	biodegradability	ecotoxicity
Environmental frame conditions (healistic worst case)	exposure scenario, e.g., – usage quantities – use pattern – sewage treatment situation	extrapolation of effect data: laboratory data → real environment (extrapolation factors)
Evaluation (hisk hharacterization)	exposure analysis: predicted environmental concentration (PEC)	effects assessment: predicted no effect concentration (PNEC)
	comparison: PEC vs. PNEC (No risk: PEC < PNEC)	

mental concentrations are therefore a prerequisite and a decisive criterion for risk evaluation.

10.4.2.1. Biodegradation

Biodegradation is the most important mechanism for the ultimate removal of organic compounds from sewages, surface waters and soils. It is a stepwise process mainly effected by aerobic microorganisms, i.e., in the presence of oxygen (Fig. 100). In a first step the parent compound is transformed to a first degradation product (primary degradation), which is subsequently degraded to a second, third, etc., intermediate, i.e., to metabolites with decreasing molecular mass and structural complexity. Ultimately, the total organic structure of the starting material has been decomposed to mineralization products (carbon dioxide, water, inorganic salts) and, in parallel, partly transformed into bacterial biomass (ultimate biodegradation).

The term "primary degradation" only refers to the first steps of the degradation process. Nevertheless, the proof of a fast and quantitative removal of the parent compound is of high ecological relevance for certain chemical classes, e.g., for surfactants. As a rule, the primary degradation of surfactants significantly reduces their surface activity (including their foaming properties) and their aquatic toxicity. For that reason, the European detergent legislation requires a minimum primary biodegradability for anionic and nonionic surfactants (see Section 10.3). Nevertheless, ultimate biodegradability is the decisive criterion for the long-term fate of a chemical and its degradation intermediates in the environment, because this will guarantee the complete re-integration of all components of the substance into the natural material cycles.

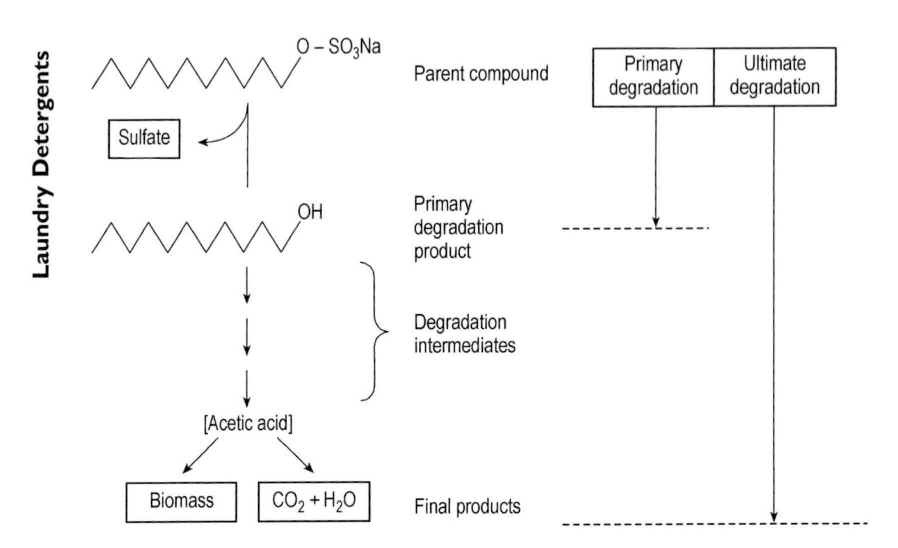

Figure 100. Terms of biodegradability evaluation exemplified by the degradation pathway of fatty alcohol sulfate

10.4.2.2. Biodegradability Standard Test Methods

Primary and ultimate biodegradability of organic chemicals can be evaluated by using suitable test methods (Table 47). The determination of the removal of the parent compound (primary degradation) requires the availability of a substance- or a substance-group-specific analytical determination method. As discussed previously (see Section 10.3), the primary biodegradation of anionic and nonionic surfactants is determined in standardized tests by measuring the removal of MBAS and BiAS, respectively. The ultimate biodegradation of chemicals can be followed in the tests by means of nonspecific analytical parameters such as carbon dioxide evolution, the BOD, or the removal of dissolved organic carbon (DOC).

Primary and ultimate biodegradability of test substances is normally evaluated by applying standardized and internationally used (OECD, ISO, EU) test procedures (Table 47). The most broadly used category of tests are the screening tests representing simple but stringent evaluation procedures. For instance, the OECD Screening Test determines the primary biodegradation of anionic and nonionic surfactants. The corresponding ultimate biodegradability screening tests are represented by the OECD tests for ready biodegradability (OECD 301 A – F) [595], e.g., the DOC die-away test (OECD 301 A), the CO_2 evolution test (OECD 301 B), or the Closed Bottle Test (OECD 301 D). If the pass level of 60 % BOD (compared to the COD) and CO_2 evolution, respectively or of 70 % COD is surpassed within the four-week test duration, it can be concluded that ultimate biodegradation of the chemical will proceed readily in the real environment [595].

It is a common feature of screening tests that a few milligrams of the test substance are diluted or suspended in a test vessel containing an aqueous mineral medium. This

Table 47. Elements of environmental safety assessment of chemicals

	Screening tests	STP-simulation tests *
Characteristics	simple discontinous flask tests test chemical is sole carbon source low bacterial inoculum stringent test conditions, no simulation test	realistic model of a municipal sewage treatment plant continuous test system/limited retention time high sludge concentration
	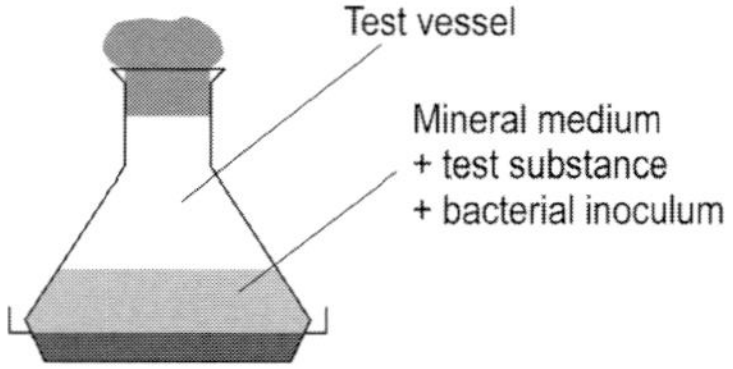 Test vessel Mineral medium + test substance + bacterial inoculum	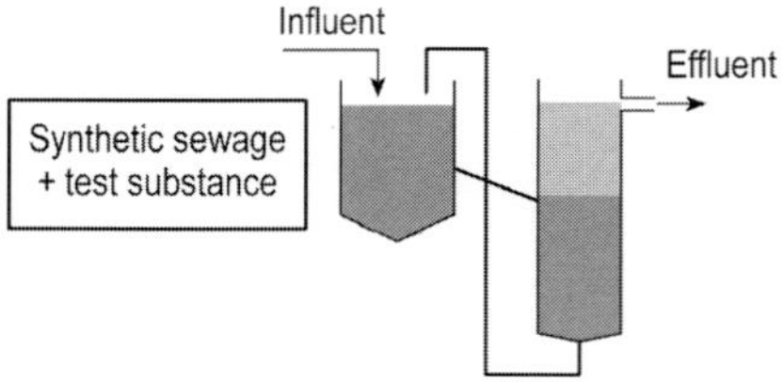 Influent Effluent Synthetic sewage + test substance
Examples	primary biodegradation: OECD Screening Test	primary biodegradation: OECD Confirmatory Test
	ultimate biodegradation: OECD tests for "ready biodegradability" (OECD 301 series)	ultimate biodegradation: Coupled Units Test

* STP = sodium triphosphate

mixture is inoculated by adding a small amount of a bacterial suspension, in most cases originating from a municipal sewage treatment plant (30 mg/L of activated sludge or ≤ 10 ml/L of plant effluent). Thus, the test inoculum contains the microorganisms ubiquitously present in wastewater treatment plants and receiving river waters. The degradation process can be analytically followed by substance-specific methods as well as by summary parameters since the test substance is the only carbon source for the degrading organisms in such a screening test system.

Due to the stringency of the screening tests, a negative or poor degradation result obtained in such a system is not necessarily a proof of lack of biodegradability in the real environment. To obtain a realistic view of the biodegradation behavior under practical conditions, model test systems are used simulating the primary and ultimate biodegradation process in the biological stage of a municipal sewage treatment plant (Table 47). In contrast to screening tests, the test chemical is continuously dosed to the model plant resulting in a hydraulic retention time of 3 – 6 h. Another important difference to screening tests is the fact that the test material is dosed together with a large surplus of highly biodegradable substances ("synthetic wastewater") to give a competitive situation which also prevails in real domestic sewage treatment plants. The bacterial concentration in a continuous activated sludge model sewage treatment plant is relatively high (ca. 3 g/L of suspended solids). Depending on the analytical test method either the primary degradation in the simulation test (requiring substance-specific analytical methods) can be determined or the substance removal based on summary analytical parameters (DOC, COD). If a nonspecific analytical method is applied, a modified test design is necessary to obtain removal data that specifically refer to the test substance. In the *Coupled Units Test* (OECD 301 A), for example, two model plants are run in parallel [595]. One is the control, which is only fed with synthetic sewage, while the influent of the second additionally contains the test substance. The test substance-related carbon removal can be calculated from the difference of the DOC concentrations in the two plant effluents compared with the carbon content of the substance in the influent of the test unit. Monitoring data on the removal of surfactants in municipal sewage treatment plants have broadly confirmed that the elimination rates obtained in model plants are a realistic or rather a conservative description of the practical situation [596].

10.4.2.3. Supplementary Biodegradation Test Methods

Apart from these internationally used standard degradation tests additional test methods exist to extend the information about the fate of chemical substances in the environment. The results from the previously discussed screening and simulation tests are, in principle, a measure of the degradation rate, i.e., the extent of degradation within the test period. Therefore, the question remains unanswered whether the test-substance-derived material which was not biodegraded within the test period may contain recalcitrant degradation intermediates. In other words, if it is possible to show that no poorly biodegradable intermediates are formed during the degradation process, unknown negative effects from the long-term existence of those materials in the environment can also be excluded. The test for *detecting recalcitrant metabolites* (metabolite test [597]) which is experimentally based on the Coupled Units Test, represents an instrument for answering this question. Thus, the test for recalcitrant metabolites can improve and supplement the conclusions drawn from positive results obtained in standard degradation tests.

In contrast to the Coupled Units Test, the effluents of the test and the control plants are reused each day as an influent after addition of a concentrate of the nutrients (control unit) and the test substance. Thus, a circuit is achieved which gives the degrading organisms the permanent chance to cope with the test compound and its intermediates under competitive conditions. Since the metabolite test is run over several weeks— corresponding to up to 100 cycles—even very small amounts of recalcitrant metabolites would accumulate and could be analytically detected. The proof of the absence of any recalcitrant metabolites was obtained for all important detergent surfactants by means of this method [598].

The issue of the completeness of the ultimate degradation in the environment has yet another dimension. Substances with a marked adsorption tendency such as surfactants or poorly water-soluble compounds may to a considerable extent reach environmental areas where no oxygen is present for the biodegradation processes of microorganisms. Examples are septic tanks, digesters, some river sediments, and soils. The *anaerobic biodegradability evaluation* is therefore an additional important aspect for those substances. A screening test method has been available for several years (ECETOC Test, ISO 11734 [599]) that measures the ultimate anaerobic biodegradation of a substance on the basis of the digester gas formation (carbon dioxide and methane). In addition, digester simulation tests with radiolabeled test materials have shown the anaerobic biodegradability of important surfactant groups [598].

10.4.2.4. Exposure Analysis

After identification of the environmental areas mainly exposed to the chemical(s) concerned (e.g., wastewater treatment plants, sludges, river waters and sediments, as well as soils are target areas of detergent chemicals) a preliminary exposure analysis may be conducted. The first tier of such a process is based on relatively simple but conservative standard exposure scenarios. Thus, the concentrations of detergent ingredients in raw wastewaters are calculated on the basis of the per capita usage of the product and the ingredient, respectively, and the per capita water consumption (see Table 44). The elimination of the chemical in sewage treatment plants is estimated by means of calculation models taking the physicochemical (adsorption behavior, volatility) and the biodegradability properties (deduced from degradation screening tests) of the substance into account. From the PEC of the chemical in the plant effluent it is possible to derive a PEC of the receiving water by applying a standard dilution factor of 10 for treated wastewater [594].

A more detailed and reliable environmental concentration prognosis can be obtained in the second tier of the exposure analysis when concrete data of the individual environmental situation are taken into account. These data are, for example, elimination data from sewage treatment plant simulation tests and information on the wastewater treatment (e.g., sewage volumes, hydraulic retention time, etc.) and river water situation (e.g., wastewater dilution ratio, flow rate, etc.). In addition, also the background concentration of the substance concerned must be considered, i.e., the PEC above the sewage outfall. This prognosis of the regional distribution is obtained by means of multimedia calculation models, for instance according to MacKay [600], which predict an average PEC in the environmental compartments of water, soil, and air.

The exposure analysis is — as is the whole environmental safety assessment — an iterative process. This means that the precision and reliability of the PEC values can be improved continuously by the improvement of the data and information based thereon. Therefore, it is understandable that the highest tier of the exposure analysis is represented by the measurement of real environmental concentrations, i.e., by environmental-monitoring data. If sufficiently specific and sensitive analytical methods are available and if the chemical concerned is already in use, such measurements can be conducted. Thus, environmental-monitoring data and PEC data from exposure models form a necessary mutual supplement to fill the environmental concentration data gap for most of the chemicals present in the environment. Due to the conservative character of the model calculations there is consensus that the data obtained from real environmental measurements take precedence over calculated values [594].

Detergent ingredients, particularly the most broadly used surfactants (LAS, alcohol ethoxylates, alcohol ether sulfates, soap) have been thoroughly investigated by means of the multi-tiered approach of exposure analysis. It could be shown that the measured concentrations in raw and treated wastewaters are considerably lower than predicted by the calculation algorithm described for the tier 1 screening phase (cf. Table 48) [601], [602]. Consequently, also the predicted river water concentrations (PEC_{river}) are higher than the measured concentrations reported for several detergent surfactants [596], [603]. The conservatism of the model calculations is also connected with the fact that the biodegradation/elimination processes of chemical compounds are only considered with regard to the removal in sewage treatment plants, while an in-stream removal after entering the receiving river is not taken into account or is based on over-conservative assumptions [604]: for instance, the half-life for LAS and other readily biodegradable surfactants in rivers was determined to be in the range of 1 – 5 h, while the calculation model used in the EU risk assessment process assumes a value of 360 h [594].

Table 48. Stepwise environmental risk assessment process exemplified on a detergent-based surfactant (LAS)

Test hierarchy	Exposure analysis		Effects assessment			PEC/PNEC
	Basis	PEC	Basis	Extrapolation factor	PNEC	
Tier 1 (screening)	calculation model: STP influent*: 12.3 mg/L Elimination in STP: 95 % Dilution factor: 10	62 μg/L	acute toxicity: lowest EC_{50} = 1.1 mg/L	1000	1.1 μg/L	56
Tier 2 (confirmatory)	STP simulation test: STP influent*: 12.3 mg/L Elimination in STP: 99 % Dilution factor: 10	12 μg/L	chronic toxicity: lowest NOEC = 0.12 mg/L	10	12 μg/L	1.0
Tier 3 (investigative)	River monitoring: 1 – 5 km downstream of STP release	$\leq$ 20 μg/L	field studies/ biocenotic tests: NOEC = 0.25 mg/L	1 – 10	250/25 μg/L	0.08/0.8

* LAS-consumption: 0.9 kg/inhabitant/year
Water consumption: 200 L/inhabitant/day

10.4.3. Assessment of Environmental Effects

The impact of chemical substances to environmental systems can be different and may also include indirect effects, for example eutrophication of water due to a nutrient function of the substance, or remobilization of heavy metal ions. However, the most important aspect to be evaluated in the environmental risk assessment is the ecotoxicity of substances, i.e., the toxic impact to organisms living in the individual environmental compartments. Of course, it is not possible to test all these organisms in the laboratory in terms of their sensitiveness to the chemical. However, a number of internationally accepted and standardized test systems exist, which cover the different trophic levels of the aquatic food chain: starting from the algae as representatives of the plants, the plant-feeding animals, represented by the daphnia while the fish stand for higher trophic levels. Finally, the bacteria close the carbon cycle in the aquatic ecosystem due to their degradation activities (Fig. 101). In the terrestrial environment, too, the ecosystem has a similar structure, consisting of producers (plants), consumers (animals), and destruents (bacteria, fungi).

As in the case of degradation tests, there is a sensible hierarchy of ecotoxicity tests, starting with relatively simple acute toxicity tests which allow a first general evaluation; these are followed by more sensitive and more expensive chronic or subchronic tests

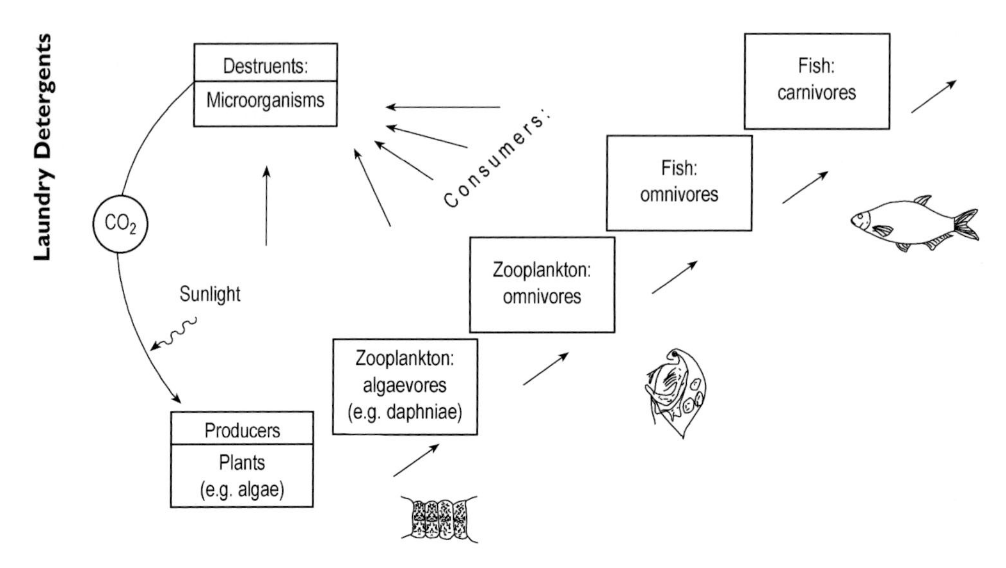

Figure 101. Scheme of the aquatic food chain

which have a higher predictive value for the practical situation. In any case, the objective of the tests is to find out a relationship between the concentration of the test substance and the ecotoxicological effect. Therefore, the test organisms are exposed to different concentrations of the chemical under defined test conditions, and the effects are evaluated by comparison with benchmarks (without addition of the test substance).

10.4.3.1. Basic Ecotoxicity Tests

The base set of ecotoxicological information of a chemical is obtained from acute tests on fish and daphnia and from an algal toxicity test (Table 49). Acute toxicity tests determine the lowest test concentration leading to death (fish) or immobilization (daphnia) of all tested organisms (LC_{100}/ EC_{100}). On the other hand, also the highest concentration showing no acute effect is experimentally determined (LC_0/EC_0). The acute toxicity is usually expressed by the LC_{50}/EC_{50} values, which can be calculated from the experimental LC_0/EC_0 and LC_{100}/EC_{100} data. The acute fish test is mainly conducted according to the pertinent OECD Guidelines [595] with zebra fish in a 96-h test. The acute daphnia test is also standardized [595] and has a duration of 48 h. The algal toxicity test differs from the previous tests, as the test criteria are based on the growth rate of the algal cells and the algal biomass formed within the 72- or 96-h incubation period [595]. Since this test not only covers the test organisms exposed at the beginning of the test but also the following generations of algae it is a chronic test determining the EC_0 and EC_{50} of algal growth (which is different from an acute algicidal effect). Despite the qualitative differences between the acute fish and daphnia tests and the chronic algae test, the LC_{50}/EC_{50} values from these three tests form the ecotoxico-

Table 49. Ecotoxicological test methods

Test	Evaluation parameter		Relevance
Acute aquatic toxicity			
Fish	lethal effects:	LC_{50}	basic data for ecotoxicological evaluation (Safety Data Sheets)
Daphnia	ability to swim:	EC_{50}	
Algae	cell growth:	EC_{50}	EU classification "dangerous for the environment"
bacteria	respiration:	EC_0, EC_{10}, EC_{50}	initial risk assessment
Chronic/subchronic toxicity			
Fish (prolonged test/early life stage)	growth:	NOEC	
Daphnia (life cycle)	reproduction:	NOEC	aquatic risk assessment
Algae	cell growth:	EC_0	
bacteria	cell growth:	EC_0, EC_{10}	
Biocenotic toxicity			
Microcosmos test	species composition:	$NOEC_{biocoen.}$	aquatic risk assessment
Model river flow			
Terrestrial toxicity			
Earthworm (acute)	lethal effects:	LC_{50}	terrestrial risk assessment
Higher plants (chronic)	growth:	EC_0	

NOEC = No observed effect concentration
LC/EC = Lethal/Effect concentration

logical criteria for the classification and labeling of substances as "dangerous for the environment" in the European Union [605]. At the same time also the environmental effects assessment for the preliminary determination of the PNEC takes reference to the lowest, i.e., most toxic LC_{50}/EC_{50} value from these tests [594]. The data from this set of three ecotoxicological characteristic values is also required when a new chemical is to be notified according to the chemical laws in EU member states [606].

Although not contained in the discussed ecotoxicological base set, the determination of the acute (e.g., bacterial respiration inhibition test [607]) or chronic (e.g., bacterial cell growth inhibition test [608]) toxicity to bacteria has a high ecological relevance for the practical situation. The chronic bacterial toxicity is measured during an 18-h incubation period using a *Pseudomonas putida* shake-flask culture and determining the rate of cell multiplication. The same bacterial species or an activated sludge culture is employed in the acute test measuring the influence of the test chemical on the oxygen consumption (respiration) rate of the bacteria. The EC_0 of this test has a high predictive value for the beginning toxicity of a substance to the purification performance of a biological sewage plant [609]. Information on the basic ecotoxicological data of a chemical product as well as the environmental classification (EU classification [605], Water Hazard Class in Germany [610]) is usually provided from the pertinent safety data sheets.

10.4.3.2. Subchronic and Chronic Ecotoxicity Tests

Information on long-term ecotoxicological effects at concentration levels fairly below their acute toxicity can be gained from standardized subchronic and chronic tests with the same test organisms also used in acute tests. This type of tests aims at determining the no observed effect concentration (NOEC), i.e., the highest tested concentration that does not exhibit any effect (vs. a benchmark). The criteria for assessing effects in the subchronic fish test (test duration 3 – 4 weeks) are not only lethality but also changes in growth, weight, behavior, etc. In the subchronic early life stage fish test with trout (OECD 204, test duration 90 d) [595] the influence of a compound on the most sensitive life stage of the fish is tested, because the development of juvenile fish from fertilized spawn is used as criteria. The 21-d daphnia reproduction test (daphnia life cycle test, OECD 202) [595] determines the number of new daphnids produced within the test period, and thus also covers the most sensitive life processes like growth, maturation, and reproduction.

10.4.3.3. Biocenotic Ecotoxicity Tests

While the discussed test systems are single-species tests, the biota in an environmental compartment form a biocenosis with many and complex interactions. Thus, the issue must be addressed whether a single-species NOEC is suitable to allow such biocenotic implications to be assessed. Laboratory tests also exist for biocenotic effects where a self-developing biocenosis of microorganisms, algae, unicellular protozoans, and small metazoans (rotifers) represents the "test organism" [611]. The possible impact of a chemical is determined by biological analysis of the qualitative and quantitative composition of the species developing within the test period. Examples of such tests are the microcosm test and the river flow simulation test [611]. Test results obtained in such complex systems indicate that biocenoses are not significantly more sensitive to toxic substance influences than single species used in standard chronic/subchronic tests [612].

10.4.3.4. Bioaccumulation

Bioaccumulation or bioconcentration describes the fact that a compound is enriched in the living organism to concentrations higher than that prevailing in the surrounding milieu. Experimentally, bioconcentration is determined in relatively costly tests with fish or other aquatic organisms. These organisms are exposed to a very low concentration of the test substance and the increase of the substance in the organism in the time course is analytically monitored. The bioconcentration factor (BCF) describes the ratio of the substance concentration in the organism and the surrounding medium after attaining a steady state. The BCF can also be calculated from the quotient of the substance uptake (k_1) and elimination (k_2) rate constants; the latter is determined in the depuration phase of the experiment after transfer of the exposed organisms to substance-free water. The most widely used bioaccumulation test is the flow-through bioaccumulation fish test (OECD 305) [595].

Because of the demanding analytical requirements with regard to the sensitivity and specific detection of a compound in the test organism, e.g., in a fish homogenate, in most cases the physicochemical properties of the test substance or quantitative structure – activity relationships (QSARs) are used for the bioaccumulation assessment. The descriptor mostly used for the bioaccumulation potential is the partition coefficient of a substance in a two-phase system of water and octanol (P_{ow}). The P_{ow} can be determined experimentally [595] or by calculation [613]. Substances having a BCF $\geq$ 100 or a log P_{ow} $\geq$ 3 are considered as significantly bioaccumulating and are to be classified as "dangerous for the environment" if their ecotoxicity is also high (EC/LC_{50} $\leq$ 10 mg/L) [595]. However, it has to be borne in mind that the P_{ow} is not generally suitable for prediction of the real bioaccumulation behavior of chemicals: it is not applicable to ionic substances and surfactants [595] and does not take into account the manifold biotransformation processes in living organisms [614].

10.4.4. Process of Environmental Risk Assessment

The exposure and effects assessment implies a hierarchy of methods starting with simple standardized basic tests up to exacting simulation tests and monitoring studies. All these investigations aim at conclusively predicting the environmental concentration (PEC) and a reliable no effect concentration (PNEC). The PEC and PNEC data generation is therefore a stepwise process. Depending on the outcome of the PEC/PNEC comparison it can be decided whether or not the PEC prognoses derived from simple but stringent (screening) tests are sufficient or whether data from more realistic (simulation) tests or monitoring data is required. A similar hierarchy of the predictive value of the data is implemented in the effects assessment by the use of application (safety) factors to derive the PNEC from acute LC/EC_{50} data and chronic NOEC data, respectively (see Table 48). It should be noted that the chronic toxicity test data of chemical substances already imply the increased body burden resulting from a potential bioconcentration so that no additional safety factors need to be applied for the PNEC derivation.

These principles form the basis of the European environmental risk assessment process described in Technical Guidance Documents [594]. Environmental risk assessment is an iterative process which can be stopped when the PEC/PNEC ratio is $\leq$ 1, which indicates the environmental safety of the chemical ("no further information/tests required"), or when all possibilities to further improve the predictive value of the data are exhausted. Risk management measures are to be taken when a PEC/PNEC ratio > 1 indicates a risk for the environment [592], [593].

According to the three-tiered process described in Section 10.4.2.4 the environmental risk of key surfactants used in detergents (LAS, alcohol ether sulfates, alcohol ethoxylates, soap) was evaluated. The LAS example (Table 48) indicates that compounds with a high environmental relevance such as detergent components often require the information even from the highest tier to reach PEC/PNEC < 1. In the end, this risk assessment showed that the most prominent surfactants in detergents are safe for the aquatic systems in spite of their high input to the environment [601], [602].

10.5. Ecological Characterization of Main Detergent Ingredients

10.5.1. Surfactants

The surfactants are the category of detergent ingredients with the highest ecological relevance due to their high consumption figures and their potential ecotoxicological impact connected with their surface-active properties. This relevance is reflected by the fact that the two most important surfactant groups, the anionics and nonionics, are subjected to legal regulations regarding their (primary) biodegradability (see Section 10.3.2). Strictly speaking, also the second main criterion for the environmental risk assessment, the ecotoxicological behavior is included therein because the primary biodegradation of surfactants and the loss of their high aquatic toxicity are normally concomitant processes. Hence, sometimes the primacy of biodegradation is emphasized since a fast and complete (ultimate) biodegradability is the best prerequisite to exclude long-term effects in the environment.

10.5.1.1. Anionic Surfactants

The oldest and largest-volume anionic surfactant worldwide is soap. However, the superiority of synthetic anionic surfactants in terms of detergency performance considerably reduced the relevance of soap in detergents in many countries. Instead, branched-chain TPS and its successor LAS have played the most important role as detergent surfactants. The poor biodegradability of TPS became evident in many rivers and sewage plants in the late 1950s and is also recognizable from the poor results in standard primary and ultimate biodegradability tests, as shown in Table 50 (see Section 10.3.1). Unlike TPS, LAS not only fulfills the legal requirements of the European detergent legislation but is also evaluated as "readily biodegradable" in the OECD 301 tests on ultimate biodegradability (Table 50). Also in the continuous activated sludge test (OECD Confirmatory Test) a high elimination result of about 95 % MBAS removal was found, which is still lower than the >99 % removal measured in real sewage treatment plants [596].

Table 50. Biodegradation data of surfactants

Surfactant	Primary biodegradation		Ultimate biodegradation			
	OECD Screen. test, % removal	OECD Confirm., % removal	Ready biodegradability test, %	Coupled Units Test, % C removal	Metabolite test, % C removal	Anaerobic biodegradability
Anionics:						
TPS (C_{12})	10–50 MBAS	20–45 MBAS	301 D: < 10 BOD/COD 301 E: 37 C removal	41 ± 9 (COD)	n.d.	no
LAS (C_{10-13})	95 MBAS	93–97 MBAS	301 E: 73–84 C removal 301 D: 55–65 BOD/COD	73–82	94.9 ± 1.2	no
C_{12-14} FA + 2EO sulfate	99 MBAS	100 MBAS	301 E: 100 C removal 301 D: 100 BOD/COD	89 ± 6	102.1 ± 3.9	yes
C_{16-18} FA sulfate	99 MBAS	98–99 MBAS	301 E: 85–88 C removal 301 D: 77 BOD/COD	96	99.9 ± 1.6	yes
C_{14-18} *sec*-alkanesulfonate	96 MBAS	97–98 MBAS	301 E: 88–96 C removal 301 D: 63–95 BOD/COD	83–96	97.4 ± 0.5	no
C_{8-18} soap	n.a.	n.a.	301 D: 79–100 BOD/COD 301 B: 80–90 CO_2 evolution	n.d.	100.8 ± 1.4	yes
Nonionics:						
C_8 FA + 6 EO	99 BiAS	n.d.	301 D: 84 BOD/COD	97 ± 3	98.5 ± 2.7	yes
C_{16-18} FA + 10 EO	99 BiAS	98 BiAS	301 D: 77 BOD/COD	90 ± 16	98.8 ± 2.7	yes
C_{12-14} FA + 30 EO	99 BiAS	n.d.	301 D: 80–83 BOD/COD	94 ± 2	95.1 ± 3.1	yes
Isononylphenol + 10 EO	6–>80 BiAS	87–97 BiAS	301 D: 5–10 BOD/COD	59 ± 22	93.6 ± 2.7	partial
C_{12-14} APG	–	99 subst.-specific	301 D: 73–88 BOD/COD 301 E: 90–93 C removal	89 ± 2	101.8 ± 2.0	yes
Cationics:						
DTDMAC	n.a.	94 DSBAS	301 D: < 5 BOD/COD	108 ± 9	n.a.	no
EQ esterquat	n.a.	93 DSBAS	301 D: 67–88 BOD/COD	89 ± 7	n.a.	yes

n.d. = not determined
n.a. = not applicable

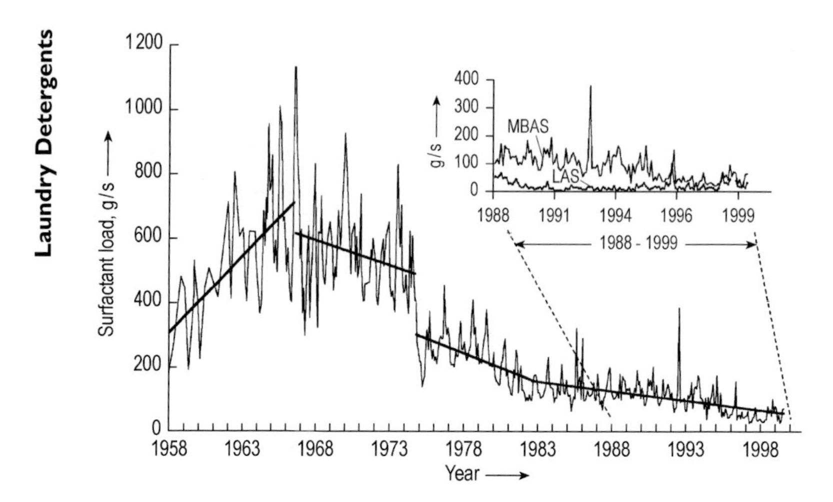

Figure 102. Changes of the anionic surfactant load (MBAS/LAS) in the river Rhine 1958 – 1999 (sampling site: Düsseldorf-Himmelgeist, Germany)

The success of the measures to improve the environmental compatibility of detergent surfactants can be impressively demonstrated by the results of a systematic monitoring program conducted by the Henkel company in the river Rhine in Germany since 1958 and in some of its tributaries [590], [591], [615] – [617]. Based on the MBAS analytical method, the usage of TPS in detergents is recognizable as a continuous and steep increase of the MBAS loads and concentrations (Fig. 102); the maximum concentration detected in the Rhine was 0.72 mg/L MBAS in the fall of 1964 while concentrations of even more than 1 mg/L were observed in its tributaries [615]. The replacement of TPS by LAS in 1964 marked the turning point: the MBAS values decreased continuously from then on in spite of the fact that surfactant consumption increased by 80 % from 1963 to 1975 [615].

A second reason for this positive development was the construction of biological sewage treatment plants. In highly populated areas such as Germany such technical measures are a prerequisite for a sufficiently effective reduction of the organic load present in sewage prior to entering receiving waters.

Based on the experiences from the detergent problem of the early 1960s, the development of new detergent surfactants has always been a search for improvements in terms of both their detergency performance and the reduction of the environmental impact. Besides the legally required primary biodegradability, the ultimate biodegradability behavior today is in the focus of the ecological safety assessment of surfactants used in laundry products. As shown in Table 50, there are a number of anionic surfactants which exhibit excellent ultimate biodegradability in the OECD 301 tests. This underlines that the high removal rates determined in model sewage treatment plants are mainly due to mineralization (carbon dioxide formation) and assimilation (biomass formation). Furthermore, it could be shown by comprehensive investigations

Table 51. Ecotoxicological data of surfactants (concentrations are given in mg/L)

Surfactant	Acute toxicity (EC/LC_{50})		Algal toxicity (EC_{50}/EC_0)	Bacterial toxicity (EC_0)	Long-term toxicity (daphnia life cycle test; NOEC)
	fish	daphnia			
Anionics:					
LAS	2.9	4 – 6	7.3/2.6	64	
C_{12-14} FA + 2 EO sulfate	7.9	79	1.8/1.0	360	0.72
C_{16-18} FA sulfate	5.2	280	30/10	35	16.5
C_{14-18} *sec*- alkanesulfonate	5	7.5	8.7 (EC_0)	390 (EC_{10})	0.6
C_{12-18} soap	46	25	10 (EC_0)	595	10
Nonionics:					
C_{12-14} FA + 9 EO	1.7	2.7	0.9/0.3	700	3.0
C_{12-14} FA + 30 EO	455	790	24/10	990	n.d.
C_{16-18} FA + 5 EO	2.7	3.9	0.3/0.1	64	n.d.
C_{16-18} FA + 14 EO	3.5	16	0.8/0.3	1000	1.0
Isononylphenol + 9/10 EO	4.6	14	2.8/1	1000	n.d.
C_{12-14} APG	3.0	7.0	12	500	1.0
Cationics:					
DTDMAC	0.6 – 3.0	1.1	0.06 (EC_{50})	10	0.38
EQ esterquat	3.0	78	1.8/0.3	90	2.7

n.d. = not determined

based on metabolic studies with radiolabeled model surfactants, or on the "test for recalcitrant metabolites", that all major anionic surfactants are biodegraded without the formation of any poorly biodegradable intermediates [618]. Surfactants exhibit a distinct adsorption behavior onto suspended solids in sewage or surface waters that cause transportation of a significant portion of these chemicals into anaerobic environmental areas (see Section 10.4.2). Thus, surfactants which have proved their ultimate biodegradability in aerobic and anaerobic environments meet a very high standard of ecological safety, as they are biodegradable under all environmental conditions. Table 50 shows that the sulfated anionic surfactants (alcohol sulfates, alcohol ethersulfates) are anaerobically degradable in contrast to the sulfonated materials (LAS, secondary alkane sulfonates, methyl ester sulfonates).

The second general essential criterion for the environmental risk assessment of chemicals is the effects assessment (see Section 10.4.3). The ecotoxicological behavior of the major detergent surfactants is very well investigated and documented involving acute, subacute/chronic, and biocenotic tests [601], [602]. Table 51 gives an overview of some ecotoxicological standard data of surfactants used for their evaluation according to the EU classification "dangerous for the environment" [605]. The comparison of the data reveals the marked aquatic toxicity of the surfactants with EC/LC_{50} values being predominantly in the range of 1 – 10 mg/L. This is a concentration range which could be achieved in surface waters if the surfactants were not biodegraded beforehand [619]. In reality, the concentrations of the most prominent detergent surfactants LAS are in the lower microgram per liter range even in small rivers receiving the effluents of biological sewage treatment plants [603], [620], while rivers receiving effluents from mechanical sewage treatment plants show LAS concentrations in the range of 500 – 600 μg/L [620]. It may therefore be concluded that the biological treatment of wastewaters is the decisive prerequisite for excluding an acute ecotoxicological impact of surfactants on surface waters' biota. Since an environmental safety assessment must take into account particularly long-term effects, predicted no effect concentrations derived from subacute and chronic toxicity data are more reliable than acute toxicity data. The no observed effect concentrations obtained in these tests with fish, daphnia, and algae are in the range of about 0.1 – 1 mg/L, i.e., one order of magnitude below the acute toxicity levels [601]; also the biocenotic NOECs determined in microcosm or mesocosm tests with anionic surfactants such as LAS and fatty alcohol sulfates [601], [612] are in the same order of magnitude. From information on the long-term ecotoxicity of key anionic surfactants a PNEC has been calculated, being 250 μg/L for LAS and 400 μg/L for alcohol ethersulfates [602]. Taking the discussed LAS concentration levels (PEC) into account it can be concluded that LAS and all other surfactants with a comparable ecological profile (readily biodegradable, ecotoxicological properties similar to LAS) will yield a ratio of PEC/PNEC < 1, i.e., are safe to the aquatic environment [602].

The bioaccumulation behavior of surfactants in general was tested in several investigations that were mainly conducted using radiolabeled test materials [621]. While in some cases a bioaccumulation in fish can be definitely excluded (e.g., alcohol sulfates), in other cases only the bioconcentration of the radioactivity in the piscine body was determined without differentiation between the parent compound and the biotransformation products. A recent and very detailed investigation of the bioconcentration behavior of several *n*-(*p*-sulfophenyl)alkanes, i.e., constituents of LAS, revealed that these chemical structures are not or only moderately bioaccumulative not the least due to the action of biotransformation processes [622].

10.5.1.2. Nonionic Surfactants

Alcohol Ethoxylates. Nonionic surfactants of the alcohol ethoxylate type are contained in almost all detergents and cleansers (see Section 3.1.2.1). Alcohol ethoxylates with a linear or single-branched alkyl chain are highly biodegradable. As shown in Table 50, they easily fulfill the legal requirements of a minimum primary biodegradability of 80 % BiAS removal and are "readily biodegradable" in the OECD tests on ultimate biodegradability. Alcohol ethoxylates are also bio-degradable under anaerobic conditions, thus providing all prerequisites for their fast and complete removal from the aquatic environment. The residue-free biodegradation of AE has been shown in the metabolite test [598] as well as in detailed investigations of the biodegradation pathways of these compounds [623], [624]. Consequently, also the elimination rates of alcohol ethoxylates in model plants (Table 50) and in real sewage treatment plants are very high (> 99 % [596]) and allow an environmental concentration (PEC) of a few micrograms per liter in receiving waters to be predicted. These figures are considerably lower than the BiAS concentrations determined in the river-monitoring programs of Henkel (10 – 100 μg/L [590], [616]). These differences are probably due to the insufficient specificity of the BiAS analytical method [625].

The ecotoxicological profile of AE has also been very well investigated. Their acute aquatic toxicity data are in the same order of magnitude as those of the anionics, although their algal toxicity seems to be more pronounced frequently (Table 51). The comprehensive set of available acute, chronic, subchronic, and biocenotic ecotoxicity data on alcohol ethoxylates was used in a recent environmental risk assessment study to derive a PNEC for alcohol ethoxylates [601]. The PEC/PNEC comparison led to the conclusion that this most prominent class of nonionic surfactants is safe to the aquatic environment [602]. Further support for this evaluation comes from comprehensive bioaccumulation studies showing that these surfactants are metabolized intensively in the piscine body [626].

Alkylphenol Ethoxylates. Another relevant group of nonionic surfactants still being used in some countries are the alkylphenol ethoxylates (see Section 3.1.2.2). In contrast to alcohol ethoxylates they show some ecological properties which have made their environmental compatibility questionable. Nonyl- and octylphenol ethoxylates have a highly branched alkyl chain which is an obstacle for microbial degradation. Thus, these surfactants are not readily biodegradable although they fulfill the legal requirement of > 80 % BiAS removal (Table 50). In winter conditions, however, even this primary biodegradation is relatively slow, so that foam problems in sewage treatment plants have been observed [627]. The major ecological reservations came from the fact that in the primary degradation process poorly biodegradable intermediates are formed which are even more toxic than the parent surfactant [628]. These intermediates (alkylphenol + 1 – 3 EO units) are poorly water soluble and are adsorbed strongly on sewage sludges. During the anaerobic treatment of these sludges in digesters alkylphenol is formed by splitting off the residual EO units. This metabolite shows an even increased aquatic toxicity [628], [629]. These findings led to a ban (e.g., in Switzerland in 1986) or voluntary renunciations (e.g., in Germany in 1987) of alkylphenol ethoxylates in laundry products. Since 1992 a PARCOM recommendation has existed to phase out alkylphenol ethoxylates in household laundry products by 1995 and in I&I cleaning products by the year 2000.

Alkylpolyglycosides. Developments towards more efficient and better environmentally compatible products are steadily in progress, also in the field of the nonionic surfactants. One interesting and promising example is the class of alkylpolyglycosides (see Section 3.1.2.6) These surfactants feature ready and complete ultimate biodegradability in aerobic and anaerobic environments (Table 50) [630]. Also, alkylpolyglycosides exhibit very low acute, subacute, and chronic toxicity data within the effect concentration range typical for surfactants (Table 51). A very conservative environmental risk assessment conducted on the basis of these data showed that this surfactant class is safe for the aquatic and terrestrial environment [630].

10.5.1.3. Cationic Surfactants

The cationics are the smallest of the three main surfactant groups (Table 44). They are not contained in usual detergents, but represent the active agent in fabric softeners (see Section 3.1.3). Until the beginning of the 1990s, ditallow dimethyl ammonium chloride (DTDMAC) was the most important cationic softener, with a market share of about 80 % [60] (see Section 3.1.3). This surfactant is almost quantitatively eliminated from sewage, as illustrated by the high elimination rates in model sewage treatment plants (columns 3 and 5 in Table 50). This is mainly due to adsorption onto sewage sludge, whereas its biodegradation is slow. As yet, cationic surfactants are not subjected to legal bio degradability requirements, although a group-specific analytical method exists to follow the primary biodegradation by measuring the decrease of the disulfine blue-active substance (DSBAS) [631].

Generally, degradation of cationics in usual screening tests is poor, maybe due to the fact that their bactericidal properties may prevent bacterial growth at the test concentrations required from the analytical point of view. However, even application of very low concentrations (< 1 mg/L) of ^{14}C-labeled DTDMAC did not result in substantial $^{14}CO_2$ formation under screening test conditions, and this indicates a poor biodegradability of DTDMAC even at concentrations that do not adversely affect bacterial growth. Results from activated sludge studies tend to be ambiguous due to the difficulty in distinguishing between elimination by adsorptive processes and real biodegradation. Analytical determinations of the residual amount of the parent DTDMAC present in the activated sludge of an OECD Confirmatory Test indicated that a major part of the elimination was due to (primary) biodegradation [632]. The strong adsorptivity of DTDMAC, its very slow aerobic biodegradability and its nonbiodegradability under anaerobic conditions resulted in relatively high concentrations of this surfactant in digested sludges [633]. These properties and findings in combination with the relatively high ecotoxicity of DTDMAC in standard tests (Table 51) triggered the development of alternative cationic surfactants exhibiting more favorable ecological characteristics.

Today, the fabric softener market in industrial countries has virtually completely switched to new cationic surfactants. The major active ingredients in softeners are now esterquats (e.g., EQ, DEQ, DEEDMAC, see Section p. 60). Esterquats are readily biodegradable under aerobic and anaerobic conditions which favors a high elimination rate in sewage treatment plants. This is attributed to genuine biodegradation (Table 50). Although the test for detecting recalcitrant metabolites is not applicable to strongly adsorbing compounds such as cationics, in-depth investigations of the degradation behavior of triethanolamine-based [60] and other esterquats [61], [634] have clearly indicated that these molecules are completely and ultimately biodegradable in the environment. The comparison of the ecotoxicological data of DTDMAC and the EQ esterquat (Table 51) points out the favorable ecological properties of the latter group of cationic surfactants. Although DTDMAC has not proved to be a risk for the aquatic environment according to the PEC/PNEC concept [635], it is even more evident that the use of esterquat cationics in fabric softeners does not have any negative impact to aquatic life.

10.5.2. Builders

As discussed in Sections 9.1.2 and 10.3.4, the complexing agent sodium triphosphate had been the main builder in almost all detergents in Europe, USA, Japan and other countries for a long time. Today, phosphate-free detergents are mainly based on builder systems consisting of the inorganic builders zeolite A and soda ash, and organic cobuilders such as polycarboxylates and/or citrate [77].

10.5.2.1. Zeolites

The major breakthrough with respect to phosphate-reduced and nonphosphate detergents was the development and introduction of inorganic water-insoluble sodium aluminum silicates of the zeolite A type. The ecological behavior of this ion exchanger had been investigated in a comprehensive research program that was started in 1973. It covered laboratory tests and field trials conducted by experts from industry, water authorities, and academia [636]. It was shown that no excessive sedimentation of the insoluble particulate matter occurred in the household sewer systems and communal sewage systems. A major part of the load of zeolite A (about 60 %) is already eliminated in the mechanical stage of sewage treatment plants (sand trap and primary settler) while total elimination is around 95 % after passing the biological stage and the secondary settler. Although the mass of the sewage sludge is increased due to zeolite, no negative effects on the purification performance of the activated sludge and on the anaerobic amenability of the sludge in the digester was found; no decreased retention of heavy metals in sewage plants has been detected in the presence of zeolite either. Investigations in the 1990s showed that zeolite A is quickly transformed into an extremely insoluble amorphous complex of basic calcium aluminum silicate phosphate under usual sewage treatment plant conditions [637].

Although only about 5 % of the zeolite A present in raw sewage will pass the sewage treatment plant, it was necessary to investigate all possible effects on the biota in receiving waters. Acute and long-term toxicity to fish and daphnia proved to be very low (EC_0 > 250 mg/L), while a growth-inhibiting effect on algae could be observed at concentrations > 10 mg/L in nutrient-poor culture media. This effect is, however, without any practical relevance because it is due to ion-exchange by zeolite A and the resulting depletion of essential trace elements in the test vessel [638]. In practice, zeolite A will never reach oligotrophic water bodies alone but only in conjunction with other nutrients in treated or untreated effluents. Long-term ecotoxicity data from fish (early life stage test) and daphnia (21-d reproduction test) revealed NOEC values of 87 mg/L and 37 mg/L, respectively [639].

Since no suitable analytical methods are available to determine zeolite concentrations in river water, calculations of the PEC were made by taking into account the consumption figures, elimination rates in sewage plants, and a river dilution factor of 10. Such a rough estimation leads to river water concentrations of 150 – 300 μg/L decreasing to 57 μg/L if a half-life of 1 – 2 months [640] is taken into account [637]. These figures prove the environmental safety of zeolite A since the PEC is far below the PNEC.

10.5.2.2. Polycarboxylates

Phosphate-free detergents require cobuilders in addition to zeolite – soda ash to achieve comparable performance to STP [77] (see Section 3.2.4). By volume, the most important organic cobuilders are polycarboxylates, i.e., high-molecular mass ($M_r \leq 100\,000$) homopolymers of acrylic acid or copolymers of acrylic acid and maleic acid (Table 44). These water-soluble polyelectrolytes form sparingly soluble coagulates with excess calcium ions and adsorb on solid surfaces [641]. Based on these properties, polycarboxylates are eliminated to a high extent (> 90 %) in mechanical-biological sewage treatment plants [641], [642]. The contribution of biodegradation to this effective elimination process is minute. Polycarboxylates are poorly biodegradable except for the low molecular constituents ($M_r \leq 1000$) as was shown by investigations using ^{14}C-radiolabeled model compounds. On the other hand, polycarboxylates in real-life concentrations have proved to be without influence on the operation of sewage plants. They exhibited no adverse effects on the biological activity of sludge bacteria, the settling behavior of the sewage sludges, and heavy metal retention in laboratory tests and in a field trial [641]. Large amounts of polycarboxylates in sewage sludges do not influence the degradation process and the polycarboxylates are not redissolved in sludge digesters [641], [642].

Similarly comprehensive studies regarding fate and effects of polycarboxylates were conducted for the environmental compartments of surface waters and soil. These studies have shown that neither significant biodegradation nor heavy metal remobilization from the sediments is to be expected at relevant river concentrations (< 0.1 mg/L) [641], [642]. Due to the low ecotoxicity of polycarboxylates concluded from results of standard acute and long-term tests (acute EC_{50} > 200 mg/L, fish embryo larvae test NOEC = 40 mg/L, algal growth NOEC = 32 to > 200 mg/L, daphnia NOEC = 1.3 – 350 mg/L) a negative influence on the biota in surface waters can be excluded. The low no-effect concentrations within the broad range of the NOEC values observed in the chronic algae and daphnia test are connected with the precipitation of the polymer in the test system, i.e., with physical effects [642]. Bioaccumulation of polycarboxylates can be excluded because of their water solubility ($\log P_{ow} < -4.2$) and their high molecular mass.

Polycarboxylates enter the terrestrial compartment when sewage sludges are used for agricultural purposes. The high-molecular polymers are retained in the top layer of the soil, as shown by lysimeter tests; the more mobile low-molecular fractions will hardly be present in real sludges because of their biodegradability and their less pronounced precipitation behavior [641], [642]. Thus, contamination of ground water by polycarboxylates can be ruled out. As expected, the ecotoxicity of these materials to terrestrial organisms is very low: the NOEC values in the plant growth test (> 200 mg/kg) and in toxicity tests with earthworms (1600 mg/kg) are orders of magnitude above the PEC_{soil} (< 0.1 mg/kg) [641], [642].

In spite of their proved environmental compatibility, the presently used polycarboxylates are still regarded as detergent components which are to be substituted as soon as efficient and biodegradable alternatives are available.

10.5.2.3. Citrates

Citrates are widely used in phosphate-free detergents and cleansers, especially in liquid formulations (Table 44). Since citric acid is a natural constituent and a common metabolite of most living organisms, it can be expected that its environmental behavior is very favorable. In fact, this compound is readily biodegraded in standard ultimate biodegradation screening tests and in continuous activated sludge tests [643]. Anaerobic biodegradation has also been proven [644], although only a small percentage of this water-soluble detergent ingredient will reach the anaerobic areas of the aqueous environment. Taking the citrate consumption in detergents and its excellent biodegradability into account, an influence of this weakly complexing agent on heavy metal remobilization in sludges and sediments can be excluded. The same positive conclusion is possible in regard to the ecotoxicological effects: citrate has low acute aquatic toxicity values of $LC_{50} \geq 500$ mg/L (fish), $EC_{50} \geq 100$ mg/L (daphnia) and $EC_{50} \geq 1000$ mg/L (algae) [643], [644].

10.5.2.4. Sodium Carbonate (Soda Ash)

The inorganic compound soda ash plays an important role as a water-soluble builder and cobuilder in phosphate-containing and nonphosphate detergents. Although detergent-based consumption figures of soda ash are high (Table 44), no environmental impact is to be expected, because the ecotoxicological effects observed at high concentrations (daphnia $EC_{50} \geq 260$ mg/L, fish $LC_{50} \geq 200$ mg/L) are due to the resulting increase in the pH value of the test medium [643], which is irrelevant under real-life conditions.

10.5.2.5. Nitrilotriacetate (NTA)

In principle, NTA is an efficient builder and cobuilder in phosphate-free detergents. Its environmental compatibility was investigated in comprehensive studies [645]. Nitrilotriacetate was found to be ultimately biodegradable in aerobic conditions, although the degradation rate depends on temperature, concentration, and other ambient conditions. Heavy-metal complexes of NTA are degraded more slowly. The ecotoxicological data of NTA characterize this compound as weakly toxic to higher aquatic organisms (fish $LC_{50} \geq 100$ mg/L, daphnia $EC_{50} \geq 80$ mg/L). The NOEC of algal growth is in the range of 10 mg/L. In spite of these ecologically favorable characteristics concern arose from the strong complexing properties of NTA and, hence, the potential of heavy-metal remobilization. Due to uncertainties in regard to biodegradation rates of NTA under real-life conditions and its heavy-metal remobilization potential, it was recommended in Germany in the 1980s to use NTA only to a limited extent and to conduct further studies. The outcome of this successive program (1987 – 1990) is that NTA is readily biodegradable and is eliminated in sewage treatment plants to a high extent (95 %) irrespective of seasonal influences [646]. Heavy-metal removal from sewage is not impaired nor is a remobilization of heavy metals from river sediments to be anticipated below a concentration of 50 μg/L of NTA. Thus, NTA today is considerably more positively evaluated than in the past. Nevertheless, NTA is only used in minute amounts in detergents (Table 44), and it seems doubtful whether this compound will play a prominent role in the future as a detergent (co)builder.

10.5.3. Bleaching Agents (see Section 3.3)

10.5.3.1. Sodium Perborate

Sodium perborate decomposes during the washing process forming hydrogen peroxide and sodium metaborate. The water-soluble metaborate is not eliminated in wastewater treatment plants and may cause a considerable increase in the boron concentration in rivers. Although large regional variations exist due to geological conditions and degree of river pollution, the ratio of anthropogenic to naturally occuring boron in more closely investigated rivers in Germany was found to be up to 50 % in the river Rhine [615], about 70 % in the Neckar river [647], and considerably more than 100 % in the river Ruhr [648]. Nevertheless, a broad systematic study on the boron concentrations in German surface waters and drinking waters conducted in the mid-1970s [649] showed that boron concentrations in most rivers are below 0.25 mg/L and less than 0.1 mg/L in drinking waters. In small and heavily polluted rivers of Germany and other European countries higher concentrations up to 1 mg/L were found [649]. In the 1990s investigations on the boron content of drinking water in Germany showed that the concentrations had not changed significantly as compared with 1974 values [650].

Accordingly, more recent data from the river Rhine and its tributaries [590] have verified that the boron concentrations have not changed (rivers Rhine and Ruhr) or have decreased considerably (rivers Main and Neckar).

The ecotoxicological behavior of borate in water and soil has been studied comprehensively [651], [652]. The ecotoxicity of borate is relatively low. This is true for fish and invertebrates although the effective concentrations vary in a very broad range, depending on the individual studies. Fish LC_{50} values between 11 and 3400 mg/L are reported. Long-term toxicity data do not indicate a pronounced ecotoxic action of borate either. NOEC values are clearly above a concentration of 1 mg/L [652]. However, literature data exists that indicate chronic effects towards certain aquatic organisms, e.g., fish, algae, or reed plants in the submicrogram per liter concentration range [643], [651], [652]. A critical re-evaluation of these previously collected data in the light of more recent investigations showed, however, that the environmentally relevant no effect concentration in fresh waters is not lower than 1 mg/L [652]. Taking the reported river water concentrations of boron into account it is evident that boron does not constitute a problem for surface waters. However, the margin between the ecotoxicological no effect concentration and the environmental concentrations in the majority of rivers is relatively small.

With respect to the terrestrial toxicity of boron, a specific phytotoxicity towards certain agricultural plants exists, such as fruit trees, tomatoes, and vineyard stock. Therefore, it is not advisable to recycle boron-containing wastewater for irrigation purposes [651].

10.5.3.2. Sodium Percarbonate

Sodium percarbonate decomposes during the washing process forming hydrogen peroxide and sodium carbonate. Sodium carbonate is a water soluble mineral and ubiquitously present in the aqueous environment, because it is in equilibrium with carbon dioxide from the atmosphere and because of the presence of other dissolved mineral carbonates. Sodium carbonate has a very low aquatic toxicity (fish LC_{50} = 200 – 740 mg/L, daphnia EC_{50} = 265 – 565 mg/L) [643] which is attributable to the increased pH value at these high test concentrations. Thus, the use of sodium percarbonate as bleaching agent in detergents will not have any significant impact on the aquatic environment.

10.5.3.3. Tetraacetylethylenediamine (TAED)

Tetraacetylethylenediamine is the most widespread bleach activator used in laundry products (Table 44). In the washing process it is almost quantitatively perhydrolyzed to diacetylethylenediamine (DAED) and peracetate, the latter ultimately forming acetate. TAED and DAED dissolve well in aqueous alkaline solutions and have a low log P_{ow} [653]. The ready ultimate biodegradability of the two compounds has been shown in several tests [643], [653]. Accordingly, the elimination in sewage treatment plants is very high ($\geq$ 95 %) [653].

The ecotoxicity of TAED and DAED is remarkably low, since no effects were observed in the majority of tests, even at the highest concentration tested. For instance, the fish LC_{50} is well above 1000 mg/L and significant effects on daphnia and algae could not be observed at 800 and 500 mg/L, respectively [653]. Based on these biodegradation and ecotoxicological data it can be concluded that no adverse ecological effects will result from the use of TAED in laundry products.

10.5.4. Auxiliary Agents

Auxiliary agents are present in small amounts in many laundry product formulations. Nevertheless, their widespread use in these products results in considerable consumption figures (see Table 44) and, consequently, in a significant entry into the environment.

10.5.4.1. Phosphonates

Phosphonate compounds with more than one phosphonate group are suitable bleach stabilizers (see Section 3.3.4), because they are effective complexing agents and have additional chemical properties that support the washing process. Phosphonates have been well investigated with regard to their environmental behavior. They are poorly biodegradable under standard test conditions, although evidence exists that the chemically very stable C–P bond in phosphonates can be microbially attacked [654]. Nevertheless, phosphonates are not persistent in the environment, because they are subject to abiotic degradation in waters and in soil [654] – [656]. The degradation rate is strongly affected by the chemical composition of waters, e.g., the presence of metal ions like Fe^{3+}, and by sunlight. The degradation of HEDP is mainly based on photochemical processes [654], [655] while degradation of nitrogen-containing phosphonates seems to be a hydrolytic process that produces decomposition products which are, at least in part, accessible to biodegradation [654], [655]. Phosphonates are partially (about 50 %) eliminated in sewage treatment plants as shown in model experiments and field trials [654], [657]. In sewage plants having a phosphate precipitation stage the elimination of phosphonates is high (> 90 %) [657].

Phosphonates are strongly complexing agents which have the potential to affect the distribution and partitioning of metal ions in sludges, sediments, and soils. Laboratory and field studies showed no mobilization of heavy-metal ions in sewage sludges at phosphonate concentrations in the 2 – 10 mg/L range [654], [657]. Similar investigations on heavy metal release from river sediments made clear that the strong adsorption of phosphonates to sludges and sediments will prevent that metal ion remobilization occurs at environmentally relevant phosphonate concentrations of < 1 mg/L in wastewater and < 10 μg/L in surface waters [654].

The aquatic toxicity of phosphonates is low. The acute fish and daphnia LC_{50} values are > 100 mg/L [654]. Also, long-term NOEC data characterizes the phosphonates as unproblematic: subchronic/chronic studies on fish and daphnia revealed NOECs > 10 mg/L, i.e., orders of magnitude above the PEC levels [654]. Similarly to other complexing agents or ion exchangers, phosphonates, too, pretend to be relatively toxic to algae in standard tests. In fact, the algal NOEC was found to be in the range of 0.1 – 10 mg/L. However, there is clear evidence that this inhibition is due to the depletion of essential micronutrients for algal growth [654]. Ultimately, bioconcentration studies on two phosphonates (HEDP, ATMP) in fish [655], [656] proved that these chemicals do not bioaccumulate significantly.

10.5.4.2. EDTA

Ethylenediaminetetraacetate played an important role as a bleach stabilizer in the past. However, issues related to the environmental properties of this strongly chelating agent have led to a considerable decrease of its use in detergents. For instance, in Germany and some other European countries the use of EDTA in detergents ceased more than a decade ago. EDTA proved to be poorly biodegradable under environmental conditions although there is evidence that microbial degradation mechanisms exist [658]. In standard tests for ready and inherent biodegradability, this complexing agent indicated insignificant degradation rates and was not eliminated in a model sewage treatment plant (4 ± 2 % DOC removal in the Coupled Units Test) [643]. The insignificant elimination of EDTA in two-stage sewage treatment plants was confirmed in monitoring studies [659]. Thus, it is obvious that EDTA present in raw wastewater will enter receiving waters almost quantitatively. Accordingly, EDTA was found in rivers of Germany, the UK, Switzerland, and Austria in significant concentrations up to 50 µg/L [658]. At the same time similar concentrations of EDTA were detected in raw drinking water at sites where it is prepared from river water. This suggested that the processes normally employed for drinking water preparation do not eliminate low EDTA levels [659]. The influence of EDTA present in surface waters on the drinking water quality, its possible effects on remobilization of heavy-metal ions from sludges and sediments, and its deficiency of being adequately eliminated raised concerns over its environmental safety. This pushed activities towards replacing EDTA by phosphonates and NTA.

From an ecotoxicological point of view no critical aspects can be deduced. EDTA is not bioaccumulating and has a very low acute toxicity to fish and daphniae. LC/EC_{50} values are fairly above 100 mg/L [643], [658]. As expected, the algal growth NOEC in standard tests is considerably lower, i.e., in the range of 10 mg/L [643] which may be explained by the already discussed deprivation of trace metals in the test medium. On the other hand, stimulation effects of EDTA on algal growth are reported which could be explained by the photodegradation of ferric complexes to form less stable ferrous complexes with an enhanced bioavailability of growth-stimulating iron [658].

10.5.4.3. Enzymes (see Section 3.4.1)

Most detergents today contain enzymes, i.e., proteases, amylases, lipases, and cellulases. Being proteins, the ultimate biodegradation of enzymes in standard tests for ready biodegradability is high [643]. Investigations on the removal of protease in a model sewage treatment plant indeed showed a loss of proteolytic activity of more than 99 %. Of course, enzymes are also accessible to anaerobic degradation. The acute aquatic toxicity of enzymes is relatively low (fish $LC_{50} = 25-350$ mg/L, daphnia $EC_{50} = 160$ mg/L) [643]. Algal toxicity values are in the range of $EC_{50} = 10-100$ mg/L. The ecotoxicological effect concentrations of enzymes are better referred to enzyme activity rather than to weight.

10.5.4.4. Optical Brighteners (see Section 3.4.5)

Fluorescent whitening agents (FWA) in laundry products are mainly anionic diaminostilbene (DAS) or distyrylbiphenyl (DSBP) derivatives. They are hydrophilic water-soluble compounds having a high affinity to textile fibers and to particulate matter such as sewage sludges. A considerable portion of optical brighteners present in wastewaters is eliminated in sewage treatment plants (about 90 % in case of the most important stilbene derivative DAS 1, and about 50 % in case of DSBP, which is the major distyrylbiphenyl compound) [660]. This elimination is only due to adsorption but not to biodegradation since FWA are not readily biodegradable and seem only to undergo slow primary biological degradation [661]. In anaerobic digesters they are not biodegraded either, so that concentrations up to 1 g per kilogram of dry matter were found [661]. The portion of FWA not eliminated in wastewater treatment plants enters receiving waters and is exposed to daylight. The photodegradation of DAS 1 and DSBP was investigated in detail by experimental work and computer simulations [660]. As the kinetics of direct photodegradation is not only a function of the chemical properties of the FWA but also of the ambient environmental conditions (spectrum and intensity of daylight) the half life of DAS 1 varies from a few days to six months, whereas DSBP has a lifetime of a few hours. The photodegradation is essentially a primary degradation that forms a large number of products from DAS 1, while the degradation of DSBP follows a relatively simple path [660]. The triazinyl derivatives formed from DAS 1 during photodegradation are poorly biodegradable but weakly toxic in the acute fish test ($LC_0 > 97$ mg/L). The DSBP degradation products are predominantly accessible to biodegradation as shown in inherent biodegradability tests [660]. Recent monitoring data of the two major FWA species obtained in Germany and Switzerland revealed results that were well comparable with individual concentrations in the microgram per liter range in sewage, in the milligram per kilogram range in sludges and sediments, and in the nanogram per liter range in river waters [660].

The aquatic toxicity of FWA is low. Tests indicate fish LC_{50} values of > 320 mg/L (DAS 1) and > 70 mg/L (DSBP), respectively. The daphnia EC_{50} value for individual

FWA is > 1000 mg/L. The algal toxicity is uncritical (EC_{50} > 80 and > 10 mg/L, respectively) as well. The low ecotoxicity of FWA, which, in addition, do not bioaccumulate [661], is confirmed by the NOEC values determined for DSBP in the prolonged fish test (> 1 mg/L) and the three-week daphnia reproduction test (> 7.5 mg/L). The comparison of the PNEC deduced from these experimental results with the measured environmental concentrations does not indicate an environmental risk.

10.5.4.5. Carboxymethyl Cellulose

Carboxymethyl cellulose is a classical soil antiredeposition agent contained in some detergents. It is a water-soluble polymer which is only very slowly biodegraded [643], [662]. The resistance to biodegradation increases with the degree of substitution of free hydroxyl groups by carboxymethyl groups [662]. Thus, it is understandable that detergent-based CMC exhibits only poor results in ready biodegradability tests. Even in inherent biodegradability tests like the Zahn – Wellens test only a 40 % elimination value was found [643]. Also in the Coupled Units model sewage treatment plant CMC showed only a partial elimination with a 25 % DOC removal [643]. On the other hand, the polymer exhibits the expected very low aquatic toxicity with EC_0 values > 1000 mg/L in the acute fish and daphnia test [643]. Based on these data CMC is not considered a detergent chemical impacting the environmental biota.

10.5.4.6. Dye Transfer Inhibitors

Dye transfer inhibitors are polymeric compounds that prevent the transfer of dyes from one laundry item to an other via the wash bath. A prominent representative of these compounds is poly(*N*-vinylpyrollidone) (PVP). Its ecological behavior is similar to that of CMC, i.e., PVP is poorly biodegradable in standard biodegradation tests and eliminated in sewage treatment plants only to a limited extent. The very low ecotoxicity of the polymer (chronic toxicity NOEC > 100 mg/L) on the other hand assures that no detrimental effects from PVP will result in the aquatic environment.

10.5.4.7. Fragrances

Fragrances are mostly present at concentrations < 1 % in almost all laundry products. Usually, they are very complex mixtures of numerous individual perfume components that differ in their chemical structure and therefore in their ecological behavior. Due to their volatility and adsorbing properties they have been considered hardly relevant to the aquatic environment. However, more recent studies have revealed that certain perfume compounds, e.g., musk xylene, can be detected in fish and river water, obviously due to a bioaccumulation process [663]. This nitro musk compound has a high $\log P_{ow}$ and is poorly biodegradable. Since this compound has also been detected in human adipose tissues and mother's milk [664], the detergent and cosmetic industry in some European countries (Germany, The Netherlands) has discontinued using this component in its products. This case exemplifies the necessity of a sufficient ecological knowledge on any detergent ingredient irrespective of whether it is contained in the product in high or very low concentrations. Presently, extensive investigation programs are under way to found a sound basis for the ecological evaluation of perfume ingredients used in detergents and cosmetics.

10.5.4.8. Foam Regulators (see Section 3.4.3)

Two key groups of chemicals used as antifoam agents are silicone fluids (polydimethylsiloxanes, PDMS) and paraffins (Table 44). Paraffins are accessible to ultimate biodegradation but the rate is strongly dependent on the chain length and the molecular mass, respectively. Polydimethylsiloxanes are not biodegradable but virtually quantitatively eliminated in sewage treatment plants due to their strong adsorptivity on sludges. PDMS that enter the terrestrial environment when sewage sludges are used for agricultural purposes are not persistent as they are abiotically decomposed in the presence of clay soils. Volatile cyclic products and linear water soluble siloxanols are formed which may undergo photolytic oxidative degradation [665]. PDMS and paraffins exhibit a low aquatic toxicity and have no bioaccumulation potential due to their polymeric properties. They will not have adverse effects on the environment.

10.5.4.9. Soil Repellents

The most important soil repellents in detergents are anionically modified polymers from terephthalic acid and poly(ethylene glycol). Due to their polymeric nature they are not readily biodegradable but, nevertheless, accessible to ultimate biodegradation under aerobic and anaerobic conditions. Based on results obtained in a static activated sludge biodegradation test system (OECD 202 B), the extent of elimination under sewage treatment plant conditions is expected to be very high. The ecotoxicity of these polymers is very low. They exhibit no acute toxic effects to fish, daphnids, and algae at the highest concentration tested (100 mg/L) which is clearly above their water solubility. Taking this data and the relatively low consumption figures (Table 44) into account it can be concluded that soil repellents pose no risk to the aquatic environment.

10.5.4.10. Dyes

Dyes are sometimes used in powder and liquid detergent products. The total usage of these complex compounds in detergents is very low (Table 44). Available ecological information on these materials from the producers' safety data sheets characterize them as poorly biodegradable and having low acute aquatic toxicity.

10.5.4.11. Sodium Sulfate

Sodium sulfate is used mainly as a filler in conventional powder detergents. The German consumption figure of this inorganic salt in detergents has decreased considerably within the last few years (1997: 100 100; 1999: 74 600 t/a [666]), mainly due to the growing market share of compact detergents. Sodium and sulfate ions are natural constituents of surface waters. Based on recent consumption figures (Table 44) and sulfate concentrations in several German rivers (30 – 160 mg/L), it can be calculated that the contribution of detergents to the sulfate load in rivers is less than 1 %. The aquatic toxicity of sodium sulfate is very low. The LC_{50} is in the range of > 1 g/L [643]. Hence, regarding the mass balance and the ecotoxicity of this compound it is obvious that sodium sulfate from detergents does not represent a significant factor influencing the aquatic environment.

11. Toxicology

Detergents are products designed for everyday use; thus, they are found in large quantity in nearly every household. The home is quite unlike the industrial workplace: one cannot assume that every user of a household product plays strict attention to warning labels or recommended safety precautions. For this reason, any product offered on the market must insofar as possible be formulated so that it represents no significant or foreseeable health hazard, regardless of whether or not the product is properly handled. This in turn means that even at the earliest stages of raw material selection, great value should be attached to chemicals known to be harmless to humans following any realistic type of exposure.

The following types of exposure must be taken into account in assessing the risk presented by detergents and their components:

Skin contact (wash bath, residues remaining on laundered items, or exposure during manufacture)
Ingestion (accidents, especially with children, and trace residues in drinking water)
Inhalation (during production or use of powder detergents)

Extensive toxicological studies are required, and the results must be carefully weighed, taking into account the amount and stability of individual substances. Such studies should examine local effects (skin irritation, development of allergic reactions on contact, skin penetration), systemic effects (both acute and chronic), and potential hazards of a more subtle nature (mutagenicity, embryotoxicity, and carcinogenicity).

This chapter begins by examining the toxicology of the main detergent constituents, after which the properties of complete detergent formulations are explored. The approach is analogous to that followed in practice by a manufacturer in the course of product development.

11.1. Detergent Ingredients

11.1.1. Surfactants

Most of the biological properties of surfactants can be understood in terms of interactions occurring between surfactant molecules and such fundamental biological structures as membranes, proteins, and enzymes. Contact between a surfactant and a membrane leads to changes in membrane permeability, and in extreme cases, can even result in membrane solubilization. The most obvious potential consequence is interference with material transport and, ultimately, cell damage. A further danger is the fact that surfactants can significantly affect the resorption of other chemicals.

Proteins form adsorption complexes with both anionic and cationic surfactants. Complex formation is often a consequence of polar interactions between the hydrophilic residue of a surfactant and charged sites on the protein molecule, but hydrophobic interactions can also play a role. Such complex formation results in protein denaturation, which in the case of an enzyme implies a reduction or even total loss of catalytic activity, corresponding to a change in metabolic function [667].

Nonionic surfactants are characterized by their lack of strongly polar functional groups. Thus, compounds of this class rarely cause protein denaturation. They can induce a limited amount of protein solubilization, but the concentrations required for the appearance of harmful effects are usually greater than those of ionic surfactants.

Transport through skin is an important consideration with respect to detergent ingredients, which tends to be quite low for anionic and cationic surfactants [668], [669]. Transport is greater for nonionic surfactants, but again the amount of material capable of entering an organism by this route is so low that it can be essentially disregarded as a potential hazard [670]. By contrast, both anionic and nonionic surfactants can be readily resorbed through the gastrointestinal tract following their ingestion, whereas the intestinal resorption of cationic surfactants is low [671]. Even absorbed surfactants are relatively harmless because they are rapidly metabolized, mainly by β- or ω-oxidation of alkyl chains and some cleavage of poly(alkylene glycol) ether linkages. Elimination occurs through the bile and the urine; significant accumulation within the body has never been demonstrated.

The ability of surface-active agents to emulsify lipids means that repeated or prolonged exposure to surfactant-containing solutions can cause damage to the lipid film layer that covers the skin surface. As a consequence, the barrier function of the lipids is impaired, leading to increased permeability and loss of moisture. This is evidenced by dryness, roughness, and flaking of the skin. Very prolonged exposure to concentrated surfactant solutions can lead to serious damage and even necrosis. Skin tolerance varies widely among the compounds making up each class of surfactants. Nevertheless, one can generalize that tolerance to surfactants tends to increase in the order cationic, anionic, nonionic materials.

Broadly speaking, all of the commercially relevant surfactants are well-tolerated in the concentration ranges applicable to detergent use [672], [673]. However, certain structure – activity relationships have been found. With nonionic surfactants, for example alcohol ethoxylates, skin irritation diminishes with an increasing degree of ethoxylation [674]. For anionic surfactants, the length of the alkyl chain is closely related to skin irritability. Thus, within a given homologous series, compounds containing saturated alkyl chains of 10 – 12 carbon atoms show the strongest effects [675].

The eye is much more sensitive than skin to damage by small amounts of surfactants. Anionic surfactant solutions with a concentration above ca. 1 % can produce minor eye irritation, although this is normally reversible [675]. Serious damage is likely only if the eye comes in direct contact with a concentrated surfactant solution and if this contact is not followed by immediate and intensive flushing with water.

Virtually all surfactants that are employed in detergents have been the subject of extensive investigation with respect to their allergenic properties. In no case has an increased risk of allergy for the consumer been demonstrated. A few isolated examples of sudden, multiple allergic reactions in connection with anionic surfactants (alkyl ether sulfates or α-olefinsulfonates) were later traced to impurities of 1,3-sultones and chlorosultones, compounds that have long been recognized as potent allergens and that can arise by bleaching of surfactants with hypochlorite at low pH [676].

The acute oral toxicity of surfactants is low; LD_{50} values normally fall in the range of several hundred to several thousand milligrams per kilogram of body weight [677]. The major detrimental effect of surfactants is damage to the mucous membranes of the gastrointestinal tract. High doses lead to vomiting and diarrhea. If a surfactant reaches the circulatory system, it can cause damage even in very low concentration as a result of interactions with erythrocyte cell membranes, ultimately resulting in destruction of the cells (hemolysis). Inhalation of dust or aerosol containing high concentrations of surfactants can interfere with pulmonary functions [678]. This interference can be partly attributed to interactions with the surface-active film of the vesicles of the lung [679].

The possibility of chronic toxicity has been a subject of intense investigation with members of all of the surfactant classes. Tests with experimental animals involving exposures for up to two years in dosage ranges of several thousand parts per million have without exception shown complete safety [672], [673], [680]. Moreover, a number of investigations have been conducted over long periods of time with human volunteers. Substantial amounts of both anionic and nonionic surfactants were administered, but no serious side effects were discovered [677].

Neither long-term oral ingestion nor continuous skin exposure has ever suggested that surfactants within the three major groups possess carcinogenic activity [672], [681]. Similarly, surfactants can be regarded as completely devoid of both mutagenic [682] and teratogenic [672], [681] characteristics.

11.1.2. Builders

The most important substances in this category are sodium triphosphate, zeolite A, and sodium nitrilotriacetate (NTA).

Pentasodium triphosphate can be regarded as innocuous. Effects of triphosphate ingestion first become acute with dosages of several grams per kilogram of body weight, and even these manifestations are simply a result of localized damage caused by the high alkalinity of concentrated phosphate solutions [683].

Zeolite A has undergone intensive toxicological investigation and has been found to possess absolutely no acute toxicity. The compound is also tolerated well locally. No adverse effects are seen with eye tissue beyond those associated with the presence of any foreign body. Studies on chronic toxicity or carcinogenicity following oral ingestion have also shown zeolite A to be completely safe when used in detergents, even considering long-term use of the products. Inhalation of the material leads neither to silicosis nor to the development of tumors. Isolated reports of alleged carcinogenic activity have been associated with synthetic zeolites, but these have been shown to be related to the presence of natural zeolites possessing a fibrous morphology. The latter are indeed capable of inducing tumors, unlike zeolite A, which consists of cubic crystals. In summary, zeolite A appears to present no health risks [684].

NTA shows little acute toxicity with respect to both inhalation and oral ingestion. Moreover, the compound produces minimal adverse effects on the skin even after repeated exposure, and it has no sensitizing properties. NTA is readily absorbed through the gastrointestinal tract, but it is then quickly eliminated without being metabolized. Numerous investigations support the conclusion that NTA is not a mutagen. Results from animal studies have shown that repeated applications of large amounts of NTA can cause changes in the urinary tract, including tumor formation, but these changes seem to be due to alterations in the metabolism of certain divalent cations by interaction with the complexing agent NTA. This effect is not considered relevant to conditions likely to be found when NTA is used in detergents. The possibility cannot be excluded that use of NTA could lead to trace amounts in drinking water. Nevertheless, concentrations to date have never been found to exceed a few micrograms per liter, and toxicological data suggest that this presents no human health risk [685] – [690].

11.1.3. Bleach-Active Compounds

Sodium perborate and sodium percarbonate show only slight acute toxicity. Extensive contact with concentrated solutions can cause irritation to the skin and mucous membranes, but this is a consequence of alkalinity rather than a specific effect of the substances themselves. The rapid hydrolysis of the bleaching agent means that any physiological effects that are observed will be those of borates or carbonates, respectively. While it is true that these substances pass readily through the mucous membranes or through damaged skin surfaces, intact skin prevents excessive amounts from being absorbed. However, the intestinal tract is capable of admitting substantial amounts of boric acid, and the substance is considered lethal to adults at a dose of 18 – 20 g [691]. Recent studies have revealed reprotoxic effects in animals treated with borate [692]. However, with respect to the exposure situation of normal use of detergents a human health risk can be excluded.

11.1.4. Auxiliary Agents

Enzymes. The proteolytic enzymes employed in detergents are quite safe from a toxicological standpoint. Evidence of local skin irritation appears to be rare, as are systemic effects following ingestion. As with all proteins, however, human allergic reactions are possible. Indeed, exposure to industrial dust in the early days of the production of enzyme-containing detergents resulted in a number of asthmatic reactions among factory workers. This risk has been eliminated by drastic reduction in the enzyme concentration of industrial dust and by the introduction of coated enzyme granulates, prills, and extrudates.

The possibility that a consumer might inhale significant amounts of enzyme from detergent dust can be essentially excluded. Thus, the only plausible and relevant form of consumer exposure is skin contact in the course of dealing with hand-washable items. At one time a definite relationship was assumed between skin reactions and use of enzyme-containing detergents, but a large body of evidence now shows the contrary. It appears safe to say that the presence of enzymes does not contribute to skin irritation by detergents, nor does it lead to increased risk of allergenic reactions [693] – [695].

Fluorescent Whitening Agents. Much information is now available that supports the conclusion that the fluorescent whitening agents employed in detergents are safe materials. All studies on toxicity, teratogenicity, mutagenicity, and carcinogenicity are consistent in affirming that such compounds present no risk. Local tolerance of the substances is also good, and they display no tendency to cause sensitization. Moreover, long-term studies involving cutaneous application have failed to reveal any carcinogenic potential. Fluorescent whitening agents bind rather tightly to proteinaceous material; as a result, they are firmly retained by the upper keratin layer of the skin, ruling out any significant percutaneous transport. The correspondingly firm binding to textile fibers implies that risks of exposure by way of clothing can also be disregarded. In other words, all available toxicological and dermatological evidence supports the belief that low-concentration detergent use of fluorescent whitening agents presents no human health risk [696], [697].

11.2. Finished Detergents

Appropriate choice of ingredients has enabled manufacturers to assure the virtual absence of health risks associated with detergent use. Potential long-term and systemic risks, such as mutagenicity, carcinogenicity, allergy, etc., all are examined and eliminated in the course of the thorough evaluations that routinely accompany raw material considerations during product development. Nevertheless, the possibility remains that finished products could reveal harmful characteristics despite the safety of the raw materials. Acute toxicity and localized reactions are both potential areas of concern, because complex mixtures occasionally show behavior in these respects different from that predicted for the sum of the components.

In fact, the usual commercial detergents display only minimal acute toxicity. Oral ingestion leads usually to vomiting, and this may be accompanied by irritation of the mucous membranes of the gastrointestinal tract and diarrhea. Low toxicity and the fact that vomiting is virtually unpreventable after ingestion of large quantities of detergent essentially exclude any serious threat of poisoning through product misuse, even with respect to children.

Skin irritation is also practically ruled out for modern detergents, provided they are used as directed. Nevertheless, the surfactants that are present and the alkalinity of detergent solutions cause natural lipids to be removed from exposed skin, which can result in minor irritation. Most reported instances of skin irritation can be attributed to extreme levels of exposure coupled with inadequate skin care.

Similarly, no risks appear to be associated with the use of laundry and clothing that has been treated with modern detergents. Only very small amounts of the detergent components (e.g., surfactants, zeolites, fabric softeners) remain on the fabric after rinsing [698]. Of these minor remnants again only traces can be transferred to the skin, amounts well under the limits of local or systemic toxicity [698], [699].

11.3. Conclusions

The constituents of modern detergents can be regarded as toxicologically well characterized and safe for the consumer. The occupational safety concerns that remain in the production process center around surfactants, alkaline components, and the local effects these elements can have on the skin and mucous membranes. Detergent enzymes present the potential for allergenic problems, but the risk is minimal if suitable protective measures are implemented.

From the standpoint of consumer protection, the standard commercial detergents are entirely safe. Even inadvertent misuse of the products can be expected to lead to only minor effects, which are readily reversible.

12. Textiles

A dominant characteristic of the current textile market is the large range of available products. Moreover, essentially all of the fiber types that are worked into textiles for everyday use lend themselves to washing with water — although not all fabrics and finished goods are equally able to withstand high wash temperatures, the alkaline medium, and vigorous mechanical action. Of course, articles that cannot be regarded as washable require dry cleaning; examples include wool suits and dresses, silk neckties, and various highly ornamented items made from a variety of different materials. Nevertheless, the overwhelming majority of fabrics and knitwear is washable and is washed — in some cases very frequently. Proper washing conditions vary from fabric to fabric, depending on a number of parameters.

Questions regarding washability and, in particular, choice of optimum washing conditions (e.g., selecting the proper cycle with an automatic washing machine), can no longer be resolved simply by determining what type of fibers is involved. Indeed, it has become increasingly necessary to take into account such factors as the types of dyes present and finishing operations to which a fabric may have been subjected.

Dyeing and Printing. *Dyes for Cellulosic Fibers.* All cellulosic fibers (cotton, linen, viscose, etc.) resemble one another in their properties and can all be dyed with the same materials and by similar procedures. What follows is a description of the major types of dyes appropriate to these fibers.

Direct dyes are transferred onto the fibers directly from a hot dye bath. Materials that have been subjected to direct dyeing are quite capable of withstanding a gentle wash cycle at 40 °C, although subsequent treatment with special chemicals can further improve both wash- and water-fastness.

If a textile color is to withstand hot-water washing (60 °C) and if a direct dye is to be used, the dye must be carefully selected and a special treatment (fixation) must be administered to increase fastness. Alternatively, dyes can be chosen that lend themselves to after-coppering. In any event, dyes must be chosen from the appropriate fastness grouping at the time of original dyeing.

If a dyed cellulosic fabric is to be subjected to particularly high demands for fastness, a dye of an entirely different type is commonly selected. Among these are vat dyes, reactive dyes, or dyes in the naphthol, phthalocyanine, or sulfur categories. Reactive dyes are capable of forming chemical bonds with cellulose molecules, which display exceptionally high degrees of water fastness.

Dyes for Wool and Silk. Only relatively few such dyes are sufficiently versatile for application to wool and silk. Wool is normally treated with acid dyes, at least in the absence of extraordinary requirements for fastness. Metal-complex dyes have been specially developed for the treatment of wool, and they possess unusually good fastness properties.

Dyes for Synthetic and Cellulose Acetate Fibers. Synthetic and cellulose acetate fibers are generally treated with disperse dyes or, in some cases, pigment dyes. The insoluble coloring agent present is subsequently fixed to the fabric with the aid of heat-setting resins.

The printing of textiles entails the use of essentially the same coloring agents that are employed in dyeing; a separate discussion of this process is therefore unnecessary. In any case, a crucial step in dying and printing of textiles is a special treatment to maximize color fastness. If the color fastness is insufficient, the dyes will severely fade upon washing. The lower the fastness, the more easily colors will fade, independent of the composition and ingredients of the detergents used. Low quality colored textiles may already fade when only water and mechanical action are applied to them.

Finishing. Consumers have come to expect fabrics emerging from washing with relatively few wrinkles. Wrinkles that remain should disappear quickly or easily be ironed off without the need for special treatment. Meeting these expectations has required manufacturers to develop special finishing methods involving the use of special resins. Appropriate water-soluble resin precondensates are applied to the raw fabric, which are induced to condense during a subsequent drying process.

With the assumption that washing is conducted under the proper conditions, resin finishing assures that the original crease angle is largely restored as the fabric dries, thereby eliminating the need for ironing. Furthermore, garments treated in this manner retain their form as they are worn. Resin-finished fabrics are frequently called permanent-press, easy-care, wash-and-wear, or no-iron fabrics.

Fabric finishes tend not to be permanent: the benefits they confer are gradually lost as a garment is subjected to repeated washing. Certain finishing characteristics can be restored by laundry aftertreatment aids, however, particularly softness, stiffness, or antistatic properties.

In some countries (e.g., the USA and Canada), flameproofing is another form of fabric finishing commonly applied to household goods. Items that have been so treated are generally identified by special labels. Strict adherence to manufacturer's washing instructions is necessary if lack of flammability is to be maintained.

A recent form of fabric finishing is the application of sun protection for special textiles such as garments for leisure, sports, and children's wear. Sun protection for fabrics is considered particularly valuable for regions with increased UV radiation due to ozone depletion in the stratosphere. Sun-protection finishings are commercially available, and are permanent and washfast [700], [701].

Washable Fabrics. Textile demand for washable fabrics in Europe (i.e., in the countries comprising the EU) for 1994 is outlined in Table 52 [702]. Three types of material can be seen to account for most of the sales: cotton (47 %), wool (9 %), and synthetic fibers (42 %). Attention is drawn to the fact that by far the largest part of synthetic fibers for domestic use is blended with natural fibers such as cotton and wool. Pure synthetic fibers are used only to a lesser extent for apparel and home furnishing

Table 52. Fiber consumption in the domestic production of the most important washable textile articles in the European Union 1994* (1000 t)

Textiles	Total	Wool	Cotton	Cellulosics	Polyamide	Polyester	Acrylics	Others
Apparel								
Workwear	65.3		41.9	1.3	1.3	19.9	0.3	0.6
Skirts	61.8	11.6	17.5	5.5	0.8	23.4	2.1	0.9
Long trousers	70.0	8.7	40.8	1.9	1.7	15.0	1.2	0.7
Sweaters	251.0	66.5	40.9	3.3	6.1	18.1	115.2	0.9
T-shirts and sweater shirts	54.9	1.9	40.7	3.8	0.4	5.3	2.6	0.2
Blouses and shirts	81.8	0.3	43.1	9.8	0.6	26.5	0.9	0.6
Underwear	66.6	0.1	54.1	1.6	4.2	5.4	0.9	0.3
Night wear	37.8	0.5	24.0	0.8	2.6	7.8	1.8	0.3
Footwear	139.0	9.3	49.4		66.5	1.0	8.6	4.2
Brassieres and foundation garments	9.0	0.3	0.8		4.8	1.3	0.2	1.6
Home furnishing								
Towels	71.8		70.3	0.9	0.2	0.1		0.3
Bed linen	181.0		132.9	4.5	0.9	39.2		3.5
Blankets	47.7	12.9	3.1	1.9	0.7	2.0	25.1	2.0
Net curtains	67.4		7.2	2.3	0.2	51.9	5.6	0.2
Table linen	21.5		12.0	1.4	0.3	3.6	1.2	3.0
	100 %	9.1 %	47.1 %	3.2 %	7.5	18.0 %	13.5 %	1.6 %

* Excluding Denmark, Ireland, and Greece.

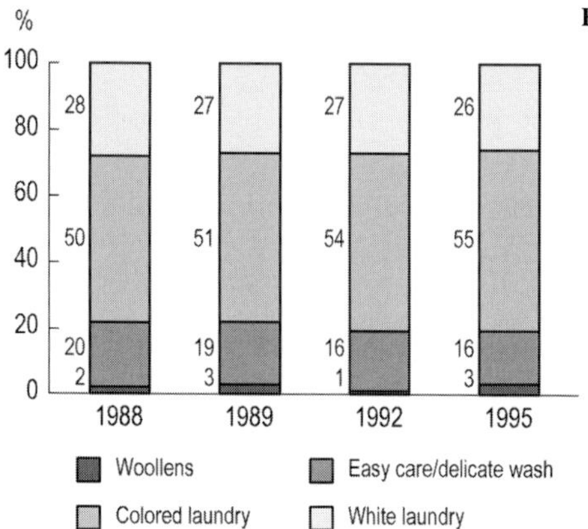

Figure 103. Proportions of wash loads in Germany (%)

textiles except upholstery and curtains. Table 52 also reveals the distribution of these fabrics over the principal categories of household washables.

In Western Europe, the proportion of white goods in a typical laundry has been quite constant since 1988. This trend is illustrated for Germany in Figure 103. These data can be extrapolated to reflect generally the situation in the European Community as a whole.

Table 53 shows the distribution of colored and white washables in the USA. The proportion of white fabrics in Japan is somewhat larger, it accounted for 35 % of total

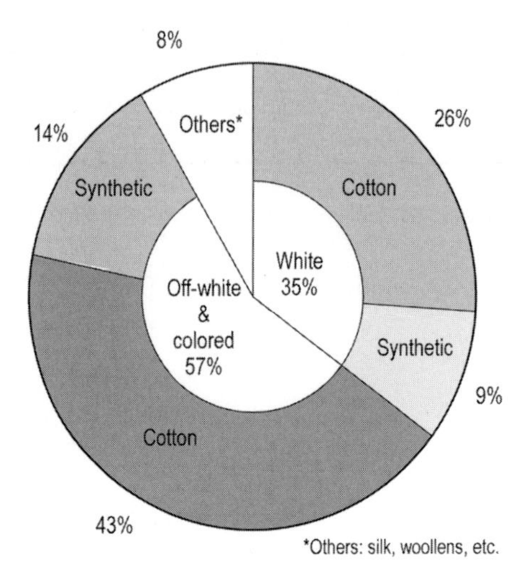

Figure 104. Distribution of white and colored washables in Japan

Table 53. Distribution of colored and white fabrics in the USA

Fabric color	Percentage			
(% white)	1993	1995	1997	1999
Only white (100 %)	18	17	18	17
Mostly white (75 %)	12	13	12	12
Mixed white (50 %)	21	19	18	18
Mostly colored (25 %)	16	17	16	17
Only colored (0 %)	33	34	36	36

laundry in 1994 (see Fig. 104). The washable fabrics comprise home-washed domestic laundry including clothing, towels, bed sheets, etc. [703].

Washing Conditions. Washable woven and knitted fabrics are normally washed either in the household, in a launderette, or by a professional laundry. The distribution among the three varies widely from country to country. In Germany, for example, more than 90 % of household laundry is done at home, usually in a horizontal-axis drum-type machine. In Europe generally, wherever the percentage of households with washing machines is high, the fraction of the laundry done at home is also high. A major reason for this pattern is the substantial reduction of effort brought by the technological advances incorporated in current automatic washers. The process has become even simpler with the advent of wash-and-wear fabrics. The finishing work associated with a load of laundry currently entails little more than folding and putting away the washed and dried articles (Fig. 105).

Establishing proper washing conditions — temperature, time, mechanical input, wash liquor ratio, and detergent — requires consideration of the characteristics of each of the materials that make up a given article of laundry. For this reason, washability is a characteristic that can only be measured with respect to finished goods. However,

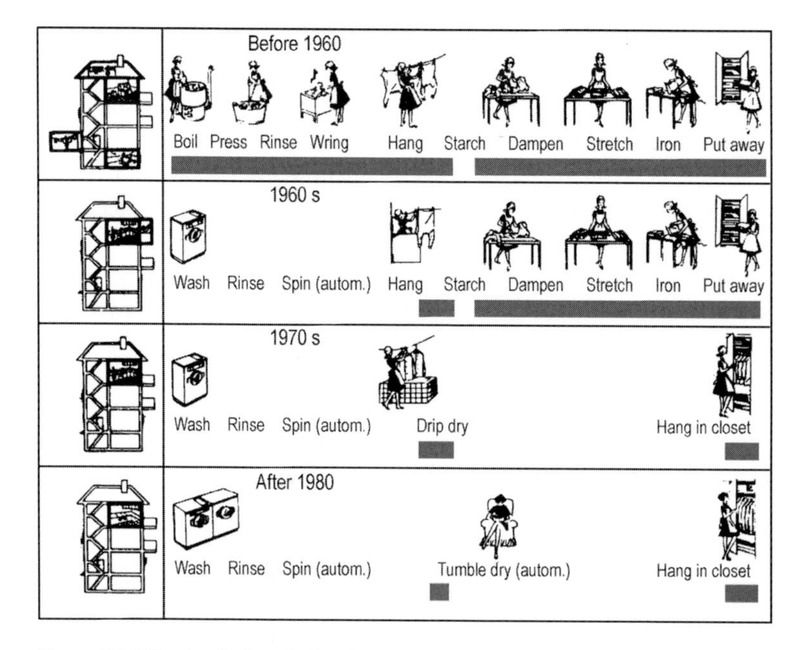

Figure 105. Effort involved in the laundry process in Germany [704]
The reduction in energy expenditure accompanying various stages of technological development through the last 50 years is symbolized by the length of the black bars under the various pictures. The more automated the process and the more nearly the fabrics show "easy-care" ("wash and wear") characteristics, the less the effort required.

conditions imposed during washing, drying, and ironing differ, depending on whether the process is carried out at home or in a commercial laundry. Differences occur in times required for washing and drying, the stress imposed in conjunction with hot-air drying, and the pressure and temperature associated with ironing.

Throughout Europe, white and colorfast cottons are usually washed with a heavy-duty detergent and a normal machine program, i.e., at 60 – 95 °C with a low wash liquor ratio and a standard amount of mechanical input. Cottons that are not colorfast are usually washed at 40 – 60°C, frequently with color detergents. Pastel coloreds often require the use of specialty detergents or heavy-duty detergents designed for colored fabrics; their main advantage is the absence of fluorescent whiteners and bleaching agents.

Woolens and silks are regarded as especially sensitive to washing conditions. In some cases, these fabrics must simply be accepted as non-washable and submitted to dry cleaning. The surface of a wool fiber has a unique "scaly" construction; this peculiar morphology contributes to wool's characteristic tendency toward *felting*. This term is used to describe the appearance of wool after mechanical action has caused individual fibers to become entangled with one another. Wool can be washed much more successfully if it is first given a special treatment known as an antifelting finish. Woolens and silks should always be washed at low temperature (maximum 30 °C), using a high wash liquor ratio and a reduced level of mechanical input. The detergent employed should be either designed for easy-care fabrics or else a product specifically intended for wool.

Synthetic fabrics made from regenerated cellulose fibers (e.g., rayon staple or artificial wool), require a wash temperature of 60 °C, but if sensitive colors are present, the temperature should be reduced to ca. 40 °C. A high wash liquor ratio is recommended, as is a limited amount of mechanical input. Cellulose acetate is generally washed at 40 °C.

Pure synthetics are best washed using a gentle program with a high wash liquor ratio. A heavy-duty detergent is considered appropriate, as is one designed for easy-care fabrics. In some cases a specialty detergent for colored fabrics is necessary (e.g., with uniformly colored pastels, when optical brighteners are to be avoided).

White polyamide fabrics can be washed at 60 °C, colored polyamide fabrics at 30 or 40 °C.

White polyester fabrics are washed at 40 or 60 °C, whereas colored polyesters can only withstand a wash temperature of 40 °C.

Knitware made of polyacrylonitrile is treated like wool, i.e., it is washed in cold water (or at a maximum of 30 °C). White polyurethane-containing undergarments can be washed at 40 – 60 °C, and colored items at either 40 or 60 °C, depending on the degree of colorfastness.

Many of the easy-care fabrics on the market consist of blends of two or more fiber types brought together as the material is woven or knitted. Such blends are treated like easy-care fabrics, using a high wash liquor ratio and gentle agitation. Recommended wash temperatures vary, depending on the nature and composition of the blend. In general, the recommended laundering conditions are those that apply to the major constituent. Wool/polyacrylonitrile or wool/polyester blends are always washed at low temperature (maximum 30 °C), whereas with permanent-press cotton/polyester blends, 60 °C is appropriate.

Items that consist of two or more types of fabric or yarn (e.g., linen tablecloths with wool embroidery, cotton shirts with polyamide piping, jackets with an easy-care exterior and a standard rayon lining, etc.) require different treatment than blended fabrics. The same applies to materials containing dyes of varying degrees of fastness. Such items must always be washed, dried, and ironed under the conditions applicable to the most sensitive fabric or color present, even if this fabric constitutes only a small portion of the whole. For example, a cotton tablecloth with a small amount of wool embroidery requires the same treatment as if it were made entirely of wool. Table 54 provides a summary of the washability of various fibers.

Care Labeling. The European Community (EC), acting in the interest of consumers, has developed a uniform and comprehensive system of care labeling for textile goods. This system has required the cooperation of textile fiber manufacturers, the textile and detergent industries, and representatives of the laundry and dry-cleaning trades. The results are illustrated in Figure 106 [705]. Labels of the approved type are permanently affixed to items of clothing prior to sale, providing the consumer with a straightforward

Table 54. Washability of textiles

Fibers	White		Colored		Pastel colored	
	Temperature, °C	Bath ratio *	Temperature, °C	Bath ratio *	Temperature, °C	Bath ratio *
Natural fibers						
Cotton	95	low	40, 60, or 95 **	low	95 or 60	low or high
Linen	95	low	40, 60, or 95 **	low	95 or 60	low or high
Wool	cold, <30	high	cold, <30	high	cold, <30	high
Silk	cold, <30	high	cold, <30	high	cold, <30	high
Chemical fibers (cellulosics)						
Rayon staple	60	high	60 or 40	high	60 or 40	high
Acetate	40	high	40	high	40	high
Chemical fibers (synthetics)						
Polyamide	60	high	30 or 40	high	30 or 40	high
Polyester	30–60	high	30 or 40	high	30 or 40	high
Polyacrylo-nitrile	cold, <30	high	cold, <30	high	cold, <30	high
Polyurethane	40–60	high	40–60	high	40–60	high

* Low: bath ratio 1 : 5 and normal mechanical input; high: bath ratio 1 : 20 to 1 : 30 and decreased mechanical input.
** Colorfast items.

guide to proper washing and handling. The symbols themselves are internationally protected trademarks, the rights to which are held by the Groupement International d'Etiquetage pour l'Entretien des Textiles (GINETEX). The German trademark rights rest with the Arbeitsgemeinschaft Pflegekennzeichen für Textilien in der Bundesrepublik Deutschland, itself a member of GINETEX. The German organization also confers the right to use these symbols. No license fees are entailed, but the symbols must be used correctly and in their entirety, as specified in the appropriate guidelines. Improper use can lead to loss of the right to employ the symbols. Care labels must be understood merely as sources of advice; they do not in any sense represent evidence of product quality. Nevertheless, any label that appears must be appropriate to the item in question and must be legible and permanently affixed. In NAFTA countries care symbols used are nearly identical to those employed in the EC. Only in a few cases do they differ somewhat.

Laundry Sorting. In Europe, all laundry is customarily sorted on the basis of the care symbols described above. The situation in the USA is rather different, primarily because of the use of different washing conditions (cf. Table 17). In particular, average wash temperatures in North America are significantly lower than in Europe, partly due to the absence of heating units in North American washing machines. Temperature recommendations applicable under these conditions are outlined in Table 55. Sorting of

Figure 106. GINETEX international care labeling symbols

Washing	
Symbol	Washing process
 	– maximum temperature 95 °C – mechanical action normal – rinsing normal – spinning normal
 	– maximum temperature 95 °C – mechanical action reduced – rinsing at gradually decreasing temperature (cool down) – spinning reduced
 	– maximum temperature 70 °C – mechanical action normal – rinsing normal – spinning normal
 	– maximum temperature 60 °C – mechanical action normal – rinsing normal – spinning normal
 	– maximum temperature 60 °C – mechanical action reduced – rinsing at gradually decreasing temperature (cool down) – spinning reduced
 	– maximum temperature 50 °C – mechanical action reduced – rinsing at gradually decreasing temperature (cool down) – spinning reduced
 	– maximum temperature 40 °C – mechanical action normal – rinsing normal – spinning normal
 	– maximum temperature 40 °C – mechanical action reduced – rinsing at gradually decreasing temperature (cool down) – spinning reduced

Washing

Symbol	Washing process
	– maximum temperature 40 °C – mechanical action much reduced – rinsing normal – spinning reduced – do not wring by hand
	– maximum temperature 30 °C – mechanical action much reduced – rinsing at gradually decreasing temperature (cool down) – spinning reduced
	– hand wash only – do not machine wash – maximum temperature 40 °C – handle with care
	– do not wash – be cautious when treating in wet stage

Chlorine–based bleaching (hypochlorite)

Symbol	Process
	– chlorine-based bleaching allowed – only cold and dilute solution
	– do not use chlorine-based bleach

Ironing

Symbol	Process
	– iron at a maximum sole-plate temperature of 200 °C

Washing	
Symbol	Washing process
	– iron at a maximum sole-plate temperature of 150 °C
	– iron at a maximum sole-plate temperature of 110 °C – steam-ironing may be risky
	– do not iron – steaming and steam treatments are not allowed

Chlorine–based bleaching (hypochlorite)	
Symbol	Process
	– dry-cleaning in all solvents normally used for dry-cleaning – this includes all solvents listed for the symbol P plus trichloroethylene and 1,1,1-trichloroethane
	– dry-cleaning in tetrachloroehtylene, monofluorotrichloromethane and all solvents listed for the symbol F – normal cleaning procedures without restrictions
	– dry-cleaning in the solvents listed in the previous paragraph – strict limitations on the addition of water and/or mechanical action and or temperature during cleaning and/or drying – so self-service cleaning allowed
	– dry-cleaning in trifluorotrichloroethane, white spirit (distillation temperature between 150 °C and 210 °C, flash point 38 °C to 60°C) – normal cleansing procedures without restrictions
	– dry-cleaning in the solvents listed in the previous paragraph – strict limitations on the addition of water and/or mechanical action and/or temperature during cleaning and/or drying – no self-service cleaning allowed

Washing	
Symbol	Washing process
	– do not dry-clean – no stain removal with solvents

Table 55. Wash water temperatures in the USA

Water temperatures	Use
Hot, 130 °F (54 °C) or above	white and colorfast pastels heavy and greasy soils
Warm, 90 – 110 °F (30 – 45 °C)	dark, bright, or noncolorfast colors moderate to light soils knits
Cold, less than 90 °F (< 32 °C)	lightly soiled fabrics colors that bleed or fade *

* Heavily bleeding coloreds are to be washed separately from other laundry items.

Table 56. Cloth sorting in the USA

Color	Amount of soil	Fabric and construction
White	heavy	permanent-press and synthetics
Colorfast	normal	towels, jeans, and denims synthetic knits
Noncolorfast	light	delicates

laundry in North America is usually limited to that required for producing entire loads subject to a common wash temperature, degree of agitation, and spin speed, as well as to the same laundry aid treatment (Table 56).

In Japan, laundry is customarily sorted into three categories, defined solely by the degree of soil: light, normal, and heavy.

Fabric Labeling. Current EC guidelines specify that all textile products offered to consumers must be accompanied by a detailed description of their content of different fibers. A "textile product" in this regulation is any item whose weight is made up of $\geq$ 80 % textile raw material.

The guidelines enumerate a very large number of categories covering the entire range of animal, vegetable, and synthetic fibers, including the newest man-made materials. In this context, the term "synthetic" is in itself not considered sufficient and definitive;

more explicit identification of material category is required (e.g., polyester, polyacrylonitrile, polyamide, polyurethane, etc.).

Fiber content is specified on the basis of a percentage of net weight, e.g.:

70 % cotton
20 % polyester
10 % silk

Constituent textile raw materials whose contribution is less than 10 % may be collected under the single heading "other fibers." Thus, a product comprised of 72 % cotton, 7 % polyester, 7 % polyamide, 7 % viscose, and 7 % acetate could be labeled as follows:

72 % cotton
28 % other fibers.

13. Washing Machines and Wash Programs (Cycles)

In contrast to the situation in the USA, the beginning of extensive mechanization of the washing process in Europe was delayed until the late 1950s. A major factor in its ultimate arrival was the development of automatic washers operating on regular home electric current, particularly those whose spin cycle was sufficiently balance-controlled so that the machine would not need to be permanently fastened to the floor.

Automation to some extent has long been present in institutional laundries. This has culminated in systems designed to operate on a production-line basis, resulting in a multifold increase in hourly output per employee and a major reduction in required man power. Nevertheless, only a small fraction of household laundry is washed commercially, primarily because the effort involved in doing laundry at home has been so dramatically reduced by the advent of the household automatic washer.

13.1. Household Washing Machines

13.1.1. Classification

Two basic types of home automatic washers exist: vertical-axis machines and horizontal-axis drum machines. In the Americas and Asia vertical-axis washers of the agitator (Fig. 107 [706]) and the related impeller (also called pulsator) types (Fig. 108 [707]) dominate. They both lack internal heating facilities and therefore must be connected to an external source of hot water if a heated wash liquor is desired. By contrast, this type of machine has lost virtually all of its former significance in Europe,

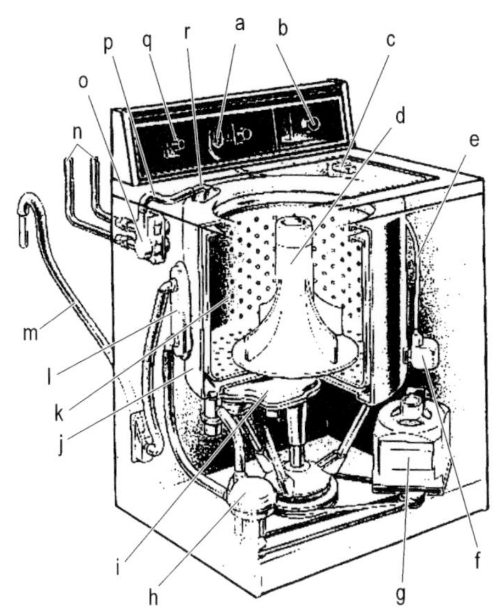

Figure 107. Agitator-type vertical-axis washer
a) Pressure switch; b) Timer, c) Lid switch; d) Agitator; e) Air-pressure tube; f) Air-pressure dome; g) Motor; h) Water pump; i) Transmission; j) Tub; k) Basket; l) Filter; m) Drain hose; n) Fill hoses; o) Water inlet mixing valve; p) Inlet hose; q) Temperature dial; r) Water inlet

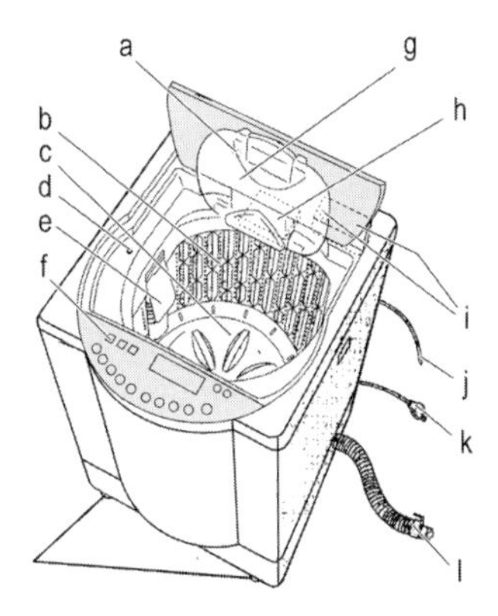

Figure 108. Impeller- (pulsator-) type washing machine
a) Water supply valve; b) Basket; c) Agitator; d) Spirit level; e) Lint filter; f) Control panel; g) Inside lid; h) Detergent and fabric softener dispenser; i) Water supply pump; j) Earth wire; k) Cord and plug; l) Drain hose; m) Leveling feet

where it has been replaced entirely by the horizontal-axis drum-type washer (Fig. 109 [708]).

Agitator Machines. Most American agitator-type washing machines mostly have an enameled metal laundry tub. Inside the tub is an open, perforated vertical axis basket, usually of a size suitable for holding up to 5 kg of laundry.

The laundry tub is filled with 50 to 75 L of water, depending on the wash program chosen and the type and weight of the load (medium or large load). Manufacturers recommend that water and detergent be introduced into the machine first, followed by the laundry, which should always be distributed as evenly as possible. Machines of this type normally have two water connections, one for hot water and one for cold. The desired wash temperature (hot, warm, or cold) is selected by proper setting of the

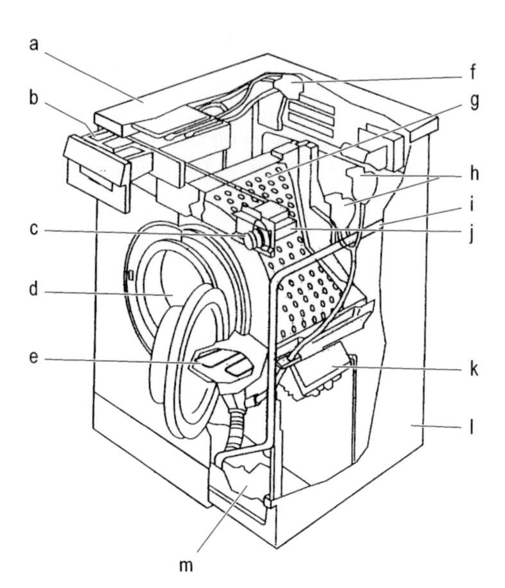

Figure 109. Horizontal-axis drum-type washer
a) Top; b) Dispenser drawer; c) Program control knob; d) Front door; e) Heating coil; f) Magnetic valves; g) Drum and tub; h) Water level control; i) Drain hose; j) Program control; k) Driving motor; l) Casing; m) Drain pump

controls, and the selected temperature is achieved by automatic mixing of water from the two input sources. Sodium hypochlorite or oxygen bleach may be added separately by hand during the wash cycle, although quite a few machines are equipped with automatic dispensers for the purpose.

After the desired water level has been reached, the agitator is set to function, and the laundry is jostled for ca. 8 – 18 min by the rhythmic motion of the agitator (Fig. 107). When the wash cycle is complete, the used wash liquor is pumped off. American machines that are set for the regular cycle then begin a series of spins and spray rinses, whereby fresh cold water is sprayed on the load during spinning to remove residual wash water. Water is again allowed to fill the tub for a brief rinse (ca. 2 min), after which it is once more drained off and a final spin drying occurs.

Impeller (Pulsator) Machines. These fully automatic washing machines are used in Japan, China, and other Asian countries. They use a rotating ribbed disk mostly mounted at the bottom of the tub as their source of mechanical input (Fig. 108). Japanese machines have a capacity of 3 to 8 kg of laundry.

Horizontal-Axis Drum Washing Machines. In automatic drum-type washers, the laundry is placed in a horizontal-axis perforated drum which rotates in a tub in alternating directions (Figs. 109 and 110). For a regular cotton wash, only the lower quarter of the drum is filled with wash liquor (bath ratio ca. 1 : 4), which means that, in contrast to agitator machines and pulsator washers, the laundry is only partly submerged. Individual laundry items are repeatedly lifted by paddles located at the inner wall of the drum, each time falling again and tossed into the wash liquor for renewed soaking and rubbing. The wash liquor is heated internally either by means of electrical

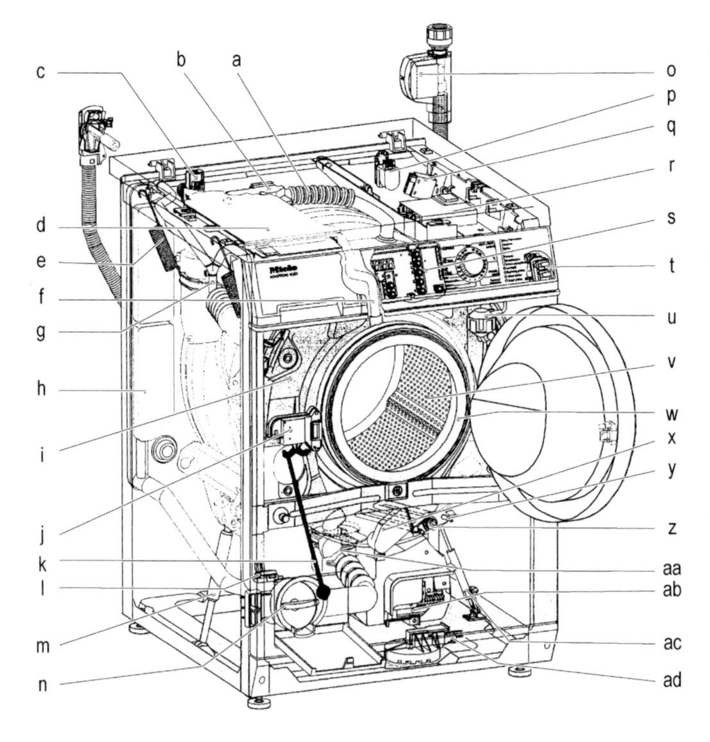

Figure 110. Front loading washing machine
a) Suds container air vent; b) Water path control system; c) Water Control System valve; d) Detergent dispenser; e) Suspension springs; f) Water intake via door glass; g) Drain pump vent/Non-return valve; h) Non-return valve; i) Counterweight; j) Door lock; k) Emergency door release; l) Drain pump; m) Hydraulic shock absorber (Only on models with loading level indication); n) Fluff filter housing; o) WaterProof System valve; p) Interference suppression filter; q) Heater relay; r) Power module; s) Control module; t) Pushbutton switch unit; u) Imbalance sensor (Only on models with loading level indication); v) Inner drum; w) Door seal; x) Air trap; y) Heating; z) NTC resistor; aa) Ball-and-seat valve; ab) Motor; ac) Load sensor (Only on models with loading level indication); ad) WaterProof System switch.

heating coils located at the bottom of the tub or by a continuous flow heater. A relatively small number of automatic drum-type washing machines are equipped with a separate hot-water inlet for use in households that have access to an economical external hot-water source. In this case the bath is often tempered initially to 30 – 40 °C by introduction of cold water, thereby preventing proteinaceous soils and stains from being thermally denatured and set in the fabric. Subsequent heating can then take place if this is required by the particular automatic cycle selected.

Laundry is loaded either from the top through an open lid in the side of the drum (top loader) or through its open front (front loader).

Detergent is introduced through a dispenser. In a front loader, the dispenser is usually a device resembling a drawer located at the front of the machine. The dispenser for a top loader is located beneath the lid of the machine. Dispensers permit automatic sequences consisting of either one or two wash cycles (i.e., a prewash and a main wash). Additional compartments are also provided for the automatic introduction of laundry

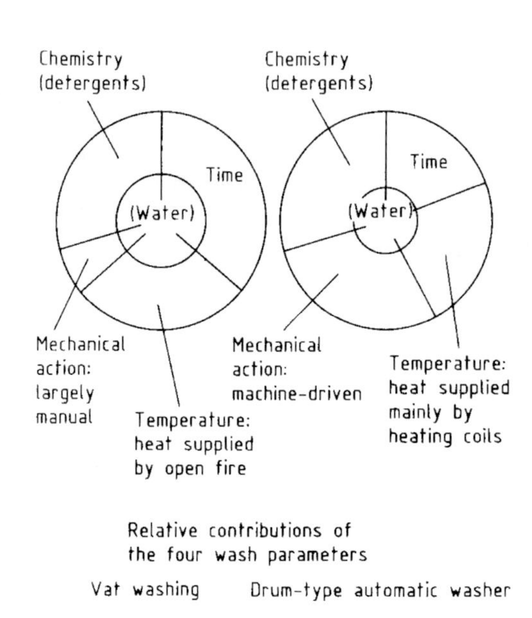

Figure 111. Circular laundry chart (Sinner's circle)

aftertreatment aids and, especially in Southern Europe, additional automatic dosage of a chlorine-containing bleach during the rinse cycles.

Fully automatic drum type washers utilize one drum for successively accomplishing three separate consecutive operations: washing, rinsing, and spin drying. The machines are equipped with special mechanical devices that compensate for load imbalances.The trend toward space-saving machines has led to the development of smaller washers and washers fitted for built-in kitchens.

Modified European-type horizontal axis drum washers have been introduced in the 1990s in the USA and Japan for reasons of improved efficiency versus current washers in terms of water and energy conservation. Both markets indicate increasing shares of these front loaders. The market shares are, however, still very low.

13.1.2. Operational Parameters

Four factors must be regulated in the washing operation: chemistry, mechanical input, wash temperature, and time.

The effect of each factor on wash performance varies, depending on the washing techniques employed. Figure 111 illustrates graphically by a Sinner's circle the rough percentage influence of the individual parameters on the overall process.

The inner circle implies that the effectiveness of the four factors is a consequence of the medium that unites them, i.e., water.

In the days when laundry was done in an open vat, the required amount of water was large, the role of mechanical action was very limited, and time was a very important factor. A washing machine provides considerably more mechanical action; thus, the wash time has been shortened.

Wash Liquor Ratio. "Wash liquor ratio" (or bath ratio) means the ratio of dry laundry (in kilograms) to the volume of wash liquor (in liters). The total amount of wash liquor required in the overall process is made up of two portions: that which is absorbed by the laundry (bound wash liquor) and that which remains in excess (free wash liquor).

Various washing processes have different wash liquor requirements. North American agitator-type washing machines typically use some 25 L of water per kilogram of laundry, i.e., a wash liquor ratio of 1 : 25, whereas Asian pulsator machines require a ratio of ca. 1 : 20 to 1 : 15. Modern horizontal axis drum-type automatic washers are characterized by a low 1 : 4 wash liquor ratio with cotton, but rather high values of up to 1 : 15 for easy-care and delicate fabrics.

Wash Liquor Level and Reversing Rhythm. A drum-type washer is programmed to rotate in a reversing fashion, i.e., first in one direction and then in the opposite. The extent of the mechanical action imparted to the laundry can be altered by changing either the rhythm of reversal or the wash liquor level (or both). A low wash liquor level (wash liquor ratio of ca. 1 : 4) and relatively rapid reversal (e.g., a reversing rhythm of 12 s of rotation/4 s of pause) causes a large mechanical effect on the laundry. In contrast, a high wash liquor level (wash liquor ratio 1 : 15) and slower reversal (e.g., reversing rhythm 4 s of rotation/12 s of pause) provides less mechanical action, hence the term "gentle cycle".

Residual Moisture. Residual moisture is defined as the amount of water remaining in the laundry after draining and spinning. Data is reported as a percentage of the mass of air-dried laundry. Separate spin drying with a vertical drum permits relatively high rates of rotation: up to ca. 2800 rpm. Corresponding residual moisture levels for a cotton laundry load are typically ca. 40 – 50 %. Automatic drum-type machines spin dry the load at rates of 400 – 1800 rpm and result in residual moisture levels for cotton of 100 – 45 %, respectively.

Wash Temperature. In Europe clothes are commonly washed over a wide temperature range of 30 – 90 °C, whereas elsewhere much lower temperatures are used (e.g., in North America up to ca. 55 °C, in Asia and South America only up to 25 °C or, in exceptional cases, 40 °C). The high wash temperatures frequently used in Europe for decades before the 1980s had been based largely on tradition and on the firmly held belief that "only clothes that have been boiled are really clean." Nonetheless, this attitude has been changing as a result of the increasing popularity of colored and permanent-press fabrics, the desire to conserve energy, and the introduction of more effective multifunctional heavy-duty detergents, designed for use at a lower temperature (cf. Section 4.1). Today, the predominant washing temperature in Europe is 40 °C.

Table 57. Predominant machine washing conditions

Machine type	Europe Front loader (drum)*	Japan Top loader (pulsator)**	USA Top loader (agitator)**
Washing time, min	20 – 60	5 – 15	8 – 18
Amount of water, L	8 – 15	30 – 60	75
Bath ratio	4 : 1	10 : 1	25 : 1
Temperature, °C	40 – 60	5 – 25	10 – 40
Water hardness	medium/high	very low	low/medium
Detergent dosage, g/L	5 – 10	0.5	1 – 2

* Horizontal-axis type.
** Vertical-axis type.

Lower temperatures have long been the rule in North America and in Japan, a fact partly explained by the hygienic function associated with use of chlorine-containing bleaches. Nevertheless, the majority of the world's population (Asia, Latin America, Africa) usually washes at tap-water temperature, i.e., cold.

Washing Times. Customary wash times also differ substantially between Europe on the one hand and North America and Japan on the other. Much of the difference is attributable to differences in washing temperatures: relatively low temperatures and the absence of built-in heating facilities in North America and in Japan, as compared to Europe, where a typical washer can heat water to 90 °C. The heating process also consumes a great deal of time. Nearly 40 min are required to raise the temperature of the wash liquor in a drum-type machine to a preset value of 90 °C, not counting the time occupied by a prewash. A wash cycle that includes a prewash entails draining about half of the first wash liquor after it has reached a temperature of about 40 °C at the end of the prewash cycle, which adds ca. 15 min to the process.

The cycles provided for most drum-type automatic washers use the temperature as their time controlling variable. Thus, a high wash temperature dictates a long wash cycle whereas a low temperature requires less time. Meanwhile it has become common for manufacturers to install optional "energy-saving cycles," where a lower wash temperature is compensated for by increasing the wash time. The resulting increase in time and mechanical input accomplishes essentially the same end as a high-temperature wash; the corresponding Sinner's circle reveals that a temperature deficit has been balanced by a gain in time and mechanical action.

Washing machines in the Americas and Asia operate considerably more rapidly. They normally require only 8 – 18 min for the wash cycle, followed by one or two brief rinse cycles. A comparison of machine washing conditions in Europe, Japan, and the USA is given in Table 57 [709].

13.1.3. Wash Programs

Since laundry habits and washing machines vary considerably in different parts of the world, discussion of washing programs under separate headings reflects the three major traditions: those found in Japan, North America, and Europe.

13.1.3.1. Japanese Washing Machines and Washing Conditions

The washing machines found in Japanese households are of the automatic vertical axis impeller/pulsator type. A declining 25 % of the machines were still semiautomatic twin-tub type washers in 1998. They feature a single housing that contains one separate tub each for washing and spin drying, with the impeller/pulsator located in the former. A characteristic of most Japanese washing machines is the absence of a provision for temperature control, since the process is predominantly conducted at tap water temperature.

The most important variables with a Japanese washing machine are the water level, the weight of the wash load, and the extent of mechanical action. Washing time is adjusted to match the degree of soil of the laundry (ca. 5 – 15 min). Table 58 depicts program selections of a typical Japanese washer.

13.1.3.2. North American Washing Machines and Washing Conditions

Although horizontal-axis drum-type machines have gained importance in the U.S. [710] and Canadian markets in the late 1990s, the automatic agitator washing machine still is by far the most common in North American households. Such machines perform a full set of laundry operations in the usual sequence, i.e., wash, rinse, and spin dry. North American machines resemble their Asian counterparts in that they lack heating coils. Therefore, they are designed for connection to domestic supply of both hot and cold water. Three wash temperatures are generally used: hot (55 °C/130 °F), warm (30 to 45 °C/90 to 110 °F), and cold (10 to 25 °C/50 to 80 °F). Rinse temperatures can also be selected, usually from among the two alternatives, warm and cold. One further variable is the water level, which is adjusted according to the quantity of laundry, e.g., medium and large load, and type of laundry to be washed. The mechanical input is also adjustable within limits by introducing changes in the agitator speed and the wash time. The following set of washing programs is typical (cf. Tables 17 and 59):

Table 58. Program selections of a typical Japanese washing machine

Program		Application
Wash Selector	Automatic program	
	Normal	For normally soiled clothes
	Gentle	For lightly soiled clothes
	Heavy	For heavily soiled clothes
	Hand Wash Dry Cleaning	For delicate fabrics with hand wash/dry mark
	Blanket	For blankets, etc.
	Optional program	
	Wash Rinse Spin/Dry	– To set preferred program (only wash, wash & rinse, only spin, only drain, etc.) – To set automatic/original wash program.
Wash Timer		To set washing time
Water Selector	Level	To set water level. If not selected, washer adjust the level automatically
	Bathtub Water	To recycle used wash water for washing/ rinsing. (Tap water is supplied to the last rinse.)
Optional Selectors*	Dosage Selector	For automatic detergent dispensing
	Basket Wash	To remove soap scum from basket using chlorine bleach. (Scum may cause malodor on clothes.)
	Spin Selector	To select spin level, depending on type of garment

* Depending on manufacturer of the washer.

Regular (heavy-duty wash)
Permanent press
Knit/gentle
Woolens
Soak
Prewash

13.1.3.3. European Washing Machines and Washing Conditions

Horizontal-axis drum-type washing machines in which the water can be heated to up to 90 °C dominate the household market in Europe. To meet the demands posed by current laundry, such machines are provided with a wide variety of wash cycles. In choosing the appropriate cycle for a given load of laundry, the first decision is whether or not a prewash is desired. Options in later stages are supported by the presence of a special third compartment in the detergent dispenser, which permits addition of aftertreatment aids to the final rinse. Machines built for the Southern European market

Table 59. Typical wash programs in North America

Wash program	Fabric	Degree of soiling	Wash time, min	Water temperature Wash	Rinse
Regular heavy	cotton, linen	heavy	12–18	warm or hot	cold
		moderate	10–12	warm	cold
		light	8–10	cold or warm	cold
Knits/gentle	synthetic	moderate	10–12	warm	cold
	delicate	light	8–10	warm	cold
Permanent press	Permanent press	heavy	12	warm or hot	cold
	cotton, cotton/synthetic fiber blends	light to moderate	8–12	warm	cold
Woolens	wool	light	4		
Soak	colorfast	heavy	30	cold or warm	none
		stained	30	cold or warm	none
Prewash	colorfast	heavy	6	cold or warm	none
		stained	6	cold or warm	none

commonly offer a fourth compartment, which allows addition of chlorine bleach to the second rinse.

Selection of the proper cycle is accomplished in one of several ways. Some machines are equipped with one control knob for establishing the program and a separate infinitely variable temperature regulation control. Alternatively, both functions can be combined in one knob while other models offer a set of pushbutton controls. Separate temperature control provides the greatest degree of flexibility.

Management of any wash cycle requires a more or less complicated switching system. Very few machines (mostly those in the lowest price category) continue to be equipped with time-based switching, an approach in which each operation is allowed to continue for a precise and predetermined length of time. The principal disadvantage of such a system is that temperature cannot be properly established without taking into account the temperature and volume of the water being introduced and the ambient temperature, all of which can vary.

Modern washing machines are mostly equipped with integrated electronic circuitry based on microprocessors, which may be regarded as microcomputers. These circuits take full functional control of the machine. So-called fully electronic washing machines of this type lack any complex mechanical switching formerly typical of automatic washers.

Modern washers frequently also have so-called fuzzy logic control governing partial processes of the washing program. Fuzzy logics are able to automatically provide, for example, variable speed of reversion of the drum depending on the amount of foam during the wash cycle [711], provide an additional rinse cycle depending on the quantity of residual foam after the regular final rinse, control the quantity of water intake and washing time depending on weight and type of wash load, etc. Fuzzy logic provides for increased fabric care and water and energy conservation.

European washing machines have been reviewed [712].

Today, most garments sold in Europe have uniform labels specifying the use of one of four wash temperatures: 30 °C, 40 °C, 60 °C, and 95 °C (see Figure 106). Table 60 gives examples of the wash cycles commonly provided to meet these needs (cf. also Table 17). Similar care labels are found in USA and Japan.

13.1.4. Energy and Water Consumption

Mechanization of the washing process has relieved much of the burden formerly resting on the person doing the wash. At the same time, however, steady increases in energy and water supply costs have made the use of a household drum-type washing machine increasingly expensive. In particular, the cost increases as the selected wash temperature is increased (Table 61), due to the fact that washers are usually poorly heat insulated.

It is not surprising that efforts to reduce the costs have begun with the 90 °C cycles, since these hold the greatest potential for energy savings (e.g., by lowering the water temperature from 90 °C to 60 °C). Some washers have herefore incorporated an optional energy conservation program. Such cycles are usually designed to function at 60 °C. Examples of typical energy conservation measures are shown in Table 62. (For information on the related trend in the USA toward washing at lower temperatures, see Fig. 68.)

With respect to conserving water and energy, European washers have made considerable progress during the 1980s and 1990s. This has been made possible through automatic control of the amount of water intake depending on the quantity of the wash load, by omitting the former intermediate cool-down rinse at the end of a hot or warm wash cycle, by considerably increasing the number of revolutions per minute for spinning, applying spray-spinning and, last but not least, by reducing the number of rinse cycles. To encourage consumers to use washing machines that operate with lower consumption of electric power and water all new washing machines manufactured in Europe since 1996 have to display the Energy Label (Fig. 112) as per Official Directive of the European Union. The label indicates the efficiency of the appliance in terms of energy and water consumption, washing performance, capacity, etc. [131].

13.1.5. Construction Materials Used in Washing Machines

Materials used in the construction of washing machines must exhibit a high degree of resistance to various mechanical, heat, and chemical influences. The following materials are particularly well-suited to the purpose:

Stainless Steel. The vast majority of drums are made of stainless steel. Chromium steel continues to be widely used for the construction of tubs, heating coils, and thermostat sensors. Wash tubs have been increasingly constructed from plastic in the 1990s.

Table 60. Typical wash programs for a European drum-type automatic washer

Designation of program	Application	Prewash temperature, °C	Main wash temperature, °C	Cycle		Water level		Number of rinse cycles	Final rinse water retainedin drum	Spin drying	
				Regular	Gentle	Low (12 – 15 L)	High (20 L)			Spin before removal	Regular
Normal ("boiling") cycle, 90°C	whites and fast coloreds (cotton, linen, and their blends)	40	85 – 90	+		+		3 – 4			+
Colored fabrics, 60 °C	nonfast colors (cotton, linen, and their blends)	40	55 – 60	+		+		3			+
White easy-care, 40 or 60 °C	white synthetics (including polyamides, polyesters, and their blends with cotton)	40	40 or 55 – 60		+	+		3	+	+	
Colored easy-care or delicate washables, 30 °C	colored synthetics, e.g., polyamides and polyesters; polyacrylonitrile, polyacrylonitrile/ wool, rayon, acetate, and cupro	30	30		+	+	+	3	+	+	
Wool, silk, 30 °C	washable woolens		30/40		+	+	+	2 – 3	+	+	+
Fabric softener	normal and colored fabrics			+		+					+

Table 61. Typical energy consumption for various wash programs with a European drum-type automatic washer

Program	Energy consumption, kWh
90 °C Boiling wash program	1.8
60 °C Colored fabrics	1.2
60 °C Synthetics	1.2
40 °C Synthetics	0.6
30 °C Woolens	0.3

Table 62. Examples of energy conservation programs for European automatic washers

Modification of program	Wash load, kg	Savings in comparison to usual 90°C-programs, %	
		Electricity	Water
No prewash cycle	4	10	11
Reduction of the washing temperature from 90 °C to 60 °C	4	30	–

Sheet Steel. Casings and lids are made from sheet steel with a protective surface (e.g., enamel paint, baked enamel, galvanization). Baked enamel provides the most permanent form of protection.

Plastics. Certain components can be made from alkali-resistant plastics, usually polypropylene. Typical applications include detergent dispensers, lint filters, paddles, operating controls, mechanical parts, and various small components. Glass fiber reinforced polypropylene has also been employed for tub construction in drum-type machines.

Rubber. Molded rubber is the material of choice for components that require a high degree of elasticity. This applies particularly to gaskets, such as those around the door, as well as to various hoses, hose connectors, and shaft seals. Synthetic rubber is used exclusively for such purposes. A drawback of rubber is its tendency to deteriorate with age. Deterioration is accelerated by exposure to grease, by which the rubber gradually acquires a soft and greasy character.

13.1.6. The Market for Washing Machines

The majority of the world population cleans laundry by hand, washing either in bowls, tubs, sinks, or in river waters. Washing machines dominate only in industrialized global regions.

The percentage of households equipped with washing machines varies considerably among the countries of the EU. Percentages for 1998 are as follows [713]:

Austria	88
France	91
Germany	95
Great Britain	93
Italy	99
Sweden	63
The Netherlands	95
Europe, average	89

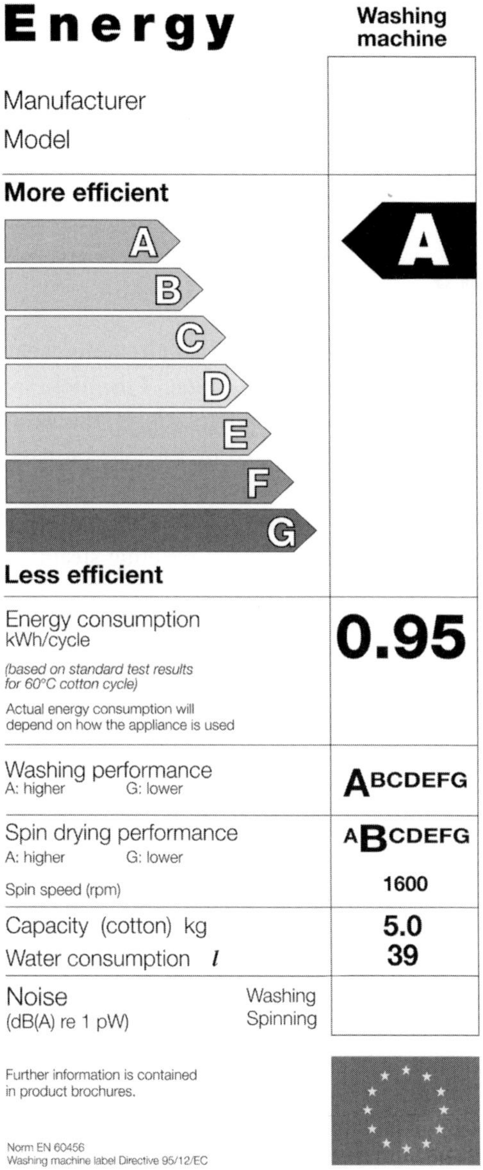

Energy Washing machine

Manufacturer

Model

More efficient

A

B

C

D

E

F

G

Less efficient

A

Energy consumption kWh/cycle *(based on standard test results for 60°C cotton cycle)* Actual energy consumption will depend on how the appliance is used	**0.95**
Washing performance A: higher G: lower	**ABCDEFG**
Spin drying performance A: higher G: lower	**ABCDEFG**
Spin speed (rpm)	**1600**
Capacity (cotton) kg	**5.0**
Water consumption *l*	**39**
Noise (dB(A) re 1 pW) Washing Spinning	

Further information is contained in product brochures.

Norm EN 60456
Washing machine label Directive 95/12/EC

Figure 112. European Union energy label

In 1998, 3.36×10^6 automatic washers were produced in Germany alone. Total European production was more than 16×10^6 units with a total consumption number of some 11×10^6 in 1998 [713]. A report and analysis on washing machines in Europe has been published. It contains detailed statistical figures for all countries of the European Union [714].

Worldwide, the USA represents the largest single washing machine market. Nevertheless, market penetration in the USA is lower than that in Europe and Japan; 80 % of American households owned a washing machine in 1998 [715]. In the same year, the USA manufactured some 7.0×10^6 washers [715].

More than 99 % of the households in Japan are equipped with a washing machine. The Japanese manufactured 4.4×10^6 units in 1999. Of these, 3.8×10^6 were fully automatic, whereas 0.6×10^6 units were of the twin-tub type.

Market penetration of washing machines in Brazil was 28 % of all households in 2000, in China it was 10 %, and in India 7 %[715].

13.2. Laundry Dryers

Laundry dryers represent a further stage in the development of laundry technology. Dryers are machines that utilize an influx of heat to cause evaporation of the moisture remaining in spin-dried laundry. Electricity is normally used to supply the heat, although some of the dryers in the USA and Europe are heated by gas. Several basic types of dryers exist, including those based on circulation of heated air, heat transfer from metal surfaces, and radiant heating. Most are drum-type machines, so-called tumblers. They can be categorized by the way they dispose of the resulting steam:

Air-Vented Dryers. Dryers in this category function on the same principle as a fan. Ambient air is drawn in by suction, passed through a heated zone, and then introduced into the laundry drum. As the air becomes enriched in moisture, it is vented through a lint filter either directly to the room or out of the building through an exhaust hose or duct.

Condenser Dryers. In these dryers the air needed for drying is drawn from the ambience and passed over a heating device before encountering the laundry. Moist air leaving the tumbler drum is cooled by an additional supply of incoming ambient air. This cooling takes place in the presence of a drip screen, which collects precipitated drops of water. Water from the drip screen collects in a condensate vessel which has to be emptied after each drying cycle. The working principle is depicted in Fig. 113 [708]. The most recent developments comprise condenser dryers that use heat generated by a heat pump [716].

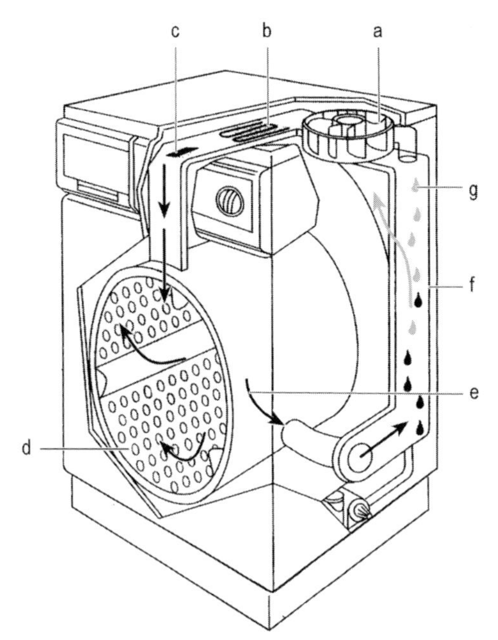

Figure 113. Condenser dryer
a) Ventilator; b) Heater; c) Drying air; d) Tumbler drum; e) Moist air exit; f) Heat echanger; g) Condensing water

Combination Washer–Dryers. The European washer–dryer is a further development of the drum-type washing machine. It is comprised of a washer combined with a water-cooled condensation dryer, and the same drum is used for washing, spinning, and drying. A drawback of these machines is the fact that the small size of the washing drum requires a full load of laundry to be divided into two batches before it can be dried. The heat necessary for drying is supplied either by radiation or by heated air. The market share of these combines in Europe has remained low. In Germany it was only some 4 % in 1999.

Market for Laundry Dryers. The market penetration for laundry dryers in the EU was 27 % on average in 1998 [713]. Percentages for individual countries vary considerably:

Austria	12
France	28
Germany	31
Great Britain	42
Italy	2
The Netherlands	44
Sweden	30

Major manufacturing countries are Great Britain (1.2×10^6 units in 1998), Germany (0.9×10^6), France (0.5×10^6) and Italy (0.3×10^6). The majority (65 %) of the dryers in Europe are air-vented dryers and the remainder (35 %) represent dryers that use condensation technology.

The situation in the USA is quite different from that in Europe. In the USA, 74 % of all households are equipped with dryers, of which some 27 % are gas-heated, the remaining 73 % being electrically heated units. U.S. production of household dryers amounted to 5.2×10^6 units in 1998 of which 23 % were gas-heated dryers [715].

In 1999, some 20 % of the Japanese households operated a laundry dryer. Approximately 45 % of the dryers have a capacity of more than 5 kg of laundry. The largest part of the remaining 55 % consists of units that have a capacity of 4 – 4.5 kg.

13.3. Washing Machines for Institutional Use

Washing machines designed for use by commercial laundries are generally designed to keep operating costs as low as possible. As a result, the more labor-intensive batch laundry systems are being increasingly replaced by more efficient continuous devices. In contrast to European households, where laundry water is heated electrically, commercial laundries worldwide usually employ high-pressure steam as their source of heat. Heat from the steam is led into the wash liquor either directly or indirectly, resulting in more rapid heating and shorter wash times.

13.3.1. Batch-Type Machines

There are many situations even today in which batch-type commercial washing machines are indispensable. This is particularly true where a frequent need exists to process numerous small batches of different kinds of laundry. Batch-type machines require a substantial amount of manual supervision; thus, their commercial use is expensive.

Conventional Drum-Type Devices. Such devices are basically scaled-up versions of household washing machines. Commercial front loaders often allow the processing of up to 400 kg of laundry, with special designs allowing even higher amounts. Reversal and rotation rates commonly produce *g*-factors during washing of 0.7 – 0.9. During spin extraction 400 *g* and higher rates are usual. The large drums require that the effect of mechanical action resulting from great fall heights and excessive weight be moderated and that loading and removing the laundry be simplified. These targets are generally accomplished by introducing into the drum a set of partitions that divide the drum into compartments. Three partitioning schemes (the Pullmann, Y, and star-pattern systems) are illustrated in Figure 114.

Machines of this type are nowadays controlled by microprocessors. Final drying is also performed in large commercial-type equipment.

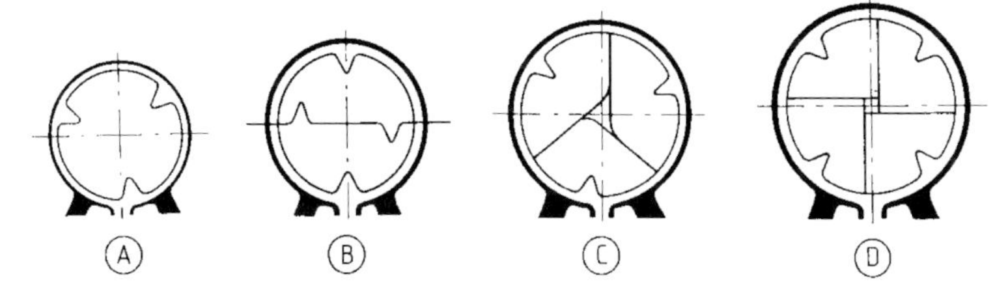

Figure 114. Four common designs for commercial washing machines
A) Open machine; internal drum diameter up to 900 mm; inner drum lacking subdivisions; B) Pullmann type machine; internal drum diameter 900 – 1400 mm; inner drum divided into two equivalent compartments by means of an axial partition; C) Y-type machine; internal drum diameter 1400 – 1700 mm; inner drum divided into three compartments by three axial partitions joined along the axis; D) Star-pattern machine, internal drum diameter > 1700 mm; inner drum divided into four compartments by four axial partitions joined along the axis

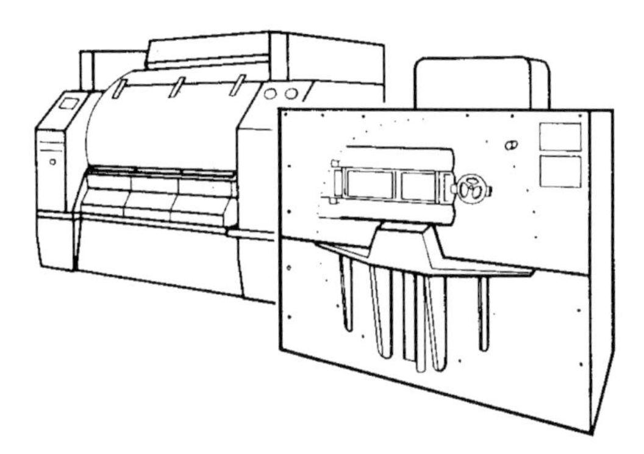

Figure 115. Schematic diagram of a wash extractor

Wash Extractors. Machines in this category, with a capacity of up to 360 kg, are capable of not only the wash step, but also a spin drying and "fluffing" of the laundered items. Spinning occurs after each wash and rinse operation, an expedient that has the effect of minimizing the number of required rinses. This in turn conserves water and reduces overall wash time to ca. 40 min (Fig. 115). Wash extractors are used in all parts of the world. The machine types, however, vary from manufacturer to manufacturer.

13.3.2. Continuous Batch Washers

The development of the continuous batch washer is a consequence of the demand for a laundry system requiring even less manual intervention. Such machines operate uninterrupted and are best suited to institutions faced with large amounts of relatively uniform laundry. Soiled laundry is loaded onto either a conveyor belt or a suspended conveyor system. As it progresses through the machine, the laundry is exposed to a series of zones designed to accomplish the equivalent of wetting, prewashing, main washing, and rinsing. Most of the zones operate on the countercurrent principle: all water encountered by the laundry flows continuously and in a direction opposite to the

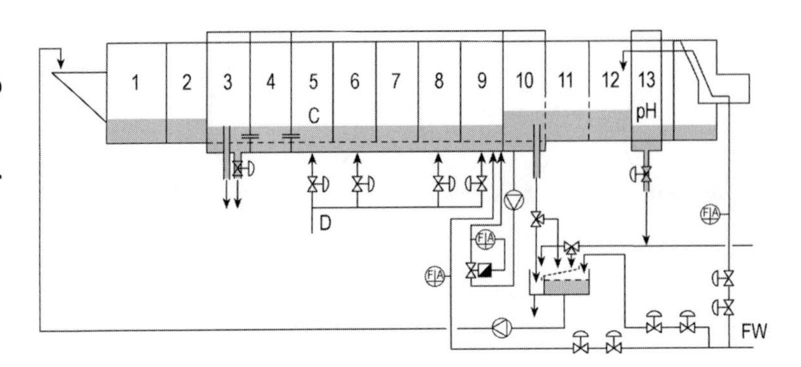

Figure 116. Continous batch washer

motion of the laundry. Each zone leads directly into the next, resulting in significant savings of energy, water, and detergent. Typical water consumption for hospital linen is 8 – 12 L/kg, sometimes even less.

Following the actual washing process, the laundry is directly loaded into an extraction press or a spin extraction to remove the water. The damp goods are then passed to steam or gas heated batch dryers. The entire process occurs without manual intervention. Continuous batch washers are found all around the globe. But especially in Europe high amounts of goods are treated in this type of machines. Figure 116 illustrates the design of a typical continuous batch washer, in this case one of the Senking type [717].

14. References

General References

K. Lindner: *Tenside-Textilhilfsmittel-Waschmittel-Waschrohstoffe,* Wissenschaftl. Verlagsges. Stuttgart, vols. I and II, 1964, vol. III, 1971.

A. M. Schwartz, J. W. Perry, J. Berch: *Surface active Agents,* **vols. I and II,** Interscience Publ., New York 1958.

Kirk-Othmer, **22,** 332 – 432.

J. Falbe: *Surfactants in Consumer Products,* Springer Verlag, Heidelberg 1987.

Kao Corporation: *Surfactants,* Chuo Toshado Printing Co., Tokyo 1983.

M. J. Rosen: *Surfactants and Interfacial Phenomena,* J. Wiley & Sons, New York – Chichester – Brisbane –Toronto 1978.

H. Stache: *Tensid-Taschenbuch,* 2nd ed., Hanser Verlag München – Wien 1981; 3rd ed., 1990.

E. H. Lucassen-Reynders: *Anionic Surfactants,* Marcel Dekker, New York 1981.

K. Laux et al.: „Tenside," in *Winnacker-Küchler,* 4th ed., **vol. 7** (1986) pp. 84 – 148.

B. Werdelmann in: *Proceedings of the World Surfactants Congress München,* **vol. I,** Kürle Verlag, Gelnhausen 1984, p. 3.

H. Andree, P. Krings, H. Verbeek: „Tenside und ihre Kombinationen in zukünftigen Waschmittelformulierungen," *Seifen Oele Fette Wachse* **108** (1982) 277–282.

K. J. Bock, H. Stache: "Surfactants" in: The Handbook of Environmental Chemistry, 3, Part B, Springer Verlag, Berlin–Heidelberg–New York 1982, pp. 163–199.

Henkel KGaA: *Fatty Alcohols. Raw Materials, Methods, Uses,* Düsseldorf 1981.

W. Kling: „Waschmittel 1928, 1968 und 2000," *Chem. Ind. (Düsseldorf)* **20** (1968) 394.

Henkel & Cie, GmbH: *Waschmittelchemie,* Hüthig Verlag, Heidelberg 1976.

H. Hauptmann: *Grundlagen der Wäschereichemie,* Spohr-Verlag, Frankfurt 1977, p. 451.

Surfactant Science Series, **vol. 1–87,** Marcel Dekker, New York.

Tensid Taschenbuch, 1st, 2nd and 3rd ed., Hanser Verlag, Munich-Vienna 1981, Leipzig 1990.

W. G. Cutler, R. C. Davis: *Detergency,* **Part I–III,** Marcel Dekker, New York 1973–1981.

„Chemie der Waschmittel und der Chemisch-Reinigung," *Chem. Ztg.* **99** (1975) 161–211.

"Proceedings-World Conference on Detergents," 1978–1998, A. Cahn (ed.), AOCS Press, Champaign IL, USA.

H. Strone, W. Löbrich, R. Senf, K. D. Wetzler: *Bildungsfibel,* **Part 1,** Deutscher Textilreinigungs-Verband, Bonn 1980.

Henkel KGaA: *Industrielles Waschen,* Rheinisch-Bergische Druckerei GmbH, Düsseldorf 1978.

A. Davidsohn, B. M. Milwidsky: *Synthetic Detergents,* 6th ed., G. Godwin Ltd., London, J. Wiley & Sons, New York 1978.

M. J. Rosen, H. A. Goldsmith: *Systematic Analysis of Surface Active Agents,* 2nd ed., J. Wiley & Sons, London 1972.

D. Hummel: *Analyse der Tenside,* Hanser Verlag, München 1962.

R. Wickbold: *Die Analytik der Tenside,* Chem. Werke Hüls AG, Marl 1976.

B. M. Milwidsky: *Practical Detergent Analysis,* Mac Nair-Dorland Comp., New York 1970.

H. König: *Neuere Methoden zur Analyse von Tensiden,* Springer Verlag, Berlin 1971.

G. F. Longman: *The Analysis of Detergents and Detergent Products,* J. Wiley & Sons, London 1975.

L. H. T. Tai (ed.): *Formulating Detergents and Personal Care Products: A Guide to Product Development,* AOCS Press, Champaign IL, USA 2000.

M. J. Rosen, M. Dahanayake (eds.): *Industrial Utilization of Surfactants,* AOCS Press, Champaign IL, USA 2000.

V. K. Pillai, D. O. Shah (eds.): *Dynamic Properties of Interfaces and Association Structures,* AOCS Press, Champaign IL, USA 2000.

F. Bohmert: *Hauptsache sauber?—Vom Waschen und Reinigen im Wandel der Zeit.* Stürtz-Verlag, Henkel KGaA, Düsseldorf 1988.

H. Eierdanz: *Perspektiven nachwachsender Rohstoffe in der Chemie.* VCH Verlagsgesellschaft, Weinheim 1996.

Fonds der Chemischen Industrie: Tenside. Folienserie 14 des Fonds der Chemischen Industrie, Frankfurt 1992.

Grugel/Puchta/Schöberl: *Waschmittel und Wäschepflege,* Falken-Verlag, Niedernhausen 1996.

Industrieverband Körperpflege und Waschmittel e.V. (IKW): Wäsche und Pflege. Informationen zu Waschmitteln, Textilwäsche und Umwelt, jährliche Berichte und weitere Materialien, Frankfurt 1996.

G. Jakobi, A. Löhr: *Detergents and Textile Washing.* VCH, Weinheim 1987.

H. Krüßmann, R. Bercovici: "Chemische Aspekte beim Waschen," *Tenside, Surfactants, Deterg.* **27** (1990) H. 1, S. 8 ff.

P. Kuzel, T. Lieser: "Bleichsysteme," *Tenside, Surfactants, Deterg.* **27** (1990) 23.

F. Lang, H. Berenhold: "Weichspüler — ein Beitrag zur modernen Wäschepflege," *Seifen, Oele, Fette, Wachse* **117** (1991) 690.

Specific References

[1] Henkel KGaA: *Fettalkohole, Rohstoffe, Verfahren und Verwendung,* Düsseldorf 1981.
[2] *Gesetz über Detergentien in Wasch- und Reinigungsmitteln,* 5. 9. 1961, BGBl. I, p. 1653.
[3] *Verordnung über die Abbaubarkeit von Detergentien in Wasch- und Reinigungsmitteln,* 1. 12. 1962, BGBl. I, pp. 698 – 706.
[4] G. Jakobi, M. J. Schwuger, *Chem. Ztg.* **99** (1975) 182 – 193.
[5] D. Nickel, C. Nitsch, P. Kurzendörfer, W. von Rybinski, *Progr. Colloid Polym. Sci.* **89** (1992) 249 – 252.
[6] ISO 2174 – 1979 (E): Surface active agents — Preparation of water with known calcium hardness.
[7] B. Werdelmann, *Soap, Cosmet. Chem. Spec.* **50** (1974) 36.
[8] M. J. Schwuger in E. H. Lucassen-Reynders (ed.): *Anionic Surfactants,* Marcel Dekker, New York 1981.
[9] D. Nickel, H. D. Speckmann, W. von Rybinski, *Tenside Surf. Det.* **32** (1995) 470 – 474.
[10] J. A. Finch, G. W. Smith in E. H. Lucassen-Reynders (ed.): *Anionic Surfactants,* Marcel Dekker, New York 1981.
[11] T. Young, *Philos. Trans. R. Soc. London* **95** (1805) 65, 82
[12] E. G. Shafrin, W. A. Zisman, *J. Phys. Chem.* **64** (1960) 519.
[13] W. Kling, H. Lange, *Kolloid Z.* **142** (1955) 1.
[14] F. Jost, H. Leiter, M. J. Schwuger, *Colloid Polym. Sci.* **266** (1988) 554.
[15] M. J. Schwuger, H. G. Smolka, *Colloid Polym. Sci.* **255** (1977) 589.
[16] R. Hofmann, D. Nickel, W. von Rybinski, *Tenside, Surfactants, Deterg.* **31** (1994) 63 – 66.
[17] C. P. Kurzendörfer, H. Lange, *Fette Seifen Anstrichm.* **71** (1969) 561.
[18] F. Schambil, M. J. Schwuger, *Colloid Polym. Sci.* **265** (1987) 1009.
[19] H. S. Kielman, P. J. F. van Steen in: *Surface Active Agents,* Society of Chemical Industry, London 1979.
[20] F. Schambil, M. J. Schwuger, *Tens. Det.* **27** (1990) 380.
[21] E. J. W. Verwey, J. T. G. Overbeek: *Theory of the Stability of Hydrophobic Colloids,* Elsevier, Amsterdam 1948.
[22] H. Sonntag, K. Strenge: *Koagulation und Stabilität disperser Systeme,* VEB Deutscher Verlag der Wissenschaften, Berlin (GDR) 1970.
[23] H. Lange in K. H. Mittal (ed.): *Adsorption at Interfaces,* ACS Symp. Ser. no. 8 (1975) 270.
[24] E. Hageböcke, *Dissertation,* Bonn 1956.
[25] M. J. Schwuger, *Ber. Bunsenges. Phys. Chem.* **83** (1979) 1193.
[26] W. von Rybinski, M. J. Schwuger, *Colloids Surf.* **26** (1987) 291 – 304.
[27] E. Smulders, P. Krings, H. Verbeek, Proceedings of the Annual Meeting of the German Society for Fat Research (DGF) 1991.
[28] D. Nickel, W. von Rybinski, E. M. Kutschmann, C. Stubenrauch, G. H. Findenegg, *Proc. 4th World Surfactant Congress* **2** (1996) 371 – 382.
[29] H. Lange, *Kolloid Z.* **169** (1960) 124; *J. Phys. Chem.* **64** (1960) 538.
[30] H. Lange in K. Shinoda (ed.): *Solvent Properties of Surfactant Solutions,* Marcel Dekker, New York 1967.
[31] H. G. Smolka, M. J. Schwuger, *Tenside Deterg.* **14** (1977) 222.
[32] H. G. Smolka, M. J. Schwuger, *Colloid Polym Sci.* **256** (1978) 270.

[33] M. J. Schwuger, H. G. Smolka, *Tenside Deterg.* **16** (1979) 233.

[34] G. Jacobi, *Angew. Makromol. Chem.* **123/124** (1984) 119 – 145.

[35] M. J. Schwuger, M. Liphard, *Colloid Polym. Sci.* **267** (1989) 336 – 344.

[36] R. Puchta, W. Grünewälder: *Textilpflege, Waschen und Chemischreinigen,* Verlag Schiele und Schön, Berlin 1973, pp. 47 – 100.

[37] M. J. Schwuger, *Chem. Ing. Tech.* **42** (1970) 433 –438.

[38] E. Smulders, P. Krings, *China Surfactant Deterg. & Cosmet.* **1** (1993) 18 – 24.

[39] A. M. Rymond, *Proceedings of the 4th World Surfactant Congress* **1** (1996) 21 – 32.

[40] G. Jakobi in H. Stache (ed.): *Tensid-Taschenbuch,* 2nd ed., Hanser Verlag, München-Wien 1981, pp. 253 – 337.

[41] M. F. Cox, T. P. Matson, J. L. Berna, A. Moreno, S. Kawakami, M. Suzuki, *J. Am. Oil Chem. Soc.* **61** (1984) no. 2, 330.

[42] H. Andree, P. Krings, *Chem. Ztg.* **99** (1975) 168 – 174.

[43] J. M. Quack, M. Trautmann, *Tenside Deterg.* **22** (1985) no. 6, 281 – 289.

[44] F. Asinger in *Chemie und Technologie der Paraffinkohlenwasserstoffe,* Akademie Verlag, Berlin 1959.

[45] H. G. Hauthal, B. Makowka, M. Pieroth, Proceedings of the 5th World Surfactants Congress (CESIO), 2000, 873 – 882.

[46] H. G. Hauthal, in H. W. Stache (ed.): Anionic Surfactants: Organic Chemistry, *Surfactant Sci. Ser.* **56** (1996) 143 – 221.

[47] H. Andree, P. Krings in Henkel & Cie GmbH (ed.): *Waschmittelchemie,* Hüthig Verlag, Heidelberg 1976, pp. 73 – 90.

[48] F. Schambil, J. Schwuger, Proceedings of the 2nd World Surfactants Congress (CESIO), 1988, 505 – 524.

[49] T. Satsuki, *INFORM* **3** (1992) 109 – 110.

[50] E. Smulders, P. Krings, *Chem. Ind.* **6** (1990) 160.

[51] A. J. Kaufmann, R. J. Ruebusch, Proceedings of the World Conference on Oleochemicals, Kuala Lumpur, 1990, 10.

[52] R. A. Peters, Proceedings of the World Conference on Oleochemicals, Kuala Lumpur, 1990, 181.

[53] H. J. Richtler, J. Knaut, 2nd World Surfactant Congress (CESIO), Paris, 1988.

[54] G. Czichocki, H. Brämer, I. Ohme, *Z. Chem.* **20** (1980) 90 – 93.

[55] G. Czichocki, A. Greiner, B. Koch, *Tenside Deterg.* **21** (1984) no. 2, 62 – 66.

[56] E. Smulders, Proceedings of the International Conference of the European Chemical Marketing Research Association (ECMRA) (1984) 305 – 348.

[57] K. Hill, W. von Rybinski, G. Stoll (eds.): *Alkyl Polyglycosides — Technology, Properties, and Applications.* Wiley-VCH, Weinheim 1997.

[58] K. Bräuer, H. Fehr, R. Puchta, *Tenside Deterg.* **17** (1980) 281 – 287.

[59] W. M. Linfield in E. Jungermann (ed.): Cationic Surfactants, *Surfactant Sci. Ser.* **4** (1970) 49.

[60] R. Puchta, P. Krings, P. Sandkühler, *Tenside Surf. Det.* **30** (1993) 186 – 191.

[61] S. T. Giolando, R. A. Rapaport, R. J. Larson, T. W. Federle, *Chemosphere* **30** (1995) 1067 – 1083.

[62] F. Lang, H. Berenbold, *Seifen, Oele, Fette, Wachse* **117** (1990) 690 – 694.

[63] H. Berenbold, *Tenside, Surfactants, Deterg.* **27** (1990) 34 – 40.

[64] G. C. Schweiker, *J. Am. Oil Chem. Soc.* **58** (1981) 58 – 61.

[65] G. Jakobi, M. J. Schwuger in Henkel & Cie GmbH (ed.): *Waschmittelchemie,* Hüthig Verlag Heidelberg 1976, pp. 91 – 120.

[66] Henkel, DE-OS 2 412 837, 1974.

[67] M. Schwuger, E. Smulders: "Inorganic Builders" in G. Cutler, E. Kissa (eds.): *Detergency: Theory and Technology, Surfactant Sci. Ser.* **20** (1987) 371 – 439.

[68] Henkel, DE-OS 2 412 838, 1975.

[69] G. Jakobi, *Angew. Makromol. Chem.* **123/124** (1984) 119.

[70] H. Krüssmann, P. Vogel, H. Hloch, H. Carlhoff, *Seifen Oele Fette Wachse* **105** (1979) 3 – 6.

[71] M. Ettlinger, H. Ferch, *Seifen Oele Fette Wachse* **105** (1979) 131 – 135, 160.

[72] W. E. Adam, K. Neumann, J. P. Ploumen, *Fette Seifen Anstrichm.* **81** (1979) 445 – 449.

[73] H. Nüsslein, K. Schumann, M. J. Schwuger, *Ber. Bunsenges. Phys. Chem.* **83** (1979) 1229 – 1238.

[74] C. P. Kurzendörfer, M. J. Schwuger, H. G. Smolka, *Tenside Deterg.* **16** (1979) 123 – 129.

[75] P. Berth, *Tenside Deterg.* **14** (1978) 176 – 180.

[76] H. Upadek, P. Krings, E. Smulders, CHIMICAoggi, January – February 1990, 61 – 67.

[77] H. Upadek, P. Krings, *Seifen Oele Fette Wachse* **117** (1991) 554 – 558.

[78] P. Zini (ed.): *Polymeric Additives for High Reforming Detergents,* Technomic, Lancaster 1995, pp. 179 – 212.

[79] P. Zini, *J. of Chem. Technol. and Biotechnol.* **50** (1991) 351 – 359.

[80] H. Upadek, B. Kottwitz, B. Schreck, *Tenside Surf. Det.* **33** (1996) 385 – 392.

[81] H. P. Rieck, *Seifen Oele Fette Wachse* **122** (1996) 376 – 391.

[82] Unilever, EP 0 384 070, 1989 (G. T. Brown, T. J. Osinga, M. J. Parkington, A. T. Steel).

[83] Joseph Crosfield & Sons, EP 0 565 364, 1993 (A. Araya).

[84] P. Kuzel, *Seifen Oele Fette Wachse* **105** (1979) 423 – 424.

[85] P. Kleinschmitt, B. Bertsch-Frank, Th. Lehmann, P. Panster, Ullmann's Encyclopedia of Industrial Chemistry, Vol. A 19 (1991) 177 – 197.

[86] V. Croud: "Oxygen Bleaches" in U. Zoller, G. Broze (eds.): *Properties, Surfactant Sci. Ser.* **82** (1999) 597 – 617.

[87] G. Whalley, *Manuf. Chem.* April 1998, 31 – 33.

[88] G. Reinhardt: " New Bleach Systems" Proceedings of the 4th World Conference on Detergents, AOCS Press Champaign IL, 1999, 195 – 203.

[89] P. Kuzel in J. Falbe (ed.): *Surfactants in Consumer Products,* Springer Verlag, Heidelberg 1987, p. 267.

[90] J. A. Church: "Hypochlorite Bleach" in U. Zoller, G. Broze (eds.): *Handbook of Detergents: Properties, Surfactant Sci. Ser.* **82** (1999) 619 – 629.

[91] M. Julémont: "Application of Hypochlorite" in U. Zoller, G. Broze (eds.): *Handbook of Detergents: Properties, Surfactant Sci. Ser.* **82** (1999) 631 – 637.

[92] W. L. Smith, Proceedings of the 4th World Conference on Detergents: Strategies for the 21st Century, AOCS Press Champaign IL, 1999, 137 – 141.

[93] H. Bloching, *Sonderheft „Haushaltswaschmittel",* Ciba-Geigy, 1975, p. 17.

[94] D. Coons in J. Falbe (ed.): *Surfactants in Consumer Products,* Springer Verlag, Heidelberg 1987, p. 274.

[95] O. Annen, H. Ulshöfer, *Melliand Textilber.* (2000) 756 – 759

[96] G. Reinhardt: "New Bleach Systems", Proc. World Conf. Deterg.: Strategies 21 Century, 4th, 1999, 195 – 203.

[97] Noury van der Lande, DE 1 162 967, 1960.

[98] J. Mathews, *J. Soc. of Dyers Colour.* **115** (1999) 154 – 155.

[99] J. Mathews, *Book. Pap.-Int. Conf. Exhib., AATC,* 1997, 462 – 470.

[100] V. B. Croud, S. J. Tompsett, Proceedings of the 4th World Surfactants Congress (CESIO), 1996, 257 – 267.

[101] M. A. Chadwick, Proceedings of the 5th World Surfactants Congress (CESIO), 2000, 1227 – 1236.

[102] V. B. Croud, I. M. George, *HAPPI Household Pers. Prod. Ind.* **34** (1997) no. 1, 82 – 92.

[103] I. M. George, V. B. Croud, *Biocides Today* (1998) 25 – 27.

[104] Procter & Gamble, US 4 412 934, 1983.

[105] A. Gilbert, *Deterg. Age* **4** (1967) July, 30 – 33.

[106] C. Render, Proceedings of the 5th World Surfactants Congress (CESIO), 2000, 1217 – 1226.

[107] R. Hage et al., *Nature (London)* **369** (1994) 637 – 639.

[108] A. Crutzen, M. L. Douglass: "Detergent Enzymes: A challenge!" in U. Zoller, G. Broze (eds.): *Handbook of Detergents, Part A: Properties,* Surfactant Sci. Ser. 82 1999, pp. 639 – 690.

[109] J. H. van Ee: "Historical Overview" in J. H. van Ee, O. Misset, E. J. Baas (eds.): *Enzymes in Detergency,* Surfactant Sci. Ser. 69 1997, pp. 1 – 10.

[110] Novo Nordisk A/S, Product Sheet, Enzyme Business B 345e-GB, 1998.

[111] J. H. Houston: "Detergent Enzymes: Market" in J. H. van Ee, O. Misset, E. J. Baas (eds.): *Enzymes in Detergency,* Surfactant Sci. Ser. 69 1997, pp. 11 – 21.

[112] H. Upadek, B. Kottwitz: "Application of Amylases in Detergents" in J. H. van Ee, O. Misset, E. J. Baas (eds.): *Enzymes in Detergency,* Surfactant Sci. Ser. 69 1997, pp. 203 – 212.

[113] H. Andree, W. R. Müller, R. D. Schmid, *J. Appl. Biochem.* **2** (1980) 218.

[114] B. Kottwitz, H. Upadek, *Seifen Oele Fette Wachse* **120** (1994) 794 – 799.

[115] G. K. Greminger Jr., *Soap Cosmet. Chem. Spec.* **2** (1978) Nov., 28 – 31, 38.

[116] G. K. Greminger, *J. Am. Oil Chem. Soc.* **55** (1978) 122 – 126.

[117] F. Lang, *Chim. Oggi* **9** (1998) 14.

[118] E. Kissa, in Detergency: Theory and Technology, *Surfactant Sci. Ser.* **20** (1986) 333.

[119] E. P. Grosselink, in Powder Detergents, *Surfactant Sci. Ser.* **71** (1998) 215.

[120] Procter & Gamble, DE-OS 2 338 464, 1974.

[121] Dow Corning, DE-OS 2 402 955, 1974.

[122] G. Rossmy, *Fette Seifen Anstrichm.* **71** (1969) 56.

[123] J. Blajely, S. Stassen, A. Bouadjaj, E. Mallen, Proceedings of the 5th World Surfactants Congress (CESIO), 2000, 1157 – 1168.

[124] R. Höfer, F. Jost, M. J. Schwuger, R. Scharf, J. Geke, J. Kresse, H. Lingmann, R. Veitenhansel, W. Erwied: "Foams and Foam Control" in Ullmann's Encyclopedia of Industrial Chemistry, Vol. A 11 (1988) 465 – 490.

[125] A. E. Siegrist, C. Eckhardt, J. Kaschig, E. Schmidt: "Optical Brighteners" Ullmann's Encyclopedia of Industrial Chemistry, Vol. A 18 (1991) 153 – 173.

[126] J. Kaschig, C. Puebla, Proceedings of the 5th World Surfactants Congress (CESIO), 2000, 1147 – 1156.

[127] E. Smulders, M. Osset Hernandez, Comun. Jorn. Com. Esp. Deterg., 28th, (1998) 13 – 32.

[128] E. Smulders, P. Krings, H. Verbeek, *Tenside, Surfactants, Deterg.* **34** (1997) 386 – 392.

[129] Th. Müller-Kirschbaum, P. Sandkühler, G. Wagner, *Naturwissenschaften im Unterricht Chemie* **12** (2001) no. 3, 11 – 15.

[130] Textile Washing Products, Euromonitor International, London, Chicago, Singapore, September 2000.

[131] Th. Müller-Kirschbaum, E. Smulders: "Facing future's Challenges — European Laundry Products on the Treshold of the 21st Century", in *Proc. World Conf. Deterg.: Strategies 21st Century, 4th* 1999, 93 – 106.

[132] R. B. McConnell, *Am. Dyest. Rep.* **67** (1978) July, 31 – 34.

[133] H. W. Bücking, K. Lötzsch, G. Täuber, *Tenside Deterg.* **16** (1979) 1 – 10.

[134] L. Hughes, J. M. Leiby, M. L. Deviney, *Soap Cosmet. Chem. Spec.* **2** (1975) Oct., 56 – 62.
[135] W. Bechstedt, W. A. Roland, *Reiniger Wäscher* **33** (1980) no. 6, 21 – 29, no. 7, 18 – 24.
[136] E. Smulders, Th. Müller-Kirschbaum, P. Sandkühler, *Chemie-Technik* **1** (1997) 96 – 100.
[137] H. Upadek, P. Krings, *Seifen, Oele, Fette, Wachse* **117** (1991) 554 – 558.
[138] Henkel KGaA, EP 0 486 592 B1, 1990.
[139] Henkel, EP 595 946, 1997.
[140] Henkel, EP 828 818, 1999.
[141] Henkel, WO 98/12299, 1998.
[142] F. Schambil, M. Böcker, *Tenside Surf. Det.* **37** (2000) 1, 48 – 50.
[143] F. Schambil, M. Böcker, G. Blasey, *Chim. Oggi,* March/April 2000 48 – 50.
[144] G. Whalley, *Manuf. Chem.* October 1998, 41 – 45.
[145] A. Syldath, K. H. Schmid, D. Kischel, Proceedings of the 5th World Surfactants Congress (CESIO), 2000, 1141 – 1146.
[146] E. Smulders, P. Krings, Proceedings of the 2nd World Surfactants Congress (CESIO), 1988.
[147] D. Nickel, Proceedings of the 5th World Surfactant Congress (CESIO), 2000, 1106 – 1111.
[148] B. Jewett, *Soap & Cosmetics* **77** (2000) no. 9, 32 – 34.
[149] H. J. Lehmann, *Chem. Unserer Zeit* **7** (1973) 82 – 89.
[150] E. P. Frieser, *Österr. Chem. Z.* **76** (1975) no. 9, 4 – 8; **76** (1975) no. 10, 1 – 6; **76** (1975) no. 11, 5 – 7; **76** (1975) no. 12, 5 – 9.
[151] K. Henning, *Chem. Lab. Betr.* **27** (1976) 46 – 50, 81 – 86.
[152] G. Gawalek: *Tenside,* Akademie Verlag, Berlin 1975.
[153] Henkel: *Waschmittelchemie,* Hüthig Verlag, Heidelberg 1976.
[154] P. Berth, G. Jakobi, E. Schmadel, J. Schwuger, C. Krauch, *Angew. Chem.* **87** (1975) 115 – 123; *Angew. Chem. Int. Ed. Engl.* **14** (1975) 94.
[155] R. Puchta, P. Krings, A. Wilsch-Irrgang, *Hauswirtschaft und Wissenschaft* **42** (1994) 70 – 75.
[156] J. Waters, K. S. Lee, V. Perchard, M. Flanagan, P. Clarke, *Tenside Surf. Det.* **37** (2000) 3, 161 – 171.
[157] W. Pietsch: *Size Enlargement by Agglomeration,* John Wiley & Sons, New York 1990.
[158] D. Jung: *Ullmann,* **vol. A8,** 1985, p. 385 ff.
[159] K. Masters: *Spray Drying Handbook,* Fifth Edition, Longman Scientific & Technical, New York 1991.
[160] Henkel KGaA, US-Patent 4 741 803, 1985.
[161] G. F. Moretti, I. Adami: Lecture presented at 6e Giornate CID, Roma 1995.
[162] J. Höhmann: *Wissensbasierte Systeme zur Prozessführung verfahrenstechnischer Anlagen am Beispiel der Sprühtrocknung,* Verlag Shaker, Aachen 1993.
[163] Henkel KGaA, EP 0 550 508 B1, 1991.
[164] Henkel KGaA, WO 95/21010, 1995.
[165] Henkel KGaA, EP 0 625 923, 1992.
[166] W. Rähse: Drying with Superheated Steam in a Spray-Drying Tower, Henkel-Referate 35, Düsseldorf 1998.
[167] W. H. de Groot, I. Adami, G. F. Moretti: *The Manufacture of Modern Detergent Powders,* H. de Groot Academic Publisher, Wassenaar 1995.
[168] Procter & Gamble, EP 0 643 130 A1, 1993.
[169] Henkel KGaA, P 38 03 966.4, 1992.
[170] Henkel KGaA, P 38 12 530.7, 1988.
[171] Procter & Gamble, WO 95/32276, 1994.
[172] Procter & Gamble, EP 0 663 439 A1, 1994.

[173] Procter & Gamble, WO 95/00630, 1995.
[174] O. T. Kraph, A. Kraglund, *Chem. Eng. (London)* **367** (1981) 149 – 153.
[175] G. F. Moretti, I. Adami: "Drying and Agglomeration Process for Traditional and Concentrated Detergent Powders", Ballestra S.p.A., Milano 1994.
[176] Henkel KGaA, DE 19844522.9, 1998.
[177] Henkel KGaA, DE 40 07601 A1, 1990.
[178] Henkel KGaA, DE 197 10 254.9, 1997.
[179] Procter & Gamble, EP 846 745, EP 846 755, EP 846 756, 1996.
[180] Unilever, EP 481 793, 1990.
[181] Unilever, EP 711 827, 1994.
[182] A. Behler, A. Syldath, Proceedings of the 5th World Surfactants Congress, 29.5 – 2.6.2000, Firenze, Vol. 1, p. 382.
[183] Korsch Pressen GmbH, Berlin, Germany.
[184] L. Spitz, *Soaps and Detergents,* AOCS Press, San Diego 1996.
[185] Henkel KGaA, DE 195 46 735, 1995.
[186] Henkel KGaA, EP 319 819, 1988.
[187] Henkel KGaA, DE 44 25 968 A1, 1994.
[188] Vomm Inc., Rozzano/Milan, Italy.
[189] Bayer AG, EP 0 163 836, 1984.
[190] Glatt GmbH, Weimar, Germany.
[191] G. Dieckelmann, H. J. Heinz: *The Basis of Industrial Oleochemistry,* Peter Pomp GmbH, Essen 1988.
[192] Fatty Alcohols, Henkel KGaA, Düsseldorf 1981 (identical with [1]).
[193] K. Winnacker, L. Küchler, H. Harnisch, R. Steiner (eds.): *Chemische Technologie,* 4. Auflage, **Bd. 7,** Carl Hanser Verlag München 1986.
[194] M. S. Showell (ed.): *Powdered Detergents,* **vol. 71,** Surfactants Science Series, 1998.
[195] Hoechst, DE-A-34 17 649, 1984.
[196] Hoechst, DE 196 01 063 A1, 1995.
[197] Hoechst, DE-A-41 42 711, 1991.
[198] Henkel KGaA, DE 444 63 63, 1994.
[199] Henkel KGaA, DE 195 33 790, 1995.
[200] Unilever/Crosfield, WO 94/26662, 1993.
[201] Unilever/Crosfield, EP 565 364, 1992.
[202] Unilever, WO 94/28098, 1993.
[203] K. Winnacker, L. Küchler, H. Harnisch, R. Steiner (eds.): *Chemische Technologie,* 4. Auflage, **Bd. 2,** Carl Hanser Verlag München 1986, p. 584 ff.
[204] W. Rähse, F.-J. Carduck, Microfiltration of Fermenter Broths, *Chem.-Ing.-Tech.* **57** (1985) 747.
[205] Biozym GmbH, Kundl, Austria.
[206] G. Sorbe, *Seifen Oele Fette Wachse* **105** (1979) 251 – 253.
[207] W. M. Linfield: *Anionic Surfactants,* **vol. 7** (Parts I and II), Marcel Dekker, New York-Basel 1976.
[208] P. Berth, M. J. Schwuger, *Tenside Deterg.* **16** (1979) 175 – 184.
[209] P. Berth, J. Heidrich, G. Jakobi, *Tenside Deterg.* **17** (1980) 228 – 235.
[210] B. Ziolkowsky, *Seifen Oele Fette Wachse* **12** (1981) 333 – 336, 339 – 343.
[211] Th. Kunzmann, *Seifen Oele Fette Wachse* **98** (1972) 179 – 181.
[212] M. J. Schwuger, H. G. Smolka, *Colloid Polym. Sci.* **254** (1976) 1062.
[213] M. J. Schwuger, H. G. Smolka, C. P. Kurzendörfer, *Tenside Deterg.* **13** (1976) 305.

[214] C. P. Kurzendörfer, M. J. Schwuger, H. G. Smolka, *Tenside Deterg.* **16** (1979) 123 – 129.
[215] H. G. Smolka, M. J. Schwuger, *Tenside Deterg.* **14** (1977) 222.
[216] H. G. Smolka, M. J. Schwuger, *Colloid Polymer. Sci.* **256** (1978) 270 – 277.
[217] M. J. Schwuger, H. G. Smolka, *ACS Symp. Ser.* **40** (1971) 696 – 707.
[218] M. J. Schwuger, H. G. Smolka, *Tenside Deterg.* **16** (1979) 233 – 239.
[219] W. Kling, *Angew. Chem.* **65** (1953) 201 – 212.
[220] W. Kling, *Parfüm. Kosmet.* **45** (1964) 1 – 5, 29 – 31.
[221] W. Hagge, *CID-Kongreß 1968,* **vol. I,** pp. 1 – 19.
[222] G. F. Longman: *Analysis of Detergents and Detergent Products,* J. Wiley & Sons, London 1975.
[223] N. Schönfeld: *Grenzflächenaktive Ethylenoxid-Addukte,* Wissenschaftl. Verlagsges., Stuttgart 1976, pp. 991 – 1045.
[224] J. Wiesner, L. Wiesnerova, *J. Chromatogr.* **114** (1975) 411 – 412.
[225] E. Kunker, *Tenside Deterg.* **18** (1981) 301 – 305.
[226] E. Heinerth, *Seifen Oele Fette Wachse* **5** (1964) 105 – 110.
[227] R. Wickbold: *Die Analytik der Tenside,* Chemische Werke Hüls AG, Marl 1976.
[228] G. Graffmann, W. Hörig, P. Sladek, *Tenside Deterg.* **14** (1977) 194 – 197.
[229] M. Teupel, *CID-Kongreß 1960,* **vol. C/III,** pp. 177 –183.
[230] G. Schwarz, *Seifen Oele Fette Wachse* **22** (1961) 715 – 717.
[231] H. König, *Fresenius' Z. Anal. Chem.* **293** (1978) 295 – 300.
[232] M. J. Rosen, H. A. Goldsmith: *Sytematic Analysis of Surface-Active Agents,* 2nd ed., J. Wiley & Sons, London 1972.
[233] Schweizerische Gesellschaft f. Analytische und Angewandte Chemie: *Seifen und Waschmittel,* Verlag H. Huber, Bonn-Stuttgart 1955, 1963.
[234] D. Hummel: *Analyse der Tenside,* Hanser Verlag, München 1962.
[235] B. M. Milwidsky, *Practical Detergent Analyses,* Mac Nair-Dorland Comp., New York 1970.
[236] British Standard 3762 : 1964, "Methods of Sampling and Testing Detergents."
[237] G. F. Longman, *Talanta* **22** (1975) 621 – 636.
[238] *Jahrbuch für den Praktiker aus der Öl-, Fett-, Seifen- und Waschmittel-, Wachs- und sonstigen chemischtechnischen Industrie 1981,* Verlag für chemische Industrie H. Ziolkowsky KG, Augsburg.
[239] J. Puschmann, *Angew. Makromol. Chem.* **47** (1975) 29 – 41.
[240] H. Specker, *Angew. Chem.* **80** (1968) 297 – 304;*Angew. Chem. Int. Ed. Engl.* **7** (1968) 252.
[241] G. Tölg, *Naturwissenschaften* **63** (1976) 99 – 110.
[242] D. C. Cullum, P. Platt, *Crit. Rep. Appl. Chem.* **32** (1991) (Recent Dev. Anal. Surfactants) 5 – 74.
[243] G. T. Battaglini, Soaps Deterg. Theor. Pract. Rev., [Soaps Deterg. Technol. Today Conf.] (1996), Meeting Date 1994, 330 – 380.
[244] B. M. Milwidsky: *Practical Detergent Analyses,* Mac Nair-Dorland Comp., New York 1970.
[245] DGF-Einheitsmethoden, Wissenschaftl. Verlagsges. Stuttgart 1950 – 1984.
[246] Official and Tentative Methods of The American Oil Chemists Society, 508 South Sixth Street, Champaign, Illinois 61 820.
[247] ISO-TC 91.
[248] G. Graffmann, H. Domels, W. Hörig, *Fette Seifen Anstrichm.* **77** (1975) 364 – 365.
[249] DIN Deutsches Institut für Normung e.V., Beuth-Vertrieb GmbH, Berlin-Köln.
[250] Jander-Blasius: *Lehrbuch der analytischen und präparativen anorganischen Chemie,* Hirzel Verlag, Stuttgart 1973.
[251] Anorganikum, VEB Deutscher Verlag der Wissenschaften, Berlin (GDR) 1967.
[252] J. Jaeger, K. Sorensen, S. P. Wolff, *J. Biochem. Biophys. Methods* **29** (1994) no. 1, 77 – 81.

[253] Fresenius-Jander: *Handbuch der Analytischen Chemie,* Springer Verlag, Berlin 1953.
[254] Deutsche Einheitsverfahren zur Wasseruntersuchung, Verlag Chemie, Weinheim 1960.
[255] H. Puderbach, *J. Am. Oil Chem. Soc.* **55** (1978) 156 – 162.
[256] I. Kawase, A. Nakae, K. Tsuji, *Anal. Chim. Acta* **131** (1981) 213 – 222.
[257] D. Millers, *Comun. Jorn. Esp. Deterg.* **21** (1990) 99 – 112.
[258] H. Bauer, *Seifen Oele Fette Wachse* **121** (1995) no. 3, 168 – 170.
[259] A. Kawauchi, M. Ishida, *J. Am. Oil Chem. Soc.* **73** (1996) no. 1, 131 – 135.
[260] F. Feigl: *Spot Tests in Organic Analysis,* 7th ed., Elsevier, Amsterdam 1966.
[261] F. Cramer: *Papierchromatographie.* 5th ed., Verlag Chemie, Weinheim 1962.
[262] E. Stahl: *Dünnschichtchromatographie.* 2nd ed., Springer Verlag, Berlin 1967.
[263] E. Heinerth, J. Pollerberg, *Fette Seifen Anstrichm.* **61** (1959) 376 – 377.
[264] R. Pribil, V. Vesely, *Chem. Analyst* **56** (1967) 51 – 53.
[265] E. Heinerth, *Fette Seifen Anstrichm.* **70** (1968) 495 – 498.
[266] J. Drewry, *Analyst (London)* **88** (1963) 225 – 231.
[267] D. Hummel: *Kunststoff-, Lack- und Gummianalyse,* vol. I, Hanser Verlag, München 1958.
[268] E. Knappe, I. Rhodewald, *Fresenius' Z. Anal. Chem.* **223** (1966) 174 – 181.
[269] K. Figge, *Fette Seifen Anstrichm.* **70** (1968) 680 – 687.
[270] G. Lehmann, M. Becker-Klose, *Tenside Deterg.* **13** (1976) 7 – 8.
[271] H. Bloching, W. Holtmann, M. Otten, *Seifen Oele Fette Wachse* **105** (1979) 33 – 38, 82 – 83.
[272] S. R. Epton, *Nature (London)* **160** (1947) 795 – 796.
[273] R. Matissek, E. Hieke, W. Baltes, *Fresenius' Z. Anal. Chem.* **300** (1980) 403 – 406.
[274] H. Brüschweiler, V. Sieber, H. Weishaupt, *Tenside Deterg.* **17** (1980) 126 – 129.
[275] M. Däuble, *Tenside Deterg.* **18** (1981) 7 – 12.
[276] J. M. Rosen, *Anal. Chem.* **27** (1955) 787 – 790.
[277] K. Bürger, *Fresenius' Z. Anal. Chem.* **196** (1963) 251 – 259.
[278] *Erfahrungsaustausch der Seifen-, Wasch- und Reinigungsmittelindustrie,* **vol. 1, pp. 9 – 12,** 1944; *Vagda-Kalender,* 5th ed., 1943, p. 87.
[279] A. Hintermaier, *Fette Seifen* 1944, Sept. 1, 367 –368.
[280] P. Friese, *Fresenius' Z. Anal. Chem.* **303** (1980) 279 – 288.
[281] J. F. Kennedy, E. H. M. Melo, *Cellul. Sources Exploit.* (1990) 371 – 376.
[282] J. F. Kennedy, E. H. M. Melo, *Eur. Pol. Paint Colour J.* **182** (1992) no. 4312, 430 – 431, 436.
[283] J. F. Kennedy, E. H. M. Melo, *Eur. Chem. Paint Colour J.* **181** (1991) no. 4289, 478, 480, 482.
[284] B. Gruening, D. Kaeseborn, H. Leidreiter, *Parfuem. Kosmet.* **77** (1996) no. 4, 244 – 248.
[285] R. Spilker, B. Menzebach, U. Schneider, I. Venn, *Tenside, Surfactants, Deterg.* **33** (1996) no. 1, 21 – 25.
[286] H. Waldhoff, J. Scherler, M. Schmitt, World Surfactants Congr., 4th (CESIO) 1, (1996) 507 – 518.
[287] W. Leonhardt, R. Peldszus, H. Wegert, *Seife Oele Fette Wachse* **113** (1987) no. 15, 511 – 513.
[288] H. P. Kaufmann: *Analyse der Fette und Fettprodukte,* Springer Verlag, Berlin 1958.
[289] R. Wickbold, *Tenside Deterg.* **9** (1972) 173 – 178.
[290] H. Hellmann, *Fresenius' Z. Anal. Chem.* **300** (1980) 44 – 47.
[291] B. Weibull: *3. Int. Kongreß für grenzflächenaktive Stoffe, Köln 1960,* **vol. 3,** Universitätsdruckerei, Mainz 1961, p. 121.
[292] B. Milwidsky, *Tenside Deterg.* **22** (1985) no. 3, 136 – 138.
[293] W. Riemann, H. Walton: *Ion Exchange in Analytical Chemistry,* Pergamon Press, Oxford 1970.
[294] P. Voogt, *Recl. Trav. Chim. Pays-Bas* **77** (1958) 889 – 901.
[295] R. Wickbold, *Seifen Oele Fette Wachse* **86** (1960) 79 – 82.

[296] N. Blumer, *Schweiz. Arch.* **29** (1963) 171 – 180.
[297] M. E. Ginn, C. C. Church, *Anal. Chem.* **31** (1959) 551 – 555.
[298] K. Bey, *Fette Seifen Anstrichm.* **67** (1965) 25 – 30.
[299] P. Voogt in H. A. Boeckenoogen (ed.): *Analysis and Characterization of Oils, Fats, and Fat Products* Interscience Publ. Inc., London 1964.
[300] R. Wickbold, *Tenside Deterg.* **13** (1976) 177 – 180.
[301] R. Wickbold, *Tenside Deterg.* **13** (1976) 181 – 187.
[302] G. Schwarz, *Fette Seifen Anstrichm.* **71** (1969) 223 – 226.
[303] E. Heinerth, *Fette Seifen Anstrichm.* **63** (1961) 181 – 183.
[304] R. Bock: *Aufschlußmethoden der anorg. und organischen Chemie,* Verlag Chemie, Weinheim 1972.
[305] Verein Deutscher Eisenhüttenleute: *Handbuch für das Eisenhüttenlaboratorium,* vol. 1, Stahleisen-Verlag, Düsseldorf 1960.
[306] *Fette Seifen Anstrichm.* **79** (1977) 203.
[307] G. Staats, H. Brück, *Fresenius' Z. Anal. Chem.* **250** (1970) 289 – 294.
[308] E. Vaeth, E. Griessmayer, *Fresenius' Z. Anal. Chem.* **303** (1980) 268 – 271.
[309] O. G. Koch, A. Koch-Dedic: *Handbuch der Spurenanalyse,* Springer Verlag, Berlin 1974.
[310] J. Drozd, *J. Chromatogr.* **113** (1975) 303 – 356.
[311] *Fette Seifen Anstrichm.* **74** (1972) 31.
[312] A. Hintermaier, *Angew. Chem.* **60** (1948) 158 – 159.
[313] V. W. Reid, G. F. Longman, E. Heinerth, *Tenside* **4** (1967) 292 – 304.
[314] S. Lee, N. A. Puttnam, *J. Am. Oil Chem. Soc.* **43** (1966) 690.
[315] R. Denig, *Fette Seifen Anstrichm.* **76** (1974) 412 – 416.
[316] R. Denig, *Tenside Deterg.* **10** (1973) 59 – 63.
[317] W. Kupfer, K. Künzler, *Fresenius' Z. Anal. Chem.* **267** (1973) 166 – 169.
[318] T. H. Liddicoet, L. H. Smithson, *J. Am. Oil Chem. Soc.* **42** (1965) 1097 – 1102.
[319] D 501 – 67, ASTM Standards of Soaps and other Detergents, Part 30, 1974.
[320] A. J. Sheppard, J. L. Iversen, *J. Chromatogr. Sci.* **13** (1975) 448 – 452.
[321] H. Schlenk, J. L. Gellerman, *Anal. Chem.* **32** (1960) 1412 – 1414.
[322] L. D. Metcalf, A. A. Schmitz, *Anal. Chem.* **33** (1961) 363 – 364.
[323] A. Seher, H. Pardun, M. Arens, *Fette Seifen Anstrichm.* **80** (1978) 58 – 66.
[324] J. D. Knight, R. House, *J. Am. Oil Chem. Soc.* **36** (1959) 195 – 200.
[325] J. Pollerberg, *Fette Seifen Anstrichm.* **67** (1965) 927 – 929.
[326] M. J. Rosen, G. C. Goldfinger, *Anal. Chem.* **28** (1956) 1979 – 1981.
[327] J. Borecky, *Mikrochim. Acta* 1962, 1137 – 1145.
[328] S. Nishi, *Bunseki Kagaku* **14** (1965) 917; *Chem. Abstr.* **64** (1966) 3858 f.
[329] R. Wickbold, *Tenside Deterg.* **12** (1975) 25 – 27.
[330] „Sauerstoffverbindungen III,"*Houben-Weyl,* **8**
[331] P. L. Buldini, J. L. Sharma, D. Ferri, *J. Chromatogr.* **654** (1993) no. 1, 129 – 134.
[332] H. Biltz, W. Biltz: *Ausführung quantitativer Analysen.* Hirzel Verlag, Stuttgart 1965.
[333] British Standard 3984: 1966 "Sodium Silicates".
[334] G. Graffmann, W. Schneider, L. Dinkloh, *Fresenius' Z. Anal. Chem.* **301** (1980) 364 – 372.
[335] J. Longwell, W. D. Maniece, *Analyst (London)* **80** (1955) 167 – 171.
[336] M. Kolthoff, V. A. Stenger: *Volumetric Analysis,* 2nd ed., vols I and II, Interscience Publ., New York 1942, 1947.
[337] G. Jander, K. F. Jahr, H. Knoll: *Maßanalyse,* De Gruyter, Berlin 1966.
[338] E. Heinerth, *Tenside* **4** (1967) 45 – 47.

[339] S. Ebel, U. Parzefall: *Experimentelle Einführung in die Potentiometrie,* Verlag Chemie, Weinheim 1975.

[340] H. Malmerig, *GIT Fachz. Lab.* **19** (1975) 400 – 404.

[341] S. Ebel, A. Seuring, *Angew. Chem.* **89** (1977) 129 – 141; *Angew. Chem. Int. Ed. Engl.* **16** (1977) 157.

[342] A. Hofer, E. Brosche, R. Heidinger, *Fresenius' Z. Anal. Chem.* **253** (1971) 117 – 119.

[343] J. Koryta, *Anal. Chim. Acta* **61** (1972) 329 – 411.

[344] *Orion Ion- and Gas-sensitive Electrodes,* Orion Research Inc., Cambridge 1973.

[345] D. L. Jones, G. J. Moody, J. D. R. Thomas, B. J. Birch, *Analyst (London)* **106** (1981) no. 1266, 974 – 984.

[346] M. Bos, *Anal. Chim. Acta* **135** (1982) no. 2, 249 – 261.

[347] T. Kobayashi, M. Kataoka, T. Kambra, *Talanta* **27** (1980) no. 2, 253 – 256.

[348] H. W. Leuchte, H. D. Kahleyss, J. Faltermann, *Fett Wiss. Technol.* **94** (1992) no. 2, 64 – 66.

[349] M. Tehrani, M. Thomae, *Am. Lab. (Shelton, Conn.)* **23** (1991) no. 19, 8, 10.

[350] G. C. Dilley, *Analyst (London)* **105** (1980) no. 1252, 713 – 719.

[351] L. I. Makovetskaya, I. L. Maislina, S. P. Volosovich, *Seife Oele Fette Wachse* **117** (1991) no. 15, 565 – 571.

[352] M. Gerlache, Z. Sentuerk, J. C. Vire, J. M. Kauffmann, *Anal. Chim. Acta* **349** (1997) no. 1 – 3, 59 – 65.

[353] B. J. Birch, R. N. Cockcroft, *Ion-Sel. Electrode Rev.* **3** (1981) no. 1, 1 – 41.

[354] R. Calapaj, L. Ciraolo, F. Corigliano, D. Di Pasquale, *Analyst (London)* **107** (1982) no. 1273, 403 – 407.

[355] G. Schwarzenbach, H. Flaschka: *Die komplexometrische Titration,* Enke Verlag, Stuttgart 1965.

[356] R. Pribil: *Komplexometrie,* VEB Deutscher Verlag für Grundstoffind., Leipzig 1960.

[357] G. Graffmann, H. Domels, M. L. Sträter, *Fette Seifen Anstrichm.* **76** (1974) 218 – 220.

[358] Merck AG: *Komplexometrische Bestimmungen mit Titriplex,* 3rd ed., Darmstadt 1966.

[359] R. Wickbold: *VI. Int. Kongreß für grenzflächenaktive Stoffe, Zürich 1972,* **vol. I,** Section A, Hanser Verlag, München 1973, p. 373.

[360] F. Öhme: *Angewandte Konduktometrie,* Hüthig Verlag, Heidelberg 1962.

[361] K. Kiemstedt, W. Pfab, *Fresenius' Z. Anal. Chem.* **213** (1965) 100 – 107.

[362] J. Cross: *Anionic Surfactants — Chemical Analysis,* **vol. 8,** Marcel Dekker, New York-Basel 1977.

[363] *Fette Seifen Anstrichm.* **73** (1971) 683.

[364] H. R. Hoffmann, W. Böer, G. W. G. Schwarz, *Fette Seifen Anstrichm.* **78** (1976) 367 – 368.

[365] "Benzalkonium (Chlorure de)," *Pharmacopée Française* 9th ed., Adapharus, Paris 1972.

[366] R. Wickbold, *Seifen Oele Fette Wachse* **85** (1959) 415 – 416.

[367] D. C. White, *Mikrochim. Acta* 1959, 254 – 259.

[368] E. E. Archer, *Analyst (London)* **82** (1957) 208 –209.

[369] D. Miller, *Comun. Jorn. Com. Esp. Deterg.* **21** (1990) 99 – 112.

[370] H. Bauer, *Seifen Oele Fette Wachse* **121** (1995) no. 3, 168 – 170.

[371] R. Hwang, *J. Test. Eval.* **14** (1986) no. 2, 128 – 130.

[372] L. Baini, G. Bolzoni, G. Carrer, E. Facetti, L. Valtora, E. Guiati, C. Pacchetti, C. Ruffo, O. Cozzoli, L. Sedea, *Riv. Ital. Sostanze Grasse* **69** (1992) no. 12, 615 – 617.

[373] AFNOR, Doc. Française A-124 (7. 7. 70)

[374] R. E. Kitson, M. G. Mellon, *Ind. Eng. Chem. Anal. Ed.* **16** (1944) 379 – 383.

[375] H. Bernhardt, (Ausschuß Phosphate und Wasser), *Zeitschr. f. Wasser Abwasser Forsch.* **7** (1974) 143 – 146.

[376] C. Burgi, R. Bollhalder, *Tenside, Surfactants, Deterg.* **31** (1994) no. 2, 86 – 89.

[377] G. van Raay, *Tenside Deterg.* **7** (1970) 125 – 132.

[378] H. Jaag: *Über proteolytische Enzyme, deren Prüfmethoden und Einsatzmöglichkeiten in der Waschmittelindustrie,* Verlag H. Lang & Cie AG, Bern 1968.

[379] T. M. Rothgeb, B. D. Goodlander, P. H. Garrison, L. A. Smith, *J. Am. Oil Chem. Soc.* **65** (1988) no. 5, 806 – 810.

[380] Boehringer, Mannheim: *Methoden der enzymatischen Lebensmittelanalytik,* 1980.

[381] E. Klein, J. Freiberg, S. Same, M. A. Carroll, *J.-Assoc. Off. Anal. Chem.* **72** (1989) no. 6, 881 – 882.

[382] N. Buschmann, S. Wodarczak, *Tenside Surfactants Deterg.* **32** (1995) no. 4, 336 – 339.

[383] F. Hui, P. Garcia-Camacho, R. Rosset, *Analysis* **23** (1995) no. 2, 58 – 65.

[384] L. K. Wang, *J. Am. Oil Chem. Soc.* **52** (1975) 339 – 344.

[385] K. Toel, H. Fujii, *Anal. Chim. Acta* **90** (1977) 319 – 322.

[386] St. Janeva, R. Borissova-Pangarova, *Talanta* **25** (1977) 279 – 282.

[387] W. Kupfer, *Tenside Deterg.* **12** (1975) 40.

[388] G. Krusche, J. Illert, H. Mandery, *Chromatographia* **31** (1991) no. 1 – 2, 17 – 20.

[389] R. Hermann, C. Th. J. Alkemade: *Flammenphotometrie,* Springer Verlag, Berlin-Göttingen-Heidelberg 1960.

[390] B. Welz: *Atom-Absorptions-Spektroskopie,* Verlag Chemie, Weinheim 1972.

[391] H. Bernd, E. Jackwerth, *Spectrochim. Acta Part B* **30 B** (1975) 169 – 177.

[392] G. Volland, G. Kölblin, P. Tschöpel, G. Tölg, *Fresenius' Z. Anal. Chem.* **284** (1977) 1 – 12.

[393] I. Ebdon, A. S. Fisher, *Anal. Spectrosc. Libr. (At. Absorpt. Spectrom.)* **5** (1991) 463 – 514.

[394] E. Jackwerth, J. Lohmar, G. Wittler, *Fresenius' Z. Anal. Chem.* **266** (1973) 1 – 8.

[395] A. Dornemann, H. Kleist, *Fresenius' Z. Anal. Chem.* **291** (1978) 353 – 359.

[396] M. H. Arab-Zavar, A. G. Howard, *Analyst (London)* **105** (1980) no. 1253, 744 – 750.

[397] J. Vilaplana, F. Grimalt, C. Romaguero, J. M. Mascaro, *Contact Dermatitis* **16** (1987) no. 3, 139 – 141.

[398] K. E. Jarvis, J. G. Williams, B. C. H. Gibson, E. Temmerman, C. De Cuyper, *Analyst (London)* **121** (1996) no. 12, 1929 – 1933.

[399] R. O. Müller: *Spektrochemische Analyse mit Röntgenfluoreszenz,* R. Oldenbourg Verlag, München 1967.

[400] E. Vaeth, E. Griessmayr, *Fresenius' Z. Anal. Chem.* **303** (1980) 268 – 271.

[401] E. Schenbeck, Ch. Jörrens, *Fresenius' Z. Anal. Chem.* **303** (1980) 257 – 264.

[402] J. Williams, *Analyst (London)* **111** (1986) no. 2, 175 – 177.

[403] H. Krischner: *Einführung in die Röntgenfeinstrukturanalyse,* Vieweg & Sohn, Braunschweig 1974.

[404] J. M. Mabis, O. T. Quimby, *Anal. Chem.* **25** (1953) 1814 – 1818.

[405] D. W. Breck: *Zeolite Molecular Sieves: Structure, Chemistry and Use,* J. Wiley & Sons, New York 1974, pp. 347 – 378.

[406] K. Lötzsch, *Seifen Oele Fette Wachse* **105** (1979) 261 – 267.

[407] F. Y. Iskander, *Appl. Radiat. Iso.* **37** (1986) no. 5, 435 – 437.

[408] J. Saint-Pierre, L. Zikovsky, *J. Radioanal. Chem.* **71** (1982) no. 1 – 2, 19 – 28.

[409] G. Zweig, J. Sherma: *Handbook of Chromatography,* CRC-Press, Cleveland, Ohio, 1972.

[410] F. Korte: *Methodicum Chimicum,* vol. 1: *Analyticum,* Part 1, G. Thieme Verlag, Stuttgart 1973.

[411] K. Heinig, C. Vogt, G. Werner, *J. Capillary Electrophor.* **3** (1996) no. 5, 261 – 270.

[412] M. W. F. Nielen, *J. Chromatogr.* **712 A** (1995) no. 1, 269 – 284.

[413] J. P. Wiley, *J. Chromatogr.* **692 A** (1995) no. 1+2, 267 – 274.

[414] J. M. Hais, K. Macek: *Handbuch der Papierchromatographie,* VEB Fischer Verlag, Jena 1963.

[415] H. Grunze, E. Thilo: *Die Papierchromatographie der kondensierten Phosphate,* Sitzungsberichte der Deutschen Akademie der Wissenschaften zu Berlin, Akademie Verlag, Berlin 1955.

[416] E. Heinerth, J. Pollerberg, *Fette Seifen Anstrichm.* **61** (1959) 376 – 377.

[417] R. E. Cline, R. M. Fink, *Anal. Chem.* **28** (1956) 47 – 52.

[418] J. C. Touchstone, M. F. Dobbins: *Practice of the Thin Layer Chromatography,* J. Wiley & Sons, New York 1978.

[419] F. Geiss: *Die Parameter der DC,* Vieweg und Sohn, Braunschweig 1972.

[420] E. Heinerth, *Fette Seifen Anstrichm.* **70** (1968) 495 – 498.

[421] A. Breyer, M. Fischl, E. Setzer, *J. Chromatogr.* **82** (1973) 37 – 52.

[422] K. Bey, *Fette Seifen Anstrichm.* **67** (1965) 217 – 221.

[423] U. Hezel, *Angew. Chem.* **85** (1973) 334 – 342; *Angew. Chem. Int. Ed. Engl.* **12** (1973) 298.

[424] J. Touchstone, T. Murawec, M. Kasparow, W. Wortmann, *J. Chromatogr. Sci.* **10** (1972) 490 – 493.

[425] D. H. Liem, *Cosmet. Toiletries* **92** (1977) 59 – 72.

[426] H. König, *Fresenius' Z. Anal. Chem.* **251** (1970) 359 – 368.

[427] R. Matissek, E. Hieke, U. Baltes, *Fresenius' Z. Anal. Chem.* **300** (1980) 403 – 406.

[428] M. Köhler, B. Chalupka, *Fette Seifen Anstrichm.* **84** (1982) 208 – 211.

[429] R. Matissek, *Tenside Deterg.* **19** (1982) 57 – 66.

[430] J. L. Jasperse, P. H. Steiger, *J. Am. Oil Chem. Soc.* **69** (1992) no. 7, 621 – 625.

[431] S. J. Patterson, E. C. Hunt, K. B. E. Tucker, *J. Proc. Inst. Sewage Purif.* 1966, 190.

[432] Ch. Buergi, T. Otz, *Tenside, Surfactants, Deterg.* **32** (1995) no. 1, 22 – 24.

[433] G. Hesse: *Chromatographisches Praktikum,* Akademische Verlagsges., Frankfurt 1968.

[434] K. Ehlert, R. Engler, *GIT Fachz. Lab.* **23** (1979) 659 – 664.

[435] P. Quinlin, H. J. Weiser, *J. Am. Oil Chem. Soc.* **35** (1958) 325 – 327.

[436] R. Wickbold, *Fette Seifen Anstrichm.* **74** (1972) 578 – 579.

[437] W. Kupfer, K. Künzler, *Fette Seifen Anstrichm.* **74** (1972) 287 – 291.

[438] M. J. Rosen, *Anal. Chem.* **35** (1963) 2074 – 2077.

[439] R. E. Kaiser: *Chromatographie in der Gasphase,* Bibliographisches Institut, Mannheim 1973.

[440] G. Schomburg, R. Dielmann, H. Husmann, F. Weeke, *J. Chromatogr.* **122** (1976) 55 – 72.

[441] H. J. Vonk et al.: *Int. Surfactants Congress, Moscow 1976,* **vol. 1, Section A,** pp. 435 – 449.

[442] G. Schomburg: *Gaschromatographie,* Verlag Chemie, Weinheim 1977.

[443] A. Kuksis, *Fette Seifen Anstrichm.* **73** (1971) 130 – 138.

[444] H. Hadorn, K. Zürcher, *Dtsch. Lebensm. Rundsch.* **66** (1970) 77 – 87.

[445] A. F. Prevot, F. X. Mordret, *Rev. Fr. Corps Gras* **23** (1976) 409 – 423.

[446] L. Julin, P. Sunila, Proc.-Scand. Symp. Lipids, 12th (1984), R. Marcuse (ed.), LIPIDFORUM, Goeteborg, 57 – 64.

[447] W. J. Carnes, *Anal. Chem.* **36** (1964) 1197 – 1200.

[448] E. Link, H. M. Hickman, R. A. Morrissette, *J. Am. Oil Chem. Soc.* **36** (1959) 20 – 23, 300 – 303.

[449] D. G. Anderson, *J. Paint. Technol.* **40** (1968) 549 –557.

[450] B. Bonney, N. F. Glennard, A. M. Humphrey, *Chem. Ind. (London)* 1973, 749 – 751.

[451] R. Schicker, *Aerosol Rep.* **13** (1974) 149 – 163.

[452] R. Schubert, L. Ketel, *J. Soc. Cosmet. Chem.* **23** (1972) 115 – 124.

[453] Hoechst, DE-OS 2 161 702, 1971.

[454] T. J. Birkel, C. R. Warner, T. Fario, *J. Am. Oil Chem. Soc.* **62** (1979) 931 – 936.

[455] C. Sommer, *Dtsch. Lebensm.-Rundsch.* **89** (1993) no. 4, 108 – 111.

[456] L. Sedea, G. Toninelli, B. Sartorel, *Tenside Deterg.* **22** (1985) no. 5, 269 – 271.

[457] F. Etzweiler, N. Neuner-Jehle, E. Senn, *Seifen Oele Fette Wachse* **106** (1980) no. 15, 419 – 427.

[458] P. Sandra, F. David, *J. High Resolut. Chromatogr.* **13** (1990) no. 6, 414 – 417.

[459] H. Engelhardt: *Hochdruck-Flüssigkeits-Chromatographie — HPLC,* Springer Verlag, Berlin-Göttingen-Heidelberg-New York 1977.

[460] J. J. Kirkland: *Modern Practise of Liquid Chromatography,* Wiley-Interscience, New York 1971.

[461] H. M. McNair, *Int. Lab.* 1980, no. 5/6, 51 – 59.

[462] L. P. Turner et al., *J. Am. Oil Chem. Soc.* **53** (1976) 691 – 694.

[463] H. Henke, *Tenside Deterg.* **15** (1978) 193 – 195.

[464] R. Murphy, A. C. Selden, M. Fisher, E. A. Fagan, V. S. Chadwick, *J. Chromatogr.* **211** (1981) 160 – 165.

[465] J. H. van Dijk et al., *Tenside Deterg.* **12** (1975) 261 – 263.

[466] L. Sedea, *Tenside, Surfactants, Deterg.* **26** (1989) no. 3, 211 – 214.

[467] J. Weiss, *Tenside, Surfactants, Deterg.* **23** (1986) no. 5, 237 – 244.

[468] J. Truchan, H. T. Rasmussen, N. Omelczenko, B. P. McPherson, *J. Liq. Chromatogr. Relat. Technol.* **19** (1996) no. 11, 1785 – 1792.

[469] D. Kirkpatrick, *J. Chromatogr.* **121** (1976) 153 – 154.

[470] B. P. McPherson, N. Omelczenko, *J. Am. Oil Chem. Soc.* **57** (1980) no. 11, 388 – 391.

[471] G. Micali, P. Curro, G. Calabro, *Analyst (London)* **109** (1984) no. 2, 155 – 158.

[472] Th. Wolf, D. Semionow, *J. Soc. Cosmet. Chem.* **24** (1973) 363 – 370.

[473] A. Nozawa, T. Ohnuma, *J. Chromatogr.* **187** (1980) 261 – 263.

[474] A. Marcomini, F. Filipuzzi, W. Giger, *Chemosphere* **17** (1988) no. 5, 853 – 863.

[475] A. Marcomini, W. Giger, *Anal. Chem.* **59** (1987) no. 13, 1709 – 1715.

[476] H. Yoshimura, T. Sugiyama, T. Nagai, *J. Am. Oil Chem. Soc.* **64** (1987) no. 4, 550 – 555.

[477] A. Marcomini, S. Stelluto, B. Pavoni, *Int. J. Environ. Anal. Chem.* **35** (1989) no. 4, 207 – 218.

[478] H. Waldhoff, J. Scherler, M. Schmitt, World Surfactants Congr., 4th (CESIO) (1996), Volume 1, 507 – 518.

[479] K. Nakamura, Y. Morikawa, I. Matsumoto, *J. Am. Oil Chem. Soc.* **58** (1981) Jan., 72 – 77.

[480] H. Ullner, I. König, C. Sander, Schwenk, *Tenside Deterg.* **17** (1980) 169 – 170.

[481] N. Chen, Jr., V. P. Nero, *J. Chromatogr.* **549** (1991) no. 1 – 2, 247 – 256.

[482] H.-U. Ehmcke, H. Kelker, K.-H. König, H. Ullner, *Fresenius' Z. Anal. Chem.* **294** (1979) 251 – 261.

[483] D. R. Zornes et al., *Petr. Eng. J.* **18** (1978) 207 – 218.

[484] W. Winkle, M. Köhler, *Chromatographia* **13** (1980) 357 – 363.

[485] K. J. Irgolic, J. E. Hobill, *Spectrochim. Acta, Part B* **42 B** (1987) no. 1 – 2, 269 – 273.

[486] H. Wagner, G. Metzner, R. R. Wagner, *Muench. Beitr. Abwasser-, Fidch.- Flussbiol. (Umweltrverträglichkeit Wasch- Reinigungsm.)* **44** (1990) 218 – 230.

[487] P. L. Buldini, J. Sharma, D. Ferri, *J. Chromatogr.* **654** (1993) no. 1, 129 – 134.

[488] J. Weiß (ed.): *Ion chromatography,* Wiley-VCH, Weinheim 1995.

[489] J. Weiß, *Tenside, Surfactants, Deterg.* **23** (1986) no. 5, 237 – 244.

[490] E. Vaeth, P. Sladek, K. Kenar, *Fresenius' Z. Anal. Chem.* **329** (1987) no. 5, 584 – 589.

[491] W. D. MacMillan, *HRC CC, J. High Resolut. Chromatogr. Chromatogr. Commun.* **7** (1984) no. 2, 102 – 103.

[492] P. Linares, M. D. Luque de Castro, M. Valcarcel, *J. Chromatogr.* **585** (1991) no. 2, 267 – 271.

[493] H. Felber, K. Hegetschweiler, M. Mueller, R. Odermatt, B. Wampfler, *Chimia* **49** (1995) no. 6, 179 – 181.

[494] T. K. Mahabir-Jagessar, S. A. M. van Stroe-Biezen, *J. Chromatogr. A* **771** (1997) no. 1+2, 1155 – 1161.

[495] E. L. Pretswell, A. R. Morrison, J. S. Park, *Analyst (Cambridge, U.K.)* **118** (1993) no. 10, 1265 – 1267.

[496] T. Wang, S. F. Y. Li, *J. Chromatogr. A* **723** (1996) no. 1, 197 – 205.

[497] J. M. Jordan, R. L. Moese, R. Johnson-Watts, D. E. Burton, *J. Chromatogr. A* **671** (1994) no. 1 – 2, 445 – 451.

[498] M. Pestemer: *Anleitung zum Messen von Absorptionsspektren im Ultraviolett und Sichtbaren,* Thieme Verlag, Stuttgart 1964.

[499] E. J. Stearns: *The Practice of Absorption Spectrophotometry,* Wiley-Interscience, New York 1969.

[500] C. N. R. Rao: *Ultraviolet and Visible Spectroscopy,* 3rd ed., Butterworths, London 1975.

[501] G. Milazzo et al., *Anal. Chem.* **49** (1977) no. 6, 711 – 717.

[502] W. J. Weber, J. C. Morris, W. Stumm, *Anal. Chem.* **34** (1962) 1844 – 1845.

[503] R. M. Kelley, E. W. Blank, W. E. Thompson, R. Fine, *ASTM Bull.* **TP 90** (1959) 70 – 73.

[504] M. Uchijama, *Water Res.* **11** (1977) 205 – 207.

[505] E. Heinerth, H. G. van Raay, G. Schwarz, *Fette Seifen Anstrichm.* **62** (1960) 825 – 826.

[506] H. Hellmann, *Tenside, Surfactants, Deterg.* **31** (1994) no. 3, 200 – 206.

[507] V. W. Reid, T. Alston, B. W. Young, *Analyst (London)* **80** (1955) 682 – 689.

[508] F. N. Stewart et al., *Anal. Chem.* **31** (1959) no. 11, 1806 – 1808.

[509] H. J. Hediger: „Infrarotspektroskopie," *Methoden der Analyse in der Chemie,* **vol. 11** Akademische Verlagsges., Frankfurt 1971.

[510] L. J. Bellamy: *Ultrarot-Spektrum und chemische Konstitution,* D. Steinkopff Verlag, Darmstadt 1966.

[511] Weissberger: *Technique of Organic Chemistry,* vol. IX: "Chemical Application of Spectroscopy," Interscience Publ. Inc., New York 1956.

[512] H. König: *Neuere Methoden zur Analyse von Tensiden,* Springer Verlag, Berlin-Göttingen-Heidelberg-New York 1971.

[513] E. Kunkel, *Mikrochim. Acta* 1977, 227 – 240.

[514] H. Günzler, H. Böck: *IR-Spektroskopie,* Verlag Chemie, Weinheim 1975.

[515] G. Socrates: *Infrared Characteristic Group Frequencies,* J. Wiley & Sons, New York 1980.

[516] Z. Huren, D. Hajman, D. Vucelic, *J. Serb. Chem. Soc.* **58** (1993) no. 9, 721 – 724.

[517] H. G. van Raay, M. Teupel, *Fette Seifen Anstrichm.* **75** (1973) 572 – 578.

[518] C. D. Frazee, R. O. Crister, *J. Am. Oil Chem. Soc.* **41** (1964) 334 – 335.

[519] H. Hellmann, *Fresenius' Z. Anal. Chem.* **294** (1979) 379 – 384.

[520] H. Hellmann, *Fresenius' Z. Anal Chem.* **293** (1978) 359 – 363.

[521] P. Friese, *Fresenius' Z. Anal Chem.* **305** (1981) 337 – 346.

[522] H. Günther: *NMR-Spektroskopie – Eine Einführung,* Thieme Verlag, Stuttgart 1973.

[523] H. König, *Fresenius' Z. Anal. Chem.* **251** (1970) 225 – 262.

[524] A. Mathias, N. Mellor, *Anal. Chem.* **38** (1966) 472 – 477.

[525] M. M. Crutchfield, R. R. Irani, J. T. Yoder, *J. Am. Oil Chem. Soc.* **41** (1964) 129 – 132.

[526] K. H. Choe, S. J. Kwon, *Taehan Hwahakhoe Chi* **27** (1983) no. 6, 419 – 423.

[527] K. H. Choe, S. J. Kwon, *Taehan Hwahakhoe Chi* **25** (1981) no. 1, 54 – 56.

[528] H. Budzikiewicz: *Massenspektrometrie – Eine Einführung,* Verlag Chemie, Weinheim 1972 (Taschentext 5).

[529] M. Köhler, M. Höhn, *Chromatographia* **9** (1976) 611 – 617.

[530] H. A. Boekenoogen: *Oils, Fats and Fat Products,* vol. 2, Interscience Publ., New York 1968.

[531] D. Jahr, P. Binnemann, *Fresenius' Z. Anal. Chem.* **298** (1979) 337 – 348.

[532] W. McFadden: *Techniques of Combined Gas Chromatography/Mass Spectrometry,* J. Wiley &Sons, London 1973.

[533] R. ter Heide, N. Provatoroff, P. C. Traas, P. J. Valois, *J. Agric. Food Chem.* **23** (1975) 950 – 957.

[534] T. Takeuchi, S. Watanabe, N. Kondo, M. Goto, D. Ishii, *Chromatographia* **25** (1988) no. 6, 523 – 525.

[535] A. L. Rockwood, T. Higuchi, *Tenside, Surfactants, Deterg.* **29** (1992) no. 1, 6 – 12.

[536] P. Rudewicz, B. Muson, *Anal. Chem.* **58** (1986) no. 4, 674 – 679.

[537] E. Stephanou, *Chemosphere* **13** (1984) no. 1, 43 – 51.

[538] M. Köhler, *Chromatographia* **8** (1975) 685 – 689.

[539] F. Ehrenberger, S. Gorbach: *Methoden der organischen Elementar- und Spurenanalyse,* Verlag Chemie, Weinheim 1973.

[540] J. Monar, *Mikrochim. Acta* **2** (1965) 208 – 250.

[541] R. Wickbold, *Angew. Chem.* **69** (1957) 530 – 533.

[542] F. Ehrenberger, *GIT Fachz. Lab.* **21** (1977) 944.

[543] R. Kaiser: *Quantitative Bestimmung Organischer Funktioneller Gruppen,* Akademische Verlagsges., Frankfurt 1966.

[544] *Technicon-Symposia, Technicon Intern. Congress 1969,* vol. II, White Plains, Mediad 1970.

[545] D. P. Lundgren, *Anal. Chem.* **32** (1960) 824 – 828.

[546] *Technicon Industrial Method 3 – 68 W,* Technicon Corporation Tarrytown/New York 10591.

[547] W. Gohla et al., *GIT Fachz. Lab.* **23** (1979) 89 – 96.

[548] H. Pfeiffer, K. H. Lange, *Fette Seifen Anstrichm.* **75** (1973) 438 – 442.

[549] C. Pipinato, L. Sedea, S. Marinaz, G. Toninelli, *Fette, Seifen, Anstrichm.* **82** (1980) no. 7, 283 – 289.

[550] E. Heinerth, *Tenside Deterg.* **7** (1970) 23 – 24.

[551] H. Grossmann, *Seifen Oele Fette Wachse* **101** (1975) 521 – 524.

[552] *Fette Seifen Anstrichm.* **77** (1975) 80.

[553] Manufacturer: Atlas Electric Devices Comp., 4114 N. Ravenswood Ave., Chicago.

[554] Manufacturer: Heraeus-Original Hanau, D-6450 Hanau.

[555] Manufacturer: United States Testing Co. Inc., Hoboken N.J.

[556] DIN 53 902, Part 1: Bestimmung des Schäumvermögens, Lochscheibenschlagverfahren.

[557] DIN 53 902, Part 2: Bestimmung des Schäumvermögens, modifiziertes Ross-Miles-Verfahren.

[558] Manufacturer: wfk-Forschungsinstitut, Campus Fichtenhain 11, D-47807 Krefeld. EMPA Eidgenössische Materialprüfungs- und Versuchsansalt, Lerchenfeldstrasse 5, CH-9014 St. Gallen.

[559] H. Stache in: *Tensid-Taschenbuch,* 2nd ed., Hanser Verlag, München-Wien 1981, pp. 460 – 461.

[560] H. Stache in: *Tensid-Taschenbuch,* 2nd ed., Hanser Verlag, München-Wien 1981, pp. 462 – 466.

[561] U. Sommer, H. Milster, *Seifen Oele Fette Wachse* **103** (1977) 295.

[562] H. Harder, D. Arends, W. Pochandke, *Seifen Oele Fette Wachse* **102** (1976) 421 – 426.

[563] H. Krüssmann, *J. Am. Oil Chem. Soc.* **55** (1978) 165.

[564] H. Harder, *Seifen Oele Fette Wachse* **94** (1968) 789 – 794, 825.

[565] H. Harder, D. Arends, W. Pochandke, *Seifen Oele Fette Wachse* **103** (1977) 180 – 183.

[566] DIN 53 990:Waschmittel zum Waschen von Textilien, Empfehlungen für die vergleichende Prüfung von Gebrauchseigenschaften.

[567] ISO TC 91-4319- 1977: Surface active agents – Detergents for washing fabrics – Guide for comparative testing of performance.

[568] A. Trius, B. Brackmann, Proceedings of the 5th World Surfactants Congress (CESIO), Firenze 2000, 56 – 72.

[569] "Higher Alcohols Market Forecast", Colin A. Houston & Associates, Inc., Pound Ridge NY 2000.
[570] M. Elton, *Soap, Cosmet., Perfum.* **73** (2000) no. 22, 51 – 56.
[571] Hauptausschuß Phosphate: *Phosphor, Wege und Verbleib in der Bundesrepublik Deutschland.* Verlag Chemie, Weinheim-New York 1978, pp. 29 – 30.
[572] E. Roland, P. Kleinschmit: "Zeolites" in *Ullmann's Encyclopedia of Industrial Chemistry,* Sixth Edition, 2001, Electronic Release, Wiley-VCH, Weinheim 2001.
[573] Statistisches Bundesamt 1998.
[574] R. Wagner, *GWF, Gas Wasserfach: Wasser/Abwasser* **119** (1978) 235 – 242.
[575] B. Böhnke, Gutachten zur Belastung der Gewässer durch Waschmittelinhaltstoffe aus Regenüberläufen und kommunalen Kläranlagen mit alleiniger mechanischer Reinigung. Industrieverband Körperpflege- und Waschmittel (IKW), Frankfurt, Dezember 1991.
[576] B. Böhnke, Gutachten zur kostenmäßigen Beurteilung des Verhaltens von Waschmitteln in kommunalen Kläranlagen. Industrieverband Körperpflege- und Waschmittel (IKW), Frankfurt, Dezember 1990.
[577] Europäisches Übereinkommen über die Beschränkung der Verwendung bestimmter Detergentien in Wasch- und Reinigungsmitteln vom 16. September 1968, BGBl 1968 I, p. 1718.
[578] Council directive 73/404/EEC, of 22nd November 1973 on the approximation of the laws of the Member States relating to detergents. OJ L347, p. 51.
[579] Council directive 73/405/EEC, of 22nd November 1973 on the approximation of the laws of the Member States relating to detergents. OJ L347, p. 53.
[580] J. Au, *Tenside, Deterg.* **18** (1981) 280 – 285.
[581] Gesetz über die Umweltverträglichkeit von Wasch- und Reinigungsmitteln (Waschmittelgesetz) vom 20.08.1975, BGBl 1975 I, p. 2255.
[582] Verordnung über die Abbaubarkeit anionischer und nichtionischer grenzflächenaktiver Stoffe in Wasch- und Reinigungsmitteln vom 30.01.1977 BGBl 1977 I, p. 244 (geändert durch Verordnung vom 18.06.1980, BGBl I, p. 706.)
[583] Verordnung über Höchstmengen für Phosphate in Wasch- und Reinigungsmitteln (Phosphathöchstmengenverordnung — PHöchstMengV) vom 04.06.1980, BGBl 1980 I, p. 664.
[584] Council Directive 82/243/EEC, of 31 March 1982 amending Directive 73/405/EEC on the approximation of the laws of the Member States relating to methods of testing the biodegradability of anionic surfactants. OJ L109, p. 18.
[585] Council Directive 82/242/EEC, of 31 March 1982 amending Directive 73/405/EEC on the approximation of the laws of the Member States relating to methods of testing the biodegradability of nonionic surfactants. OJ L109, p. 1.
[586] R. Wickbold, *Tenside* **9** (1972) 173 – 177.
[587] D. Brown: "Introduction to surfactant biodegradation", in D. R. Karsa, M. R. Porter (eds.): *Biodegradability of surfactants,* Blackie Academic & Professional, London 1995, p. 1.
[588] Proposed method for the determination of the biodegradability of surfactants used in synthetic detergents. OECD, Paris 1976.
[589] Hauptausschuß Phosphate, Phosphor, Wege und Verbleib in der Bundesrepublik Deutschland, VCH, Weinheim, Germany 1978.
[590] P. Gerike, K. Winkler, W. Jakob, *Tenside Surf. Det.* **26** (1989) 270 – 275.
[591] P. Gerike, K. Winkler, W. Schneider, W. Jakob, J. Steber, *Tenside Surf. Det.* **28** (1991) 86 – 89.

[592] Commission Directive 93/67/EEC on the assessment of risks to man and the environment of substances notified in accordance with Directive 67/584/EEC on dangerous substances. Off. J. Europ. Comm. L227 of Sept. 8, 1993.

[593] Council Regulation (EEC) No. 793/93 of 23 March 1993 on the evaluation and control of the risks of existing substances. OJ L84. p. 1.

[594] Technical Guidance Documents in support of the Commission Directive 93/67/EEC on risk assessment for new notified substances and the Commission Regulation (EC) 1488/94 on risk assessment for existing substances, ECB, Ispra April 19th, 1996.

[595] OECD Guidelines for testing of chemicals, Volume 1, OECD, Paris 1993.

[596] J. Waters, T. C. J. Feijtel, *Chemosphere* **30** (1995) 1939 – 1956.

[597] P. Gerike, W. Holtmann, W. Jasiak, *Chemosphere* **13** (1984) 121 – 141.

[598] J. Steber, *Textilveredlung* **26** (1991) 348 – 354.

[599] R. Birch, C. Biver, R. Campagna, W. E. Gledhill, U. Pagga, J. Steber, H. Reust, W. J. Bontinck, *Chemosphere* **19** (1989) 1127 – 1550.

[600] ECETOC, Technical report No. 61: "Environmental exposure assessment", European Centre for Ecotoxicology and Toxicology of Chemicals (ECETOC), Brussels, November 1992.

[601] T. Feijtel, E. J. van de Plassche, Environmental risk characterization of 4 major surfactants used in The Netherlands, RIVM report No. 679101, National Institute of public health and environmental protection, Bilthoven 1995.

[602] AISE/CESIO: "Environmental risk assessment of detergent chemicals", Proceedings of the A.I.S.E./CESIO Limelette III Workshop on 28 – 29 November 1995. AISE Brussels 1996.

[603] R. Schröder, *Tenside Surf. Det.* **32** (1995) 492 – 497.

[604] J. Steber: "Biodegradation kinetic data for environmental risk assessment on surface waters and soil," in S. G. Hales, T. Feijtel, H. King, K. Fox, W. Verstraete (eds.): Biodegradation kinetics, Generation and Use of Data for Regulatory Decisions Making, Proceedings of the SETAC-Europe workshop 1996, SETAC Europe, Brussels, p. 81.

[605] Council Directive 92/32/EEC of 30.04.1992 amending for the seventh time, Directive 67/548/EEC on the approximation of the laws, regulations and administrative provosion relating to the classification, packaging and labelling of dangerous substances. OJ L154, p. 1.

[606] Verordnung über Anmeldeunterlagen und Prüfnachweise nach dem Chemikaliengesetz (ChemG Anmelde- und PrüfnachweisV), Änderung vom 31.5.1989, BGBl 1989 I, p. 1074.

[607] DIN 38412, L127: "German standard methods for the examination of water, waste water and sludge; bio-assays (group L); determination of the inhibitory effect of waste water on the oxygen consumption of Pseudomonas putida", November 1992.

[608] EN ISO 10712, "Water quality — Pseudomonas putida growth inhibition test (Pseudomonas cell multiplication inhibition test), December 1995.

[609] W. Guhl, P. Gode, *Vom Wasser* **72** (1989) 165 – 173.

[610] Allgemeine Verwaltungsvorschrift zum Wasserhaushaltsgesetz über die Einstufung wassergefährdender Stoffe in Wassergefährdungsklassen vom 17. Mai 1999. Bundesanzeiger, Jahrgang 51, Nr. 98a, ausgegeben am 29.05.1999.

[611] W. Guhl, *Z. Angew. Zool.* **74** (1987) 385 – 409.

[612] W. Guhl, P. Gode, *Tenside Surf. Det.* **26** (1989) 282 – 287.

[613] W. J. Lyman, E. F. Reehl, D. H. Rosenblatt (eds.): *Handbook of chemical property estimation methods,* McGraw Hill Book Comp., New York 1982.

[614] ECETOC, Technical Report No. 67: "The role of bioaccumulation in environmental risk assessment: The aquatic environment and related food webs". European Centre for Ecotoxicology and Toxicology of Chemicals (ECETOC). Brussels, March 1995.

[615] W. K. Fischer, K. Winkler, *Vom Wasser* **47** (1976) 81 – 129.

[616] P. Gerike, K. Winkler, W. Schneider, W. Jakob, *Tenside Surf. Det.* **26** (1989) 21 – 26.

[617] R. Schröder, *SOFW-J.* **121** (1995) 420 – 427.

[618] J. Steber, H. Berger: "Biodegradability of anionic surfactants", in D. R. Karsa, M. R. Porter (eds.): *Biodegradability of surfactants,* Blackie Academic & Professional, London 1995, p. 134.

[619] P. Berth, P. Gerike, P. Gode, J. Steber, *Tenside Surf. Det.* **25** (1988) 108 – 115.

[620] P. Schöberl, *Tenside Surf. Det.* **32** (1995) 25 – 35.

[621] J. Tolls, P. Kloepper-Sams, D. T. H. M. Sijm, *Chemosphere* **29** (1994) 693 – 717.

[622] J. Tolls, M. Haller, I. de Graaf, M. A. T. C. Thijssen, D. T. H. M. Sijm, *Environ. Sci. Technol.* **31** (1997) 3426 – 3431.

[623] J. Steber, P. Wierich, *Appl. Environ. Microbiol.* **49** (1985) 530 – 537.

[624] J. Steber, P. Wierich, *Water. Res.* **21** (1987) 661 – 667.

[625] H. Hellmann, *Vom Wasser* **64** (1985) 29 – 42.

[626] C. S. Newsome, D. Howes, S. J. Marshal, R. A. van Egmont, *Tenside Surf. Det.* **32** (1995) 498 – 503.

[627] Shell Chemicals U.K.Ltd., *Chim. Oggi* **8** (1983) 19 – 22, 53.

[628] E. Stephanou, W. Giger, *Environ. Sci. Technol.* **16** (1982) 800 – 805.

[629] M. Tschui, P. H. Brunner, *Vom Wasser* **65** (1985) 9 – 19.

[630] J. Steber, W. Guhl, N. Stelter, F. R. Schröder, *Tenside Surf. Det.* **32** (1995) 515 – 521.

[631] H. A. Painter: "Biodegradability testing", in D. R. Karsa, M. R. Porter (eds.): *Biodegradability of surfactants,* Blackie Academic & Professional, London 1995, p. 65.

[632] P. Gerike, *Tenside Surf. Det.* **19** (1982) 162 – 164.

[633] H. Hellmann, *Z. Wasser Abwasser Forsch.* **22** (1989) 131 – 137.

[634] J. Waters, H. H. Kleiser, M. J. How, M. D. Barratt, R. R. Birch, R. J. Fletcher, S. D. Haigh, S. G. Hales, S. J. Marshal, T. C. Pestell, *Tenside Surf. Det.* **28** (1991) 460 – 468.

[635] D. J. Versteeg, T. C. J. Feijtel, C. E. Cowan, T. E. Ward, R. A. Rapaport, *Chemosphere* **24** (1992) 641 – 662.

[636] Umweltbundesamt: Die Prüfung des Umweltverhalten von Natrium-Aluminium-Silikat Zeolith A als Phosphatersatzstoff in Wasch- und Reinigungsmitteln, Materialien 4/1979, E. Schmidt-Verlag, Berlin 1979.

[637] C. P. Kurzendörfer, P. Kuhm, J. Steber: "Zeolites in the environment" in M. J. Schwuger (ed.): *Detergents in the environment,* Marcel Dekker Inc., New York 1997, p. 127.

[638] P. Gode, *Z. Wasser Abwasser Forsch.* **16** (1983) 210 – 219.

[639] A. W. Maki, K. J. Macek, *Environ. Sci. Technol.* **12** (1978) 573 – 580.

[640] H. E. Allen, S. H. Cho, T. A. Neubecker, *Water Res.* **17** (1983) 1871 – 1879.

[641] H.-J. Opgenorth: "Polymeric materials polycarboxylates", in O. Hutzinger (ed.): *The Handbook of environmental chemistry,* **Vol. 3, Part F,** Springer-Verlag Berlin Heidelberg 1992, p. 337.

[642] ECETOC, Joint Assessment of Commodity Chemicals No. 23: "Polycarboxylate polymers as used in detergents." European Centre for Ecotoxicology and Toxicology of Chemicals (ECETOC) Brussels 1993.

[643] P. Schöberl, L. Huber, *Tenside Det. Surf.* **25** (1988) 99 – 107.

[644] H. L. Hoyt, H. L. Gewanter: "Citrate" in O. Hutzinger (ed.): *The Handbook of environmental chemsitry,* **Vol. 3, Part F,** Springer-Verlag Berlin Heidelberg 1992, p. 229.

[645] H. Bernhardt: *Studie über die aquatische Umweltverträglichkeit von Nitrilotriacetat,* Verlag Hans Richarz, Sankt Augustin 1984.

[646] "Aquatische Umweltverträglichkeit von Nitrilotriessigsäure (NTA)", Kernforschungszentrum Karlsruhe GmbH, Projektträger Wassertechnologie und Schlammbehandlung (PtWT) 1991.

[647] G. Müller, U. Nagel, I. Purba, *Chem. Ztg.* **102** (1978) 169 – 178.

[648] F. Dietz, *GWF, Gas Wasserfach: Wasser/Abwasser* **116** (1975) 301 – 308.

[649] G. Graffmann, P. Kuzel, H. Nösler, G. Nonnenmacher, *Chem. Ztg.* **98** (1974) 499 – 504.

[650] B. Wiecken, S. Wübbold-Weber, *SOFW-J.* (1994).

[651] K. Raymond, L. Butterwick: "Perborate" in O. Hutzinger (ed.): *The Handbook of environmental chemistry,* **Vol. 3, Part F,** Springer-Verlag Berlin Heidelberg 1992, p. 287.

[652] W. Guhl, *SOFW-J.* **18** (1992) 1159 – 1168.

[653] P. A. Gilbert: "TAED-Tetraacetylethylenediamine" in O. Hutzinger (ed.): *The Handbook of environmental chemistry,* **Vol. 3, Part F,** Springer-Verlag Berlin Heidelberg 1992, p. 319.

[654] W. E. Gledhill, T. C. J. Feijtel: "Environmental properties and safety assessment of organic phosphonates used for detergent and water treatment applications" in O. Hutzinger (ed.): *The Handbook of environmental chemistry,* **Vol. 3, Part F, Detergents,** Springer-Verlag Berlin Heidelberg 1992, p. 261.

[655] J. Steber, P. Wierich, *Chemosphere* **15** (1986) 929 – 945.

[656] J. Steber, P. Wierich, *Chemosphere* **16** (1987) 1323 – 1337.

[657] G. Müller, J. Steber, H. Waldhoff, *Vom Wasser* **63** (1984) 63 – 78.

[658] K. Wolf, P. A. Gilbert: "EDTA-Ethylenediaminetetraacetic acid" in O. Hutzinger (ed.): *The Handbook of environmental chemistry,* **Vol. 3, Part F,** Springer-Verlag Berlin Heidelberg 1992, p. 243.

[659] AIS Environmental Safety Working Group: An assessment of the implications of the use of EDTA in detergent products. AIS, Brussels, April 1987.

[660] J. Kaschig, R. Hochberg, M. Zeller: "Concerning the environmental fate of fluorescent whitening agents", in: Proc. of the 4th World Surfactants Congress 3 (CESIO), 1996, 155.

[661] J. B. Kramer: "Fluorescent whitening agents" in O. Hutzinger (ed.): *The Handbook of environmental chemistry,* **Vol. 3, Part F,** Springer-Verlag Berlin Heidelberg 1992, p. 351.

[662] J. G. Bartelaan, C. G. van Ginkel, F. Balk: "Carboxymethylcellulose (CMC)" in O. Hutzinger (ed.): *The Handbook of environmental chemistry,* **Vol. 3, Part F,** Springer-Verlag Berlin Heidelberg 1992, p. 329.

[663] G. Rimkus, M. Wolf, *Chemosphere* **30** (1995) 641 – 651.

[664] H. F. Geyer, G. Rimkus, M. Wolf, A. Attar, C. Steinberg, A. Kettrup, *UWSF-Z. Umweltchem. Ökotox.* **6** (1994) 9 – 17.

[665] N. J. Fendinger, R. G. Lehmann, E. M. Mihaich, "Polydimethylsiloxane" in O. Hutzinger (ed.): *The Handbook of environmental chemistry,* **Vol. 3, Part H,** Springer-Verlag Berlin Heidelberg 1997, p. 181.

[666] Industrieverband Körperpflege und Waschmittel, Frankfurt, 2000.

[667] M. J. Schwuger, F. G. Bartnik in Ch. Gloxhuber (ed.): *Anionic Surfactants, Biochemistry, Toxicology, Dermatology,* Surfactant Sci. Ser., 10, 1981, pp. 1 – 49.

[668] J. G. Black, D. Howes in Ch. Gloxhuber (ed.): *Anionic Surfactants, Biochemistry, Toxicology, Dermatology,* Surfactant Sci. Ser., 10, 1981, pp. 51 – 85.

[669] R. B. Drotman in V. A. Drill, P. Lazar (eds.): *Cutaneous Toxicity,* Academic Press, New York, 1977, pp. 96 – 109.

[670] R. B. Drotman, *Toxicol. Appl. Pharmacol.* **52** (1980) 38.

[671] B. Isomaa, *Food Cosmet. Toxicol.* **13** (1975) 231.

[672] A. Siwak, M. Goyer, J. Perwak, P. Thayer in K. L. Mittal, E. J. Fendier (eds.): *Solution Behavior of Surfactants,* **vol. I,** Plenum Publ. Corp., New York 1982, p. 161.

[673] R. A. Cutler, H. P. Drobeck in E. Jungermann (ed.): *Cationic Surfactants,* Surfactant Sci. Ser., 5, 1970, p. 527.

[674] D. L. Opdyke, C. M. Burnett, *Proc. Sci. Sect. Toilet. Goods Assoc.* **44** (1965) 3.

[675] W. Kästner in Ch. Gloxhuber (ed.): *Anionic Surfactants, Biochemistry, Toxicology, Dermatology,* Surfactant Sci. Ser., 10, 1981, pp. 139 – 307.

[676] Ch. Gloxhuber, M. Potokar, S. Braig, H. G. van Raay et al., *Fette Seifen Anstrichm.* **76** (1974) 126.

[677] R. D. Swisher, *Arch. Environ. Health* **17** (1968) 232.

[678] W. B. Coate, W. M. Busey, W. H. Schoenfisch, N. M. Brown et al., *Toxicol. Appl. Pharmacol.* **45** (1978) 477.

[679] W. Kissler, K. Morgenroth, W. Weller, *Prog. Respi. Res.* **15** (1981) 121.

[680] M. S. Potokar in Ch. Gloxhuber (ed.): *Anionic Surfactants, Biochemistry, Toxicology, Dermatology,* Surfactant Sci. Ser., 10, 1981, pp. 87 – 126.

[681] K. Oba in Ch. Gloxhuber (ed.): *Anionic Surfactants, Biochemistry, Toxicology, Dermatology,* Surfactant Sci. Ser., 10, 1981, pp. 327 – 403.

[682] Y. Yam, K. A. Booman, W. Broddle, L. Geiger et al., *Food Chem. Toxicol.* **22** (1984) no. 9, 761 – 769.

[683] Ch. Gloxhuber, *Fette Seifen Anstrichm.* **74** (1972) 49.

[684] Ch. Gloxhuber, M. Potokar, W. Pittermann, S. Wallat et al., *Food Chem. Toxicol.* **21** (1983) no. 2, 209 – 220.

[685] G. A. Nixon, *Toxicol. Appl. Pharmacol.* **18** (1971) 398 – 406.

[686] W. R. Michael, J. M. Wakim, *Toxicol. Appl. Pharmacol.* **18** (1971) no. 2, 407 – 416.

[687] J. A. Budny, J. D. Arnold, *Toxicol. Appl. Pharmacol.* **25** (1973) no. 1, 43 – 53.

[688] G. A. Nixon, E. V. Buehler, R. J. Nieuwenhuis,*Toxicol. Appl. Pharmacol.* **21** (1972) no. 2, 244 – 252.

[689] National Cancer Institute, NCI Techn. Rep. Ser. no. 6, DHEW Publication no. (NHI) 77 – 806 (1977).

[690] P. S. Thayer, C. J. Kensler, A. D. Little, *CRC Crit. Rev. Environ. Control* **3** (1973) 335 – 340.

[691] Ch. Gloxhuber, *Med. Welt* **19** (1968) 351 – 357.

[692] ECETOC Technical Report No. 63: "Reproductive and General Toxicology of some Inorganic Borates and Risk Assessment for Human Beings", Brussels Febr. (1995), p. 1 – 91.

[693] Ch. Gloxhuber, J. Malaszkiewicz, M. Potokar, *Fette Seifen Anstrichm.* **73** (1971) 182 – 189.

[694] G. J. Schmitt, *Z. Hautkr.* **49** (1974) 901.

[695] P. H. Andersen, C. Bindslev-Jensen, H. Mosbech, H. Zachariae, K. E. Andersen, *Acta Derm.-Venereol* **78** (1998) 60 – 62.

[696] F. Coulston, F. Korte (eds.): *Fluorescent Whitening Agents,* Georg Thieme Verlag, Stuttgart 1975.

[697] A. W. Burg, M. W. Rohovsky, C. J. Kensler, *CRC Crit. Rev. Environ. Control* **7** (1977) 91 – 120.

[698] W. Matthies, A. Löhr, H. Ippen, *Dermatosen Beruf Umwelt* **38** (1990) 184 – 189.

[699] G. J. Schmitt, *Tenside Deterg.* **16** (1979) 226 – 228.

[700] R. Teichmann, C. Vogel, G. Busch, W. Körber, *Melliand Textilber.* **80** (1999) 638 – 639.

[701] J. Rieker, T. Guschlbauer, *Textilveredlung,* **34** (1999) no. 11/12, 4 – 6.

[702] CIRFS (Comité International de la Rayon et des Fibres Synthétiques, Paris): *Data on Synthetic Fibers,* 1996, pp. – 51.

[703] S. Mizobuchi, S. Sekiguchi, Household Science Laboratories, Lion Corporation, Tokyo 2000.

[704] R. Weber, *Textilveredlung* **15** (1980) 380 – 385.

[705] ISO 3758, EN DIN 23758.

[706] Brochure No. 3951059 Rev. B, Whirlpool Co., Benton Harbor MI, USA 1999.

[707] Brochure 9/2000, National Matsushita Electric Industrial Co., Tokyo 2000.

[708] Hauptberatungsstelle für Elektrizitätsanwendung (HEA) e.V., Bilderdienst, Frankfurt 1992.

[709] A. Cahn, *HAPPI* **32** (1995) no. 6, 77 – 86.
[710] J. Motavalli, *E/The Magazine* **8** September—October 1997no. 8, 77 – 86.
[711] M. Liphard, A. Griza, *Tenside Surf. Det.* **34** (1997) no. 6, 410 – 416.
[712] H. G. Hloch, Proceedings of the 50th International Detergency Conference (wfk), Luxembourg 1999, 31 – 42.
[713] Appliance, Nov. 1999, 61 – 62.
[714] European Energy Network, Working Group for Efficient Appliances, Danish Energy Agency, Copenhagen 1995.
[715] Appliance, Sept. 1999, 77 – 80.
[716] AEG, Brochure of Washers and Dryers, 2000.
[717] Senking Wäschereianlagen, Hildesheim, Germany, 1997.

15. Persil – Nine Decades of Research for Detergent Users

"You will see that is not easy
to manufacture a modern laundry detergent
and that it demands exceptional efforts
to keep a product up to standard
and to improve it still further."
Fritz Henkel, 1910

Market leadership comes from the determination to seek continual improvement – something realized by company founder Fritz Henkel. In 1907 his company launched an internationally unparalleled innovation on the market and one which would revolutionize the lives of millions of people. This was the first automatic laundry detergent, which achieved dazzlingly white laundry in a single boiling cycle, without physical effort or time-consuming bleaching on the grass. This was Persil. The fact that what was then a unique branded product still ranks among today's leading laundry detergents is attributable above all to Henkel's consistent innovation management.

It all started with a passion for chemistry. Fritz Henkel, the son of a teacher from Hesse, showed an early interest in the secrets of nature and what could be done with them. Nevertheless, instead of studying natural sciences, he embarked on a commercial apprenticeship at a paint factory in Elberfeld, and advanced there to the position of Prokurist (holder of a special commercial power of attorney). In 1876, the 28-year-old boldly decided to go his own way and founded his own detergent factory in Aachen with three employees.

Two years later, Henkel & Cie., a factory for chemical products, moved to Düsseldorf because of its better traffic links. With his heavy-duty laundry detergent consisting of waterglass and soda, he soon made a name for himself. "Henkel's Bleaching Soda" counteracted the yellowing of textiles caused by the iron contained in the water, and gave housewives white washing. Detergents hitherto had failed on this count. By 1899, Fritz Henkel was achieving sales of a million Reichsmarks with this innovation.

At the turn of the century, Fritz Henkel's sons joined the company. Fritz junior, the older of the two, took charge of the commercial side of the business together with his father in 1893. His brother, Dr. Hugo Henkel, following his study of chemistry, injected new impetus in research from 1905 onward. Together with the scientist Dr. Hermann Weber, he succeeded in mixing perborate, silicate, bleaching soda and soap to yield an epoch-making laundry detergent, the name of which was derived from two of the main chemical ingredients, *per*borate and *sil*icate.

Contributing the oxygen for bleaching, perborate was thus an ideal alternative to the chlorine and grass bleaching widespread at that time. Combined with the soaps, the silicate acted as a soil-release agent during boiling. Henkel's Persil was thus the first automatic detergent, a milestone in practical research. And a veritable miracle for the housewife during Germany's period of industrialization and economic expansion (Gründerzeit), after centuries of arduously punishing the laundry on the washboard and time-consuming grass bleaching.

It is hardly surprising, then, that the sales figures rocketed almost instantaneously. Persil embarked on its conquest of Europe. By 1908, Henkel was producing 4,700 metric tons; seven years later, in 1915, seven times this quantity left the Düsseldorf site. The competition was not slow to recognize the advantages of "chemical sun-bleaching", and sent its own products by the names of "Mach's allein", "Bleichin", "Schneeflocken", and "Fix und Fertig" into battle. For Henkel, this was the chief incentive not to rest on its laurels, and to initiate its systematic detergent research.

Henkel's chemists turned laundry washing into a science and developed new test procedures, laid down standards and tested new substances.

With the start of the 1st World War, raw materials became hard to come by. The German government banned the production of detergents containing soap, prompting Henkel to start production of a "war-time soap powder" in 1916. Not until 1920 when supplies of raw materials were back to normal did Henkel announce on posters and in advertisements that "Automatic Persil is back again – in peace-time quality." In fact, the quality of the laundry detergent had been improved still further. A newly developed powder spray-drying process in spray towers made the time-consuming mixing of Persil's ingredients over a large surface area redundant and gave the detergent a finer structure. Operations soon got back to normal. In 1921, Henkel produced 34,000 tons of Persil and beat its own output record of 1915. At the end of the 20s, Henkel's researchers made another astonishing breakthrough in discovering the soil-releasing action of phosphates and used this discovery to improve the Persil formula.

This formula was upgraded again in 1934, with diphosphate and magnesium silicate being employed for the first time. The advantage of this was that the new formulation prevented scale deposits on the laundry and made bleaching even more gentle. Persil thus managed to maintain its market leadership – even though it now had to compete with 170 other soap powders for the consumer's favor. In 1939, annual output peaked at an incredible 103,392 tons. The only problem was that Persil was still based exclusively on natural fats and oils, even though synthetic detergents had been in existence – even from Henkel – since the beginning of the 30s. These replaced the natural ingredients of soaps – mainly palm kernel oil and cotton oil – with synthetic substances. To secure supplies of fats in spite of the restrictive import policy, Henkel even set up its own whaling fleet in 1935 and undertook three Henkel expeditions to the Antarctic Ocean from 1936 to 1939.

The 2nd World War put an end not only to whaling, but also to Persil production. Once again the ban was imposed by the authorities. From now on there were just standard-formula detergents.

It was to be another 11 years before green, white and red flags on the streetcars in the newly created Federal Republic announced the return of the detergent to the shops. Yet again, Henkel's researchers had optimized the formula. Optical brighteners now made the laundry even whiter. The packets, sorely missed for so long, were sold out in a flash. The market launch of "Persil D" in 1951, which contained a synthetic surfactant and a new additive for hard water, helped Persil to withstand the increasingly tough competition from abroad.

In the coming years, the company did all it could to make up the eleven years of lost ground in Persil research and production. In 1959, Henkel ushered in a new era with "Persil 59 – The best Persil there has ever been". The slogan certainly didn't overshoot the mark. The new Persil 59, with its appealing new fragrance, contained synthetic substances which were largely insensitive to hard water and thus highly wash-active, and also acted as foam boosters. Having fallen behind because of the war, Henkel was now well and truly back at the forefront of international laundry detergent research.

At the beginning of the 60s, fully automatic washing machines made their entry into the modern home. Henkel research reacted quickly and came up with Persil 65, the first machine detergent with an inbuilt foam inhibitor. Foam development was controlled by additives in relation to temperature and ensured Persil's suitability for front- and top-loaders, and laundry boilers, and for washing by hand. Persil was now faced not only with new textiles and the growing proportion of colored fabrics, but also with increasingly ominous environmental problems. Here, too, Henkel spearheaded developments.

In 1970 it introduced enzymes for clean laundry without boiling. Even protein stains were no longer problematical. Additives launched in 1973 gave gentle treatment to textiles made of synthetics. And in the same year, chemists at Henkel brought forth a new fully serviceable substitute for the environmentally questionable phosphates. This was Zeolite A, known by the brand name of "Sasil®", which represented the decisive step on the road to environmentally compatible laundry detergents.

In 1981 Persil introduced a totally new formulation, offering not only greater wash power, better color protection, superior water softening and better graying inhibitors, but also anticipated coming legislation on phosphates. In 1986 Henkel brought out the first non-phosphate Persil. And in May 1987 the company entered the market for liquid detergents with its Persil Liquid.

From then on changes to the formulations came thick and fast. Henkel consistently responded to the new demands made of laundry detergents, resulting from the ever expanding diversity of textiles and from the consumer's changing washing habits. Henkel's laboratories were confronted again and again with the central challenge of optimizing Persil fabric care still further and treating the environment gently. Thanks to its staff, who continually applied all their expertise, creativity and ingenuity in attempts to break out of the conventional conceptual mold, the Company has succeeded in the 90s in cementing its position in the top flight of detergent research.

For everything depends ultimately on research. In 1990 Persil introduced Persil perfume-free, i.e. unscented. The same year saw another major advance in the ecological field in the shape of Henkel's Persil SUPRA. This innovative product was a concentrated detergent which has a 50 percent higher yield than classical powder and saves on packaging material at the same time.

In 1991 Henkel presented another unparalleled innovation on the detergent market. This was Persil COLOR, the first heavy-duty detergent with color protection, which washes colored fabrics stain-free from 40°C and at the same time prevents color transfer.

The next innovations arrived only twelve months later: Persil SUPRA Liquid and Persil COLOR Liquid in the eco light pack with new surfactant combinations featuring 50 percent more active wash ingredients. The following year Persil SUPRA was improved. By employing the enzyme lipase for the first time, the concentrate now had a stain-releasing system for 40°C washing, which breaks down natural fats and oils in the wash liquor and thus makes them much easier to rinse out.

1993 saw the next important step towards improved environmental compatibility. Persil Plantaren® is the first quality powder detergent containing surfactants derived mainly from renewable resources such as rapeseed, palm, coconut and palm kernel oil. These are totally biodegradable, even in the absence of oxygen (i.e. in anaerobic conditions). At the same time, Henkel introduced Persil SUPRA and Persil COLOR in the 2 kg refill pack.

In 1994 the Düsseldorf Company gave birth to a totally new generation of detergents in the form of Persil Megaperls®. This product represented a genuine milestone in consisting of pearls rather than powder. In a patented production process, the detergent ingredients are compressed under exceptionally high pressure into compact beads. With the European patent in its pocket, Henkel once again demonstrated its innovative powers.

In 1996 Henkel again responded to the trend towards colored textiles and the growing proportion of laundry washed at low temperatures. Thanks to the surfactant APG® (alkyl polyglycoside) employed for the first time in a powder, the special detergent "Persil for Fine Fabrics" generates a highly stable care foam which cushions fine fabrics. In time for the anniversary year of 1997, Henkel achieved its next breakthrough in the form of "Persil Power Gel" and "Persil COLOR Gel". These two innovations embody not only the transformation of liquid detergents into gels, but also the enhancement of wash power at low temperatures. The gel dissolves stains better even at 30°C.

The following year saw the dawn of the detergent tablet age. Henkel was one of the first suppliers of heavy-duty laundry detergent tablets on the European market. Depending on the degree of soiling and water hardness, only one, two or three tablets are required. These are metered quite simply via the machine's detergent dispenser. After further development work, Persil COLOR tablets were also launched in 1999.

Following a relatively long period of development and now that sufficient raw materials had become available, Henkel changed the bleaching system in Persil Megaperls® from sodium perborate to sodium percarbonate in the same year. Percarbonate gives new scope to formulation development as it combines bleaching, alcalinity, and builder properties.

However, this by no means exhausts the list of innovations and successes in 1999. In February of that year, Persil Megaperls® Sensitive was launched on the German market.

Developed and tested in cooperation with the German Allergy and Asthma Society (DAAB), Persil Sensitive is a heavy-duty detergent without colorants or perfumes. This new formulation with its extra-low irritation potential combines the accustomed Persil washing power with suitability for users with particularly sensitive skin.

In the year 2000 the entire Persil product range was enriched with an improved system of graying inhibitors.

To mark Henkel's 125th anniversary in 2001, Persil Sensitive was launched as a gel in the middle of the year. This means that consumers (and lovers) of bleach-free liquid detergents now also have a highly skin-compatible product available.

In spite of the swift expansion of the Persil family particularly in the last few years, the brand has remained consistently true to itself. Gentle care and optimum cleanliness still have absolute priority.

The fact that the consumers still appreciate this commitment to quality is demonstrated by sales figures. Persil continues to be the most frequently sold laundry detergent in Germany today.

16. Index

acetate fibers 210
acid dyes 209
acidimetry 149
activators, bleaches 80 f
active oxygen loss 108
additives
- detergent ingredients 38 ff, 84 f, 120
- ecologics 198
- toxicology 207
adhesion 21 f
adsorption 26, 30, 35
aftertreatment aids 115
agents *see individual types like*: chelating, complexing, sequestering, bleaching etc.
agglomeration 126 f, 139
agitator washing machines 221 f
air vented dryers 234
alcohol ethoxylates 2, 26
- detergent ingredients 42,52
- ecologics 190
- toxicology 205
alcohol sulfates 42, 138
alcohol sulfonates 101, 160
alcohols 109
algal toxiticity test 182
alkalies
- builders 62 f
- industrial detergents 121
alkaline hydrolysis 149
alkalitriphosphate 20
alkanesulfonates 42, 47
alkyl chains 39
alkyl ether sulfates (AES) 42, 50 f
alkyl poly(ethylene glycol) ethers 40
alkyl sulfates 2
alkylamine oxides 55
alkylbenzenesulfonates 2, 16
- detergent ingredients 42, 45
- economics 160, 186 ff
- heavy-duty detergents 101
- powder detergents 138
alkyldimethlbenzylammonium chlorides, detergent ingredients 59
alkylphenol ethoxylates (APE)
- detergent ingredients 42, 55
- ecologics 191
- economics 160
alkylpolyglycolides (APG) 6 ff, 16, 26
- detergent ingredients 43, 56
- ecologics 191
- powder detergents 141
alkylsulfonates 40
allergic reactions 203
aluminum, photobleaching 83
ambroxan 97
amine oxides 90
amorphous silicates 142
amphoteric surfactants 39, 61
amylase 5, 87
- detergent ingredients 85 f
- heavy-duty detergents 101
- powder detergents 143
anaerobic biodegradability 179
analcime 31
analytical methods, compositions 145–154
anionic surfactants
- detergent ingredients 39, 45 f
- heavy-duty detergents 109
- oil/grease 15
- powder detergents 112, 138
- regulatory limitations 169
- soil removal 24
- toxicology 204
- wastewater 167
antiredeposition agents 4, 34 ff, 88f
- ecologics 201
- heavy-duty detergents 101
- industrial detergents 121

aquatic food chain, ecologics 182
atomic absorption spectrometry 151
auxiliary agents *see*: additives

Bacillus species 140
bacterial inoculum 170
bar extrusion 123
base detergents, industrial 120
batch type washing machines 236
bath ratio 225
behenate soap 5
benzyl acetone 97
betaines 40 f, 90
bioaccumulation 184, 190
biocenotic toxicity 183 ff
bioconcentration factor (BCF) 184
biodegradability 2
biodegradation tests 170 ff, 175 ff
bis(benzoxazole) 93
bleach activators 3, 5, 75, 80 f, 206
bleach catalysts/stabilizers 83
bleachable dyes 11
bleaches 1, 74 ff, 142
bleaching agents
– detergent ingredients 38 ff
– ecologics 196 ff
– heavy-duty detergents 101
– industrial detergents 121
– laundry aids 113 ff
– powder detergents 138
blending 129
block flow diagram, powder production 124
blood 10
blueing agents 92
boosters 113 ff
Born repulsion 22
builders 1 f
– combinations 74
– detergent ingredients 38, 61 ff
– ecologics 193
– economics 160 f
– heavy-duty detergents 101, 109
– powder detergents 141 f
– toxicology 206

calcium binding capacity 63, 69 f
calcium carbonate 32
calcium containing soils 29 ff
calcium phosphate 32
calcium salts 7
capillary zone electrophoresis 151
carbohydrates 11
carbon black 10 f, 24, 31
carbonates 11
carboxyl groups 29
carboxymethyl cellulose (CMC)
– antiredeposition 88
– ecologics 201
carboxymethyltartronic acid 65
carcinogenicity 203 ff
care labeling 214 ff
catalysts 83
cationic bridges 29
cationic surfactants 4
– detergent ingredients 39, 57 f, 112
– ecologics 187, 192 f
– fabric softeners 116
– soil removal 24
– toxicology 204
– wastewater 167
cellulase 6
– detergent ingredients 85 f
– heavy-duty detergents 101
cellulose ethers 36
celluosic fibers 209
chelating agents 11
chemothermal sterilization 121
chloramine T 121
chlorine, industrial detergents 121
chlorine bleaching 217
chromium steel 230
chronic toxicity 184, 205
citrates
– ecologics 193, 195
– heavy-duty detergents 104, 107

– wastewater 167
citric acid 65
coagulation, soil removal 22, 28
cobuilders, heavy-duty detergents 101
coconut oil 139
coffee 11
colored laundry 111
coloring agents 97
colors 107, 209
column chromatography 151
compact heavy-duty detergents 103 f
complex formation, toxicology 204
complexing agents 2
– bleaches 83
– builders 63 f
– industrial detergents 121
– soil removal 25, 31
complexometry, detergent ingredients 150
composition analysis 145–154
compositions *see*: formulations
condenser dryers 234
conductometry 150
construction materials, washing machines 230 ff
consumer tests 157
consumption
– detergents 99
– fabric softeners 164
– surfactants 158
continuous batch washers 237
corn starch 117
corrosion inhibitors 92
cotton 215
coumarin 93
crutchers 125
cycles, wash programs 220–238
cyclohexyl salicylate 97
cyclovertal 97

Daphnia tests 183
decane 16
decomposition methods, detergent ingredients 148
densifying processes, powder detergents 123, 129 ff
density 155
Derjaguin-Landau-Verwey-Overbeek theory (DLVO) 21
desmine 31
desorption processes, calcium containing soils 30
detergent ingredients 38–98
– analytical methods 146 ff
– ecologics 186 ff
– economics 158 ff
– toxicology 204 ff
dialkyldimethylammonium chlorides 58
diaminostilbenes 200
dicarboxylic acid 107
dimethyl ammonium chloride 192
diperoxydodecanedioic acid (DPDDA) salts 77
disinfectant additives 121
disintegration time, tablet detergents 136
dispersion, soil removal 34
dissolution, tablet detergents 136
dissolved organic carbon (DOC) 176
distearyldimethylammonium chloride (DSDMAC) 57, 117
distryrylbiphenyl 93
distyrylbiphenyl derivatives 200
n-dodecyl sulfate (SDS)
– adsorption 35
– oil/grease 15, 26
n-dodecyl sulfonate 67
domestic laundry 155
dosing units, industrial detergents 120
drum type washing machines 236
dry densification 130
dry milling, powder detergents 123
dryers 234
– laundry aids 113, 116, 119

dye transfer inhibitors
– detergent ingredients 96, 112
– ecologics 201
– heavy-duty detergents 107
dyes
– bleachable 11
– detergent ingredients 97
– ecologics 202
– textiles 209 ff
– wastewater 167

ecology 165–203
economic aspects 157–165
ecotoxicity tests 183 ff
effervescents 107
egg 10
electrolytes 18
electrophoretic mobility 24
elemental analysis, detergent ingredients 153
embryotoxicity 203 ff
emulsification, oil/grease 16
energy consumption, washing machines 230
environmental aspects 165–203
enzymes 3
– detergent ingredients 84 f
– ecologics 200
– heavy-duty detergents 101, 109
– industrial detergents 121
– powder detergents 112, 143
– toxicology 207
– wastewater 167
esterquats (EQ) 60, 117
ethanol extract 146
ethoxy groups 28
ethoxylates 2, 26
– detergent ingredients 42, 52 f
– economics 160
– powder detergents 140
– toxicology 205
ethylene 140
ethylenediaminetetraacetate (EDTA) 199
ethylenediaminetetraacetic acid 65
European washing machines 228
exposure assessment 174, 179
extraction, detergent ingredients 148
extractors, institutional uses 237
extrusion
– enzyme production 145
– heavy-duty detergents 103, 106
– powder detergents 123, 132 f

fabric labeling, textiles 219
fabric softeners 116, 164 f
fat analysis data 153
fat saponification 1
fats 10
fatty acid alkanolamides (FAA) 43, 55, 90
fatty acid ethoxylates 140
fatty alcohol ether sulfates 42, 50
fatty alcohol ethoxylates 140
fatty alkyl sulfate (FAS) 16
faujasite 31
fermenters 143
fibers, soil removal 23
fillers 97, 101
finishing 210
fixations 209
flame photometry 151
flameproofing 210
flocculation, soil removal 12, 21
floramat 97
flow packers, tablet detergents 137
flowability 155
fluorescent whitening agents 4, 92 ff
– ecologics 200
– industrial detergents 121
– toxicology 207
foam regulators 3 f, 90 f
– ecologics 202
– heavy-duty detergents 101
food chain, ecologics 182

formulations
– bleaches 115
– boosters 115
– fabric softeners 118
– heavy-duty detergents 101 ff, 108 f
– powder specialty detergents 112
– stiffeners 118
fragrances 4
– detergent ingredients 96, 112
– ecologics 201
– fabric softeners 117
– heavy-duty detergents 101, 109
fruits 11

gas chromatography 151
gel formation, tablet detergents 135
gelatin 36
gelling effect 107
GINETEX labeling symbols, textiles 216 ff
granulation
– heavy-duty detergents 103, 106
– powder detergents 123
grass 10
gravimetric analysis 149
grease 10, 13

hand washing 111
hardness, water 7, 9
heavy-duty detergents 99 ff
hedione 97
hematite 26
hemolysis, toxicology 205
herbavert 97
heterocoagulation, soil removal 12
hexadecyltriglycol ether 15
horizontal drum washing machines 220 f
household laundry products 98–120
household washing machines 220 ff
hydrogen peroxide 114
hydrophilic-lipophilic balance (HLB) values 52 f
hydrophilic groups 27
hydrophobic residues 41 f
hydrophobicity/hydrophilicity 33
hydroxyethanediphosphonate (HEDP) 83
hypochlorite bleaching 75, 78, 217

imidazolinium salts 59
impeller washing machines 220 f
industrial detergents 120 ff
ingestion/inhalation 203
inorganic components 146
inorganic salts 10
institutional detergents 120 ff
institutional uses, washing machines 236
ion exchangers 2 f
– builders 68 f
– calcium containing soils 30
– detergent ingredients 148
IR spectrometry 152
iron oxides 31
ironing 217
isopropyl myristate (IPM) 16

Japanese washing machines 227

kaolin
– builders 66
– calcium containing soils 31
keratin 10

laundry aids 112
laundry dryers 234
laundry sorting, textiles 215
laws/legislation, ecologics 167 ff
layered silicates 141
lime soaps 1
linalol 97
linear alkylbenzenesulfonates (LAS) *see*: alkylbenzenesulfonates
linen 215
lipase 6, 87

– detergent ingredients 84 f
– heavy-duty detergents 101
– powder detergents 143
liquid chromatography 152
liquid detergents 108, 111

magnesium salts 7
manganese complexes 83
manufacturing processes, powder detergents 124 ff
market
– dryers 235
– washing machines 232 f
mass spectrometry 153
mechanical input, washing machines 212, 224
metabolite test 179
metal complex dyes 209
metal complexes 64
metal oxides 10
N-methylglucamides (NMG) 5 f, 140
micelle concentration 54
microfiltration 144
milk 10
miscellaneous titrimetric analysis 150
mixer processes, powder detergents 125f
mobility, electrophoretic 24
monoperoxyphthalic acid 77
mutagenicity 203 ff

naphthol 209
natural fibers 215
nitrilotriacetic acid (NTA)
– builders 61, 65
– ecologics 196
NMR spectrometry 153
no effect concentration (NOEC) 185, 194
nonionic granulation 123
nonionic surfactants
– detergent ingredients 39, 52 ff
– ecologics 186 f, 190 f
– heavy-duty detergents 109
– powder detergents 112, 139
– regulatory limitations 169
– toxicology 204
– wastewater 167
nonphosphate heavy-duty detergents 105
nonylphenol ethoxylates 2
North American washing machines 227

odor removers 113, 119 f
OECD test 170
oils 10, 13
oily/greasy soils 13
α-olefinsulfonates (AOS) 42, 48
oligomerization 140
optical brighteners 4, 92,112
– ecologics 200
– heavy-duty detergents 101, 109
– wastewater 167
oral toxicity 205
organic components 146
oxidative bleaching 75
oxygen 121
oxygen loss 108

packaging volume 106
palm kern oil 139
pancreatic enzymes 4
paper chromatography 151
paraffins
– heavy-duty detergents 101
– powder detergents 140
– soil removal 28
– wastewater 167
particulate soils 21
peracetic acid-hydrogen peroxide mixture, disinfectant 121
permanent stiffeners 118
peroxide bleaching 75
peroxygen bleaches 142
perspiration, soil types 10

phase behavior, detergent
penetration 18
phosphate content, maximum 172
phosphates 167
phosphonates 198
photobleaching 83
phthalocyanine dyes 209
physical chemistry, washing
process 7–38
pigments 11, 23
plastics 14, 232
poly(acrylic acid) 69, 73
polyamide 13
polycarbonates 167
polycarboxylates
– cobuilders 3
– ecologics 193 f
– heavy-duty detergents 104
– soil removal 32
polyester fabrics 214
polyester fibers 18
poly(ethylene glycol) 89
poly(α-hydroxyacrylic acid) 69, 73
polymeric stiffeners 118
polytetrafluoroethylene 13
poly(tetramethylene-1,2-dicarboxylic
acid) 69, 73
polyurethane 215
poly(vinyl acetate) 117
poly(vinylpyridine *N*-oxide) 96
poly(*N*-vinylpyrrolidone) 96
postaddition processes, powder
detergents 135
potassium octanoate 18
potato starch 117
potentiometry 149
powder characteristics 155
powder detergents, production 122–145
powder heavy-duty detergents 100
powder specialty detergents 111
precipitation 144
predicted environmental
concentration (PEC) 185, 190
pretreatment, laundry aids 113
printing, textiles 209
production
– enzymes 144
– heavy-duty detergents 101
– powder detergents 122–145
protease 4, 84
– heavy-duty detergents 101
– powder detergents 143
proteins 10, 204
Pseudomonas putia 183
pulsator washing machines 220 f

qualitative analysis, detergent
ingredients 147
quantitative analysis, detergent
ingredients 149 f
quantitative structure-activity
relationship (QSAR) 185
quaternary ammonium
– disinfectants 121
– fabric softeners 116

radiometric methods 151
raw materials, powder detergents 138
rayon 215
refreshing products 113, 119
release agents 88 f
removal processes, soil types 11 ff
repellent agents 22, 35 ff, 88 f
residual moisture 225
reversing rhythm, washing
machines 225
rice starch 117
roller compaction 123, 131
rolling-up processes 13 f
rubber 232

sample preparation 147
Sandelice 97
savinase 86
sebum 10, 19
sedimentation, soil removal 12

semiesters 138
separation methods 151
sequestering agents 67
sewage load, laundry contribution 166 ff
Shell Higher Olefins Process (SHOP) 140
silicates
– heavy-duty detergents 104
– powder detergents 112, 141
– soil types 11
– wastewater 167
silk 209
Sinner's circle, washing machines 224
size exclusion chromatography 150
skin contact, toxicology 203
slurries 125
soaps
– detergent ingredients 42, 44
– economics 162
– heavy-duty detergents 109
soda 63
soda ash 1 f
– ecologics 193, 195
– powder detergents 125 f
– wastewater 167
sodalite 31
sodium alkylbenzenesulfonates 32
sodium aluminium silicate *see*: zeolite A
sodium carbonate 1, 3, 61
– heavy-duty detergents 101
– laundry aids 114
sodium citrate 5
sodium diphosphates 65 f, 68
sodium hypochlorite 114
sodium nitrilotriacetate (NTA) 206
sodium perborate 1, 3, 76, 142
– bleaches 80
– ecologics 196
– heavy-duty detergents 101
– laundry aids 114
– powder detergents 112, 142
– toxicology 206
sodium percarbonate
– bleaches 80
– ecologics 197
– powder detergents 143
– toxicology 206
sodium *p*-nonanoyloxybenzenesulfonate (NOBS) 81, 101
sodium polycarboxylates 66, 114
sodium silicate 1, 3
– builders 61, 114
sodium sulfates
– laundry aids 114
– ecologics 203
– heavy-duty detergents 101
sodium triphosphate 2
– builders 65, 71
– ecologics 177
– economics 160 f
– industrial detergents 121
– interfacial tension 20
– powder detergents 141
– toxicology 206
soil antiredeposition 3, 34 ff, 88 ff
soil removers, laundry aids 113
soil repellents 3, 35 f
– ecologics 202
– heavy-duty detergents 101
– wastewater 167
soil types 7 ff
solubilization
– oil/grease 17
– soil removal 34
solubilizers 109
soybean oil 139
spaghetti extrusion 123, 132
specialty detergents 110, 120
spectrometric determination 150
spheronizer densification 130
spray drying 106, 125
stabilizers
– powder detergents 143
– bleaches 83
– ecologics 198

– heavy-duty detergents 101, 109
stain removers 113
stainless steel, washing machines 230
standardization 155
starch 117
steam drying 127
sterilization 121
stiffeners 113, 116 ff
stilbenes 93, 101
structure determination, detergent ingredients 152
subchronic ecotoxicity tests 184
sublation 148
sugar 10
α-sulfo fatty acid methyl esters (MES) 42, 49
sulfobetaines 90
sulfuric acid semiesters 138
supercompact heavy-duty detergents 103 f
supercritical chromatography 152
superheated steam drying 127
surface tension 7, 12 ff
surfactants 1 ff
– bioaccumulation 190
– detergent ingredients 38 ff
– ecologics 186 f
– economics 158 f
– heavy-duty detergents 101, 109
– industrial detergents 121
– oil/grease 15
– powder detergents 112, 138 f
– regulatory limitations 169
– soil removal 24
– toxicology 204 ff
– wastewater 167
see also: anionic, ionic, nonionic, and other individual types
swelling agents 107
synthetic fibers 210, 214
synthetic polymeric stiffeners 118
synthetic surfactants 1, 3

tableting
– dry densification 135
– heavy-duty detergents 106
– powder detergents 123
tallow oil 139
tea 11
temperatures 99
– bleaches 82
– textiles 212 ff
– washing machines 224
terephthalic acid 89
terrestrial toxicity 183
test methods 154–157
tetraacetylethylenediamine (TAED)
– bleaches 81
– ecologics 197
– heavy-duty detergents 101
– wastewater 167
tetrapropylenebenzenesulfonate (TPS) 2, 45, 187
textile fibers 4, 7, 33, 209–220
thermal sterilization 121
thin-layer chromatography 151
threshold effect 66
tower process, powder detergents 125
toxicology 182, 203–209
transport properties 204
triglycerides 139
turbidity 28

urea 10
UV spectrometry 152

van der Waals-London attraction 22
vegetables 11
vertical axis washing machines 220 f
Vertofix Coeur 97
viscosity 12
volumetric analysis, detergent ingredients 149

wash cycles 155
wash liquor ratio 225

wash programs 220–238
washable fabrics 210
washing conditions, textiles 212
washing machines 220–238
washing process, physical chemistry 7–38
wastewate 165
water consumption, washing machines 230
water influences 7
water softeners 113
wax 10
wet granulation
– heavy-duty detergents 103
– powder detergents 123, 131
wetting 13 ff, 33, 39
wine 11
wool 111, 209
world soap production 162

X ray diffraction 151
xenobiotics 174 f

Young equation 13

zeolite A 2 f, 5
– builders 68, 70
– calcium containing soils 30
zeolite P 74
zeolites
– adsorption 35
– builders 61, 68, 73
– ecologics 193
– economics 161
– heavy-duty detergents 101, 104
– industrial detergents 121
– laundry aids 114
– powder detergents 112, 127, 141
– wastewater 167